Sample Preparation-Quo Vadis: Current Status of Sample Preparation Approaches

Sample Preparation-Quo Vadis: Current Status of Sample Preparation Approaches

Editors

Victoria Samanidou
Irene Panderi

MDPI • Basel • Beijing • Wuhan • Barcelona • Belgrade • Manchester • Tokyo • Cluj • Tianjin

Editors

Victoria Samanidou
Laboratory of
Analytical Chemistry
Aristotle University
of Thessaloniki
Thessaloniki
Greece

Irene Panderi
School of Pharmacy
Laboratory of Pharmaceutical
Analysis
National and Kapodistrian
University of Athens
Athens
Greece

Editorial Office
MDPI
St. Alban-Anlage 66
4052 Basel, Switzerland

This is a reprint of articles from the Special Issue published online in the open access journal *Molecules* (ISSN 1420-3049) (available at: www.mdpi.com/journal/molecules/special_issues/ molecules_sample_preparation).

For citation purposes, cite each article independently as indicated on the article page online and as indicated below:

LastName, A.A.; LastName, B.B.; LastName, C.C. Article Title. *Journal Name* **Year**, *Volume Number*, Page Range.

ISBN 978-3-0365-1310-2 (Hbk)
ISBN 978-3-0365-1309-6 (PDF)

Contents

About the Editors

Victoria Samanidou

Victoria F. Samanidou is a professor of analytical chemistry at the School of Chemistry of the Aristotle University of Thessaloniki, Greece, and the director of the same laboratory. She obtained her doctorate (PhD) in chemistry in 1990 from the same department. She has published more than 165 original research articles in peer-reviewed journals and 60 reviews and chapters in scientific books. Her H-index is 40 (Scopus, October 2021).

Irene Panderi

Irene Panderi is a pharmacist with a PhD in pharmaceutical analysis. She is a professor and director of the Laboratory of Pharmaceutical Analysis at the Faculty of Pharmacy, National and Kapodistrian University of Athens in Greece, and the director of the postgraduate program "Pharmaceutical Analysis—Quality Control". She has over 20 years of experience in leading research in the fields of pharmaceutical analysis and analytical chemistry, developing novel analytical methods for clinical testing, toxicology studies, doping control, and quality control of drugs.

Article

Determination of Intact Parabens in the Human Plasma of Cancer and Non-Cancer Patients Using a Validated Fabric Phase Sorptive Extraction Reversed-Phase Liquid Chromatography Method with UV Detection

Anthi Parla [1], Eirini Zormpa [1], Nikolaos Paloumpis [1], Abuzar Kabir [2], Kenneth G. Furton [2], Željka Roje [3], Victoria Samanidou [4], Ivana Vinković Vrček [5] and Irene Panderi [1,*]

1 Laboratory of Pharmaceutical Analysis, Division of Pharmaceutical Chemistry, School of Pharmacy, National and Kapodistrian University of Athens, Panepistimiopolis, Zografou, 15771 Athens, Greece; anthiparla@hotmail.com (A.P.); eirini.zorba1@gmail.com (E.Z.); paloumpisnick@gmail.com (N.P.)
2 Department of Chemistry and Biochemistry, Florida International University, Miami, FL 33199, USA; akabir@fiu.edu (A.K.); furtonk@fiu.edu (K.G.F.)
3 Department for Plastic, Reconstructive and Aesthetic Surgery, University Hospital Dubrava, 10 000 Zagreb, Croatia; zroje@kbd.hr
4 Laboratory of Analytical Chemistry, Department of Chemistry, Aristotle University of Thessaloniki, 54124 Thessaloniki, Greece; samanidu@chem.auth.gr
5 Institute for Medical Research and Occupational Health, Ksaverska cesta 2, 10 000 Zagreb, Croatia; ivinkovic@imi.hr
* Correspondence: irenepanderi@gmail.com; Tel.: +30-210-72 74 820

Citation: Parla, A.; Zormpa, E.; Paloumpis, N.; Kabir, A.; Furton, K.G.; Roje, Ž.; Samanidou, V.; Vinković Vrček, I.; Panderi, I. Determination of Intact Parabens in the Human Plasma of Cancer and Non-Cancer Patients Using a Validated Fabric Phase Sorptive Extraction Reversed-Phase Liquid Chromatography Method with UV Detection. *Molecules* **2021**, *26*, 1526. https://doi.org/10.3390/molecules26061526

Academic Editor: Farid Chemat

Received: 31 January 2021
Accepted: 8 March 2021
Published: 11 March 2021

Publisher's Note: MDPI stays neutral with regard to jurisdictional claims in published maps and institutional affiliations.

Abstract: Parabens have been widely employed as preservatives since the 1920s for extending the shelf life of foodstuffs, medicines, and daily care products. Given the fact that there are some legitimate concerns related to their potential multiple endocrine-disrupting properties, the development of novel bioanalytical methods for their biomonitoring is crucial. In this study, a fabric phase sorptive extraction reversed-phase liquid chromatography method coupled with UV detection (FPSE-HPLC-UV) was developed and validated for the quantitation of seven parabens in human plasma samples. Chromatographic separation of the seven parabens and *p*-hydroxybenzoic acid was achieved on a semi-micro Spherisorb ODS1 analytical column under isocratic elution using a mobile phase containing 0.1% (*v/v*) formic acid and 66% 49 mM ammonium formate aqueous solution in acetonitrile at flow rate 0.25 mL min^{-1} with a 24-min run time for each sample. The method was linear at a concentration range of 20 to 500 ng mL^{-1} for the seven parabens under study in human plasma samples. The efficiency of the method was proven with the analysis of 20 human plasma samples collected from women subjected to breast cancer surgery and to reconstructive and aesthetic breast surgery. The highest quantitation rates in human plasma samples from cancerous cases were found for methylparaben and isobutylparaben with average plasma concentrations at 77 and 112.5 ng mL^{-1}. The high concentration levels detected agree with previous findings for some of the parabens and emphasize the need for further epidemiological research on the possible health effects of the use of these compounds.

Keywords: fabric phase sorptive extraction (FPSE); parabens; *p*-hydroxybenzoic acid esters; human plasma; bioanalysis

1. Introduction

Parabens are widely employed preservatives for extending the shelf life of foodstuffs, medicines, and daily care products [1]. These compounds are chemically stable without imparting any smell or taste and exhibit antimicrobial activity against a broad range of microorganisms. Such a combination of properties makes it difficult to find alternative preservatives to satisfactorily replace parabens. After ingestion, they are rapidly metabolized in the human intestine and liver into the relatively inactive metabolite, *p*-hydroxybenzoic acid, and its sulfuric acid and glucuronic conjugates, with less than 24 h biological half-life [2]. After dermal application, these compounds are partly metabolized to *p*-hydroxybenzoic acid by the skin enzymes [3], and the shorter alkyl chains derivatives cross the stratum corneum more easily than the longer chain derivatives [4]. Most studies showed that the estrogenic potential of the main metabolite of parabens, *p*-hydroxybenzoic acid, is weaker than that of the intact ingredients [5]. Although parabens are not mutagenic [6,7], there are concerns related to their potential multiple endocrine-disrupting action that are suspected to cause various health effects [8–11]. In the early 2000s, some studies indicated that the estrogenic activity of parabens increases with increasing the length of the linear alkyl chain group, with branching in the alkyl chain group or by the addition of a benzyl ring [12,13]. Further research has shown that despite the rapid metabolism rate, the concentration levels of the parent paraben esters in various human samples are not insignificant with the percentage of intact paraben esters excreted in urine to be dependent on the route of exposure [14,15]. The analysis of human breast tissue samples revealed that at least one intact paraben was present in 99% of the analyzed samples with a total median concentration at 85.5 ng/g [16]. In 2007, the Food and Agriculture Organization of the United Nations (FAO) and World Health Organization (WHO) Joint Expert Committee on Food Additives (JECFA) recommended the exclusion of propylparaben from use as a food preservative due to its adverse effects and allowed only the use of a group acceptable daily intake (ADI) of 0 to 10 mg/kg body weight (bw) for the sum of methylparaben and ethylparaben [17]. In the European Union, the allowed limit for propylparaben and butylparaben in cosmetics is 0.14% when used individually or together, while a safe concentration has been established for mixtures of parabens in cosmetics where the sum of the individual concentrations should not exceed 0.8% (as acid) [18]. Additionally, since 2014, isopropylparaben, isobutylparaben, benzylparaben, phenylparaben, and pentylparaben have been banned from use in cosmetics [19]. Recently, the Scientific Committee on Consumers Safety (SCCS) considered the concerns related to the potential endocrine-disrupting properties of propylparaben and concluded that the compound is safe when used as a preservative in cosmetic products up to a maximum concentration of 0.14% [20]. The concentrations of some parabens such as methylparaben, propylparaben, and the sum of parabens in umbilical cord plasma were associated to the levels of androgens in the same biofluid, and the widely used propylparaben was found to be negatively associated to the testosterone levels [21]. Recently, intact parabens have been detected in endometrial carcinoma tissues at a higher extent than in the normal endometrium, and propylparaben, isobutylparaben, and butylparaben were detected most frequently in all the tissue samples [22]. In another study contacted in Japan, the urine levels of parabens in pregnant women were measured in significantly high quantities indicating widespread exposure to parabens among these subjects [23]. Children and especially infants are vulnerable to the exposure of endocrine-disrupting chemicals (EDCs) in the environment due to their immature metabolic pathways [24]. Parabens are metabolized to *p*-hydroxybenzoic acid at different rates, and as a result, their exposure doses cannot be accurately assessed based only on their concentrations in human urine. Therefore, the development of novel bioanalytical methods for biomonitoring these compounds in human plasma has also been highlighted [25].

The literature survey revealed several chromatographic methods undertaken for the analysis of parabens in various biofluids [9,15,26–29]. These procedures were based on the use of liquid–liquid extraction (LLE) [30,31], solid phase extraction (SPE) [32], and several others sample preparation techniques [33–37]. Despite the extensive research in the quantitation of parabens in various biofluids, only a few published methods refer to the determination of the isomers of both propylparaben and butylparaben. These methods include a gas chromatographic-mass spectrometric (GC-MS) method for the quantitation of seven parabens in human breast milk [29,31] and liquid chromatographic methods coupled to diode array detection (HPLC-PDA) and the fabric phase sorptive extraction (FPSE) technique for the analysis of seven paraben residues in human whole blood, plasma, urine, and breast tissue samples [38–40].

In recent years, the development of novel sample preparation procedures following the philosophy of Green Analytical Chemistry (GAC) is a matter of growing interest among the analytical and bioanalytical scientists. To this purpose, we thought that it would be of interest to develop a method for the quantitation of seven parabens, namely methylparaben (MPB), ethylparaben (EPB), isopropylparaben (iPPB), propylparaben (PPB), butylparaben (BPB), isobutylparaben (iBPB), and benzylparaben (BzPB) in human plasma samples using a novel, eco-friendly, and efficient fabric phase sorptive extraction (FPSE) technique. FPSE simplifies the analytes extraction from complex matrices and reduces the solvent consumption. This technique utilizes a variety of fabric substrates chemically coated with sol–gel hybrid organic–inorganic sorbents, resulting in an efficient and eco-friendly sample pretreatment technique [41,42]. Since its discovery in 2014 by Kabir and Furton, FPSE has been applied to several analytes in variable samples including biological, environmental, and food samples, and by using various analytical techniques such as liquid chromatography-mass spectrometry (LC-MS), gas chromatography-mass spectrometry (GC-MS), and high-performance liquid chromatography with photodiode array detection (HPLC-PDA) [43–45]. In this work, the assay was based on the use of a small fraction (50 µL) of human plasma followed by an improved FPSE protocol using a sol–gel Carbowax® 20M polar sorbent that exhibited improved sensitivity and reversed-phase HPLC-UV analysis on a semi-micro reversed phase analytical column. The method adequately separated the targeted analytes from their main metabolite, *p*-hydroxybenzoic acid [46], and it is in accordance with the green analytical chemistry philosophy. Finally, the proposed method was successfully applied to the analysis of human plasma samples taken from women subjected to malignant and benign plastic, reconstructive, and aesthetic breast surgery.

2. Results and Discussion

2.1. Method Development

2.1.1. Chromatography Optimization

The chemical structures and the main physicochemical parameters that affect both chromatography and the FPSE procedure were calculated by ChemBioDraw ver. 13.0 (Perkin Elmer Informatics, Billerica, MA, USA) and are presented in Table 1. Several reversed-phase analytical columns have been tested for the chromatographic separation of the selected parabens, including Hypersil Gold® C18 (150.0 mm × 2.1 mm, 5 µm), Spherisorb® ODS1 (150.0 mm × 2.0 mm, 3 µm), and XTerraMS® C18 (150.0 mm × 3.0 mm, 5 µm). Among these columns, both Hypersil Gold® C18 and XTerraMS® C18 did not allow the adequate separation of the isomers. On the other hand, Spherisorb® ODS1 gives adequate separation for the selected seven parabens within a reasonable chromatographic runtime, and therefore, it was selected for this study. Consequently, chromatography was optimized for the adequate separation of the targeted analytes from matrix interferences within a chromatographic runtime of less than 24 min.

Various combinations of ammonium formate aqueous solution mixed with acetonitrile or methanol with an altered content of each factor were examined to discover the optimal mobile phase. It was observed that acetonitrile, compared to methanol, decreases the retention of the targeted analytes and allows for their adequate separation from matrix interferences. The effect of the percentage of acetonitrile in the mobile phase, φ_{ACN}, was evaluated in experiments where φ_{ACN} was varied from 30 to 40%, whereas ammonium formate concentration was kept constant at 31.5 mM in whole mobile phase and the percentage of formic acid was constant at 0.1% *v/v*. As it can be observed in Figure 1a, the increase in the percentage of acetonitrile yielded to a linear decrease in the retention factor (log*k* values) of the targeted analytes, as it was expected based on the retention mechanism of the reversed phase chromatography. Consequently, 34% acetonitrile was chosen as the optimum φ_{ACN} content in the mobile phase, as it allows adequate separation of all the analytes within less than 22 min.

Table 1. Chemical structures and physicochemical parameters of the seven parabens under study.

Chemical Structures/Chemical Names	Physicochemical Parameters [1]
p-hydroxy benzoic acid	Log P: 1.2 CLogP: 1.5572 pKa: 4.109, 9.685
Methylparaben	Log P: 1.46 ± 0.47 CLogP: 1.9846 pKa: 9.007
Ethylparaben	Log P: 1.83 ± 0.49 CLogP: 2.5136 pKa: 8.981
Isopropylparaben	Log P: 2.12 ± 0.47 CLogP: 2.8226 pKa: 8.955
Propylparaben	Log P: 2.29 ± 0.47 CLogP: 3.0426 pKa: 8.970
Isobutylparaben	Log P: 2.69 ± 0.47 CLogP: 3.4416 pKa: 8.959
Butylparaben	Log P: 2.71 ± 0.47 CLogP: 3.5716 pKa: 8.965
Benzylparaben	Log P: 3.20 ± 0.47 CLogP: 3.8126 pKa: 8.931

[1] Physicochemical parameters were calculated by ChemBioDraw ver. 13.0.

With a constant acetonitrile content in the mobile phase at 34% and formic acid content at 0.1%, the concentration of ammonium formate was altered from 25 to 80 mM. Figure 1b indicates that an increase in the concentration of ammonium formate slightly increases the retention factors (log$k$ values) of all parabens without affecting the separation of the isomers. In all tested mobile phases, the selected parabens are eluted in order of increased lipophilicity (Table 1). Thus, MPB, which is less lipophilic (LogP 1.46), is firstly eluted followed by EPB (LogP 1.83), iPPB (LogP 2.12), PPB (LogP 2.29), iBPB (LogP 2.69), BPB (LogP 2.71), and BzPB (LogP 3.20). It was also found that the column backpressure decreased, and analytes separation from matrix interferences was improved by increasing the ammonium formate concentration in the mobile phase. The best results achieved when a mobile phase consisting of 66% 49 mM ammonium formate aqueous solution in acetonitrile with 0.1% *v/v* formic acid was used for the separation of the targeted analytes. Column equilibration was achieved within 1 h, and the proposed isocratic LC method allows for adequate separation without the need for column re-equilibration.

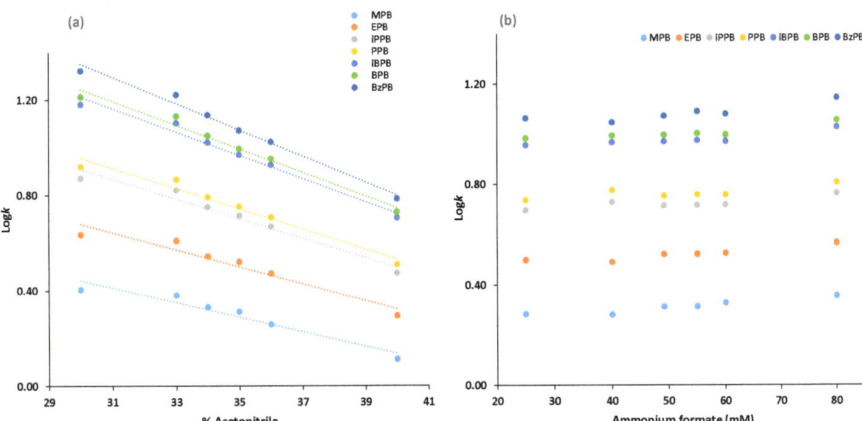

Figure 1. Plots of log*k* values of the targeted analytes versus (**a**) the percentage of acetonitrile and (**b**) ammonium formate concentration in the mobile phase.

Figure 2 shows a chromatogram of a mixed standard solution of the seven parabens at 100 ng mL^{-1} and *p*-hydroxybenzoic acid at 200 ng mL^{-1} prepared in 60% 5 mM ammonium formate aqueous solution in acetonitrile (dilution solvent) and detected at 257 nm. Under the optimum chromatographic conditions, the seven parabens are well separated within less than 22 min, and their main metabolite, *p*-hydroxybenzoic acid, is eluted at 2.16 min and therefore, it does not interfere in their analysis.

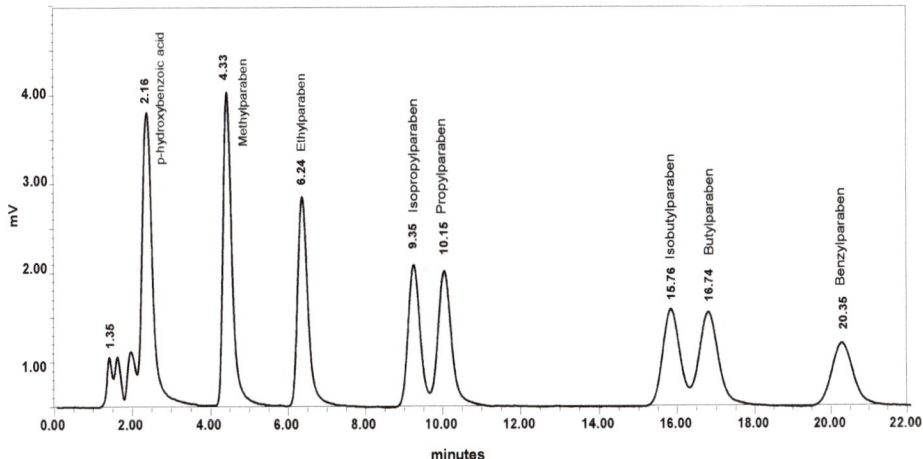

Figure 2. HPLC-UV chromatogram of a mixed working standard solution of the seven parabens at 100 ng mL^{-1} and *p*-hydroxybenzoic acid at 200 ng mL^{-1}. Chromatographic conditions: Spherisorb® ODS1 column; mobile phase: aqueous ammonium formate solution at 49 mM/acetonitrile (66:34, *v/v*) with 0.1% formic acid, column temperature 25 °C, flow rate 0.25 mL min^{-1} at 257 nm.

2.1.2. Optimization of the FPSE Procedure

Human plasma contains various ingredients, mainly proteins such as albumin, globulin, and fibrinogen, electrolytes, hormones, vitamins, lipids, and other substances. Due to this complex composition, an appropriate sample preparation procedure for removing matrix interferences is crucial prior to the chromatographic analysis. In addition, the optimization of a sample preparation procedure for the analysis of the selected parabens in human plasma can be challenging due to their varied polarity, logP values range from 2.08 for MPB to 3.28 for BzPB.

FPSE is a new, innovative, and promising technique for sample preparation. This technique uses a natural or synthetic fabric membrane as a substrate that is chemically coated in the form of a very thin but spongy coating of sol–gel organic–inorganic hybrid sorbent [47]. In this study, a sol–gel Carbowax 20M coated FPSE membrane was used for sample preparation of the human plasma samples [38,39]. This membrane consists of a biocompatible organic poly(ethylene glycol) polymer, an inorganic precursor, and a hydrophilic natural polymer cellulose fabric that synergistically complement each other to determine the overall selectivity of the extraction device. The FPSE membrane has been previously used for the analysis of parabens in various biofluids [35], including human plasma. However, in this study, the procedure was modified and optimized for maximum recovery of the analytes using a small volume of human plasma samples, intending to achieve increased sensitivity for the analysis of samples collected from women subjected to malignant and benign plastic, reconstructive, and aesthetic surgery of breasts as well as evaluate the presence of parabens. In this regard, several parameters were thoroughly investigated including the type and the volume of the extraction solvent, the extraction time, the desorption solvent, the desorption time, and the reconstitution solvent mixture. To achieve maximum extraction efficiency, the selected parameters (factors) were studied and evaluated at each stage of the technique, each time changing one factor and keeping the rest stable. In all the experiments, a small amount of human plasma samples (50 µL) was used spiked with the targeted analytes at a concentration of 250 ng mL^{-1}, and the sol–gel Carbowax 20M FPSE membrane was cut into a size of 2.25 cm^2.

The type and the volume of the extraction solvent are critical parameters in any FPSE procedure, and they were optimized to deliver the maximum percentage recovery for each analyte. As it is shown in Figure 3a, when an aliquot of 0.35 mL of water was used to extract 50 µL of a human plasma sample, the percentage recovery of the analytes ranged from 14.9% for MPB to 59.1% for iPPB. The seven parabens under study are in a neutral state under acidic conditions since their pKa values are around 8.9 (Table 1). Thus, to increase the interaction of the targeted analytes with the neutral extraction sorbent of the sol–gel Carbowax 20M-coated FPSE membrane, the extraction solvent should be acidified. Consequently, the aqueous extraction solvent was acidified by the addition of formic acid. It was found that the percentage recovery for MPB, which is the less lipophilic compound, increased radically and reached 47.5% after acidification of the extraction solution with 0.2 mL of a 0.1% formic acid aqueous solution.

To further improve the percentage recovery of the targeted analytes, various experiments have been performed where the total extraction volume varied from 0.4 to 6 mL, with varied content of 0.1% formic acid aqueous solution. As it is shown in Figure 3b, the maximum percentage recoveries for all the analytes were attained with an extraction volume of 0.8 mL acidified with 200 µL of 0.1% formic acid aqueous solution.

During the extraction and desorption (back-extraction) procedure, the screw-capped glass vials were placed in a vertical rotating mixer with a reciprocal rotation speed set at 12 rpm and reciprocal rotation tilt angle range set at 20°. Three different time periods were tested for extraction, and as it is illustrated in Figure 4a, a 20-min extraction time gave the highest percentage recoveries for all the analytes. As it is illustrated in Figure 4b, the optimum percentage recovery for the targeted analytes is attained using 0.8 mL methanol as desorption (back-extraction) solvent. Three different time intervals have been tested for desorption—10, 20, and 30 min—and it was found that within 20 min, optimum percentage recovery is attained. After desorption, the methanolic eluent was evaporated to dryness under a gentle stream of nitrogen at 30 °C.

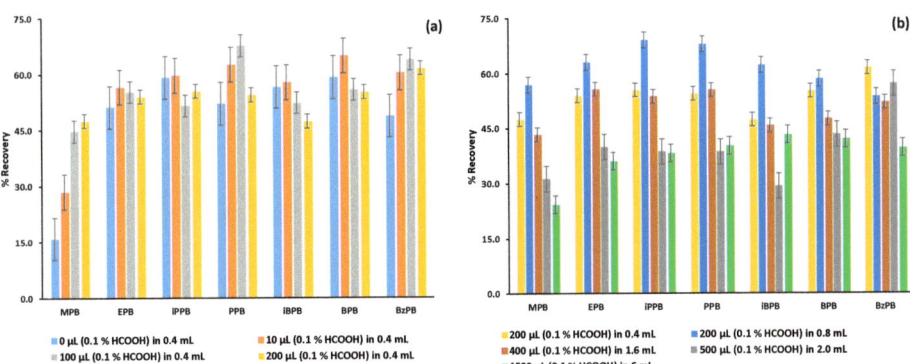

Figure 3. Influence of (**a**) the type of extraction solvent and (**b**) the extraction volume on the percentage recovery of the seven parabens; MPB for methylparaben, EPB for ethylparaben, iPPB for isopropylparaben, PPB for propylparaben, BPB for butylparaben, iBPB for isobutylparaben, and BzPB for benzylparaben.

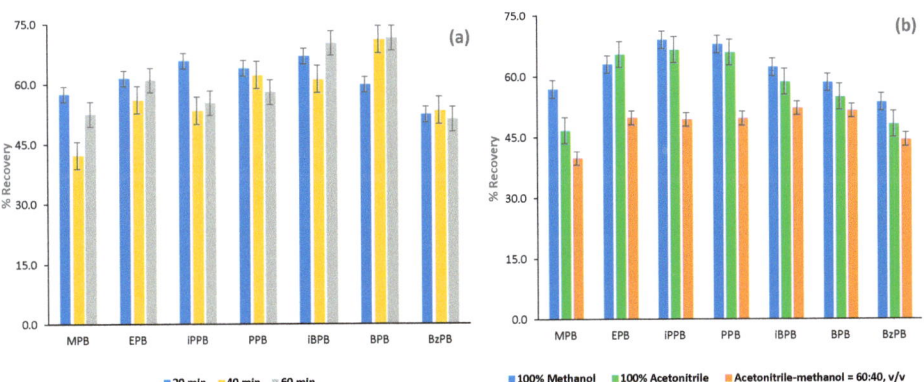

Figure 4. Influence of (**a**) the extraction time and (**b**) the type of the desorption solvent on the percentage recovery of the seven parabens; MPB for methylparaben, EPB for ethylparaben, iPPB for isopropylparaben, PPB for propylparaben, BPB for butylparaben, iBPB for isobutylparaben, and BzPB for benzylparaben.

In the early steps of method development, we have noticed that the reconstitution solvent affects the chromatographic response (peak area) of the analytes (Figure 5a). To this regard, various solvent mixtures have been tested for reconstitution of the samples after evaporation of the methanolic back-extraction solvent, including water, acetonitrile–water mixture 60:40, v/v and acetonitrile with various concentrations of ammonium formate aqueous solution 60:40, v/v. As it is clearly shown in Figure 5b, the optimum solvent mixture for reconstitution is acetonitrile/5mM ammonium formate aqueous solution 40/60, v/v as it gives the highest percentage recovery for the analytes with the lowest percentage recovery, namely MPB and BzPB.

The residues were reconstituted with 150 µL of the reconstitution solvent and then filtered through a syringe filter prior to HPLC-UV analysis. Various syringe filters such as nylon membrane, hydrophobic polytetrafluorethylene (PTFE), and hydrophilic PTFE have been tested for the final filtration step prior to the HPLC-UV analysis. A 13 mm hydrophilic PTFE membrane syringe filter with 0.22 µm pore size gave a percentage recovery greater than 98.0% for all the analytes, and therefore, it was selected as the optimum for the analysis of the seven parabens under study.

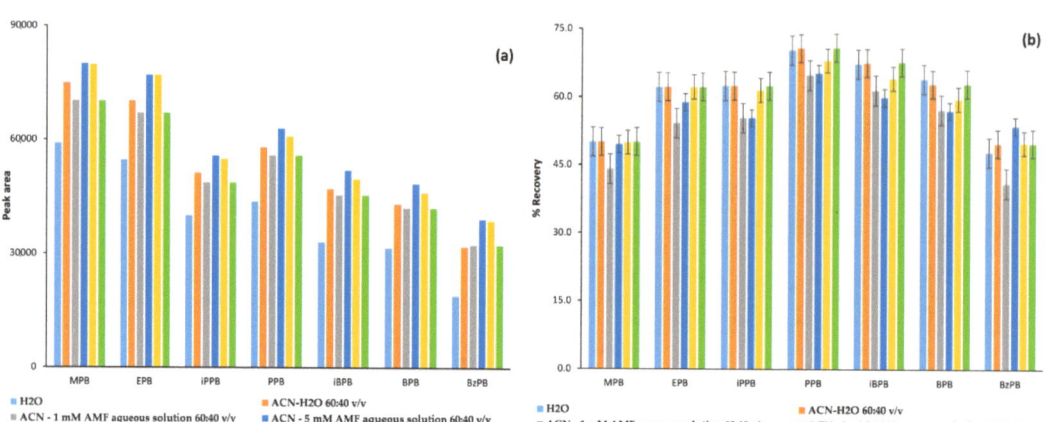

Figure 5. Influence of the reconstitution solvent on (**a**) the chromatographic response and (**b**) the percentage recovery of the seven parabens; MPB for methylparaben, EPB for ethylparaben, iPPB for isopropylparaben, PPB for propylparaben, BPB for butylparaben, iBPB for isobutylparaben and BzPB for benzylparaben.

2.1.3. Mechanism of Extraction in FPSE Membrane

Classical sample preparation techniques often utilize highly viscous polymeric sorbents such as poly(dimethylsiloxane). The sorption properties of these viscous sorbents toward the target analyte(s) are described by (a) solubility (S) and the partition coefficient (K). These two parameters define the relative concentration of the analyte at equilibrium between the polymeric sorbent and the sample matrix [48]. Due to the high viscosity of the polymer, the mass transfer of analyte between the sorbent and the sample matrix is relatively slow. However, the extraction of analytes in the FPSE membrane is governed by adsorption. Sol–gel sorbents are inherently porous, possessing sponge-like architecture containing many functional moieties that are highly affinitive toward the target analytes. When the FPSE membrane is inserted into the aqueous solution, the analytes approach toward the FPSE membrane via diffusion and interact with the sorbent via different intermolecular interactions such as London dispersion, dipole–dipole interaction, and hydrogen bonding. The planer geometry of the FPSE membrane, the sponge-like porous architecture of the sol–gel sorbent, and the built-in pores of the fabric substrate allow rapid permeation of the aqueous sample through the FPSE membrane. The continuous passage of the same sample through the FPSE membrane ensures fast extraction equilibrium and exhaustive/near-exhaustive extraction in a relatively short period.

At the end of extraction, the FPSE membrane is introduced into a small volume of organic sorbent such as methanol. The solvent solvates both the FPSE membrane as well as the analytes. As a result, the interaction between the sorbent and the analytes are shattered, and the analytes are solvated with methanol. Due to the high porosity and sponge-like internal structure of sol–gel sorbent, a small volume of methanol can quantitatively scavenge the analytes from the FPSE membrane very fast. Both the extraction and desorption processes are presented in Figure 6.

2.1.4. Green Attributes of Fabric Phase Sorptive Extraction

Fabric phase sorptive extraction was invented as a new generation green sample preparation technology. Galuszka et al. [49] compiled 12 principles of green analytical chemistry. Surprisingly, FPSE meets 10 out of 12 principles. One major green attribute of FPSE is the substantial reduction of steps involved in the overall sample preparation workflow. As such, FPSE has not only simplified the sample preparation process but also significantly reduced organic solvent consumption, eliminated sample pretreatment and post-treatment steps, and supported field deployability. FPSE is the only sample preparation technique that allows the deployment of custom membrane size based on the volume of sample to be analyzed. For biological samples, FPSE allows whole blood analysis without converting it into plasma or serum. Sample preparation without any sample pretreatment ensures minimal analyte loss during the sample preparation and improves the quality and confidence in analytical data.

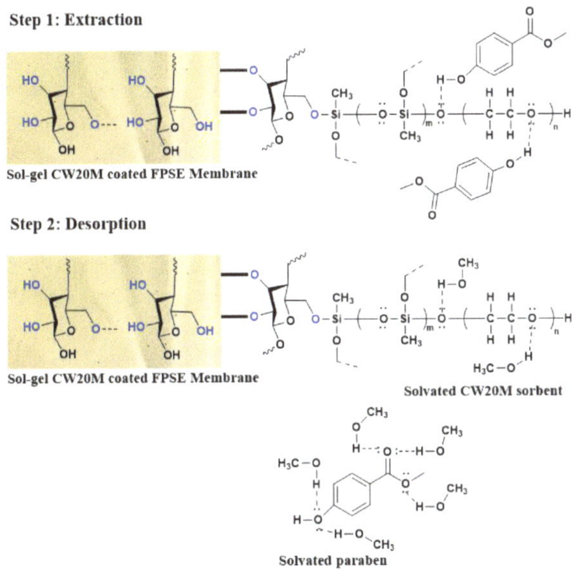

Step 1: Extraction

Step 2: Desorption

Figure 6. Schemes demonstrating extraction and desorption processes in fabric phase sorptive extraction.

2.2. Statistical Analysis of Method Validation Data

2.2.1. Selectivity and Specificity

All the chromatograms obtained from the analysis of five blank plasma samples (negative control samples) contained no co-eluting peaks greater than 5% of the peak area of the targeted analytes at 20 ng mL^{-1} showing the selectivity of the chromatographic method. In addition, the carry-over test met the predefined requirements, as no interfering peaks with responses greater than 5% of the peak areas at 20 ng mL^{-1} of each analyte were detected in blank human plasma samples analyzed after a high concentration calibration standard spiked in human plasma.

The developed FPSE-HPLC-UV method selectivity is further demonstrated in Figure 7, where a chromatogram of a blank plasma sample is superimposed with a chromatogram of a calibration plasma sample at 200 ng mL^{-1} for each analyte. MPB, EPB, iPPB, PPB, iBPB, BPB, and BzPB were eluted at 4.46, 6.39, 9.41, 10.33, 16.35, 17.36, and 20.88 min, respectively.

2.2.2. Linearity, Precision, and Accuracy

A weighted linear regression analysis with a weighting factor of $1/y^2$ was adopted due to data heteroscedascity and because of the better results regarding other weighting factors ($1/x$, $1/x^2$) and unweighted linear regression, which was also tested. Data are presented in Table 2 and indicate that linear relationships were attained between the responses of the targeted analytes with regard to the corresponding concentrations. Back-interpolated concentrations in the calibration curves were less than 15.8% of the nominal concentration at lower limit of quantitation (LLOQ) levels, which is in agreement with international guidelines [50].

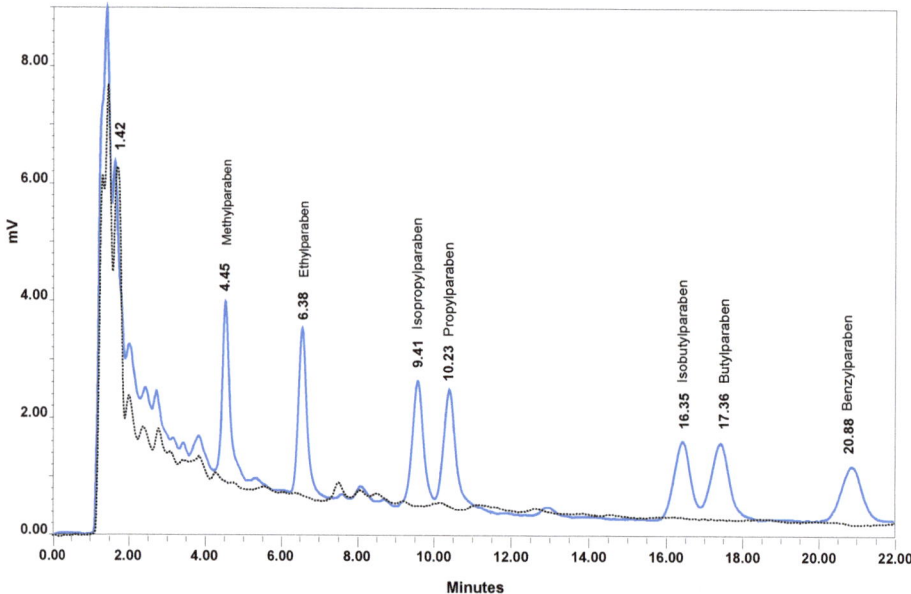

Figure 7. HPLC-UV chromatogram of a blank human plasma sample (black dotted line) overlaid with a chromatogram of a calibration plasma sample spiked with the targeted analytes at 200 ng mL^{-1} (blue line).

Table 2. Linearity data for the quantitation of the seven parabens in human plasma samples as assessed by the fabric phase sorptive extraction reversed-phase liquid chromatography method coupled with UV detection (FPSE-HPLC-UV) method.

Compound [1]	Matrix	Regression Equations [2]	r [3]	Standard Deviation		S$_r$ [4]
				Slope	Intercept	
MPB	Water	$S_{MPB} = 113.9 \times C_{MPB} - 795$	0.998	2.6	92	0.05
	Human plasma	$S_{MPB} = 61.2 \times C_{MPB} - 598$	0.996	2.4	78	0.08
EPB	Water	$S_{EPB} = 122.1 \times C_{EPB} - 805$	0.9994	1.8	65	0.04
	Human plasma	$S_{EPB} = 70.4 \times C_{EPB} - 477$	0.998	2.2	70	0.06
iPPB	Water	$S_{iPPB} = 96.4 \times C_{iPPB} - 774$	0.991	4.1	145	0.08
	Human plasma	$S_{iPPB} = 63.1 \times C_{iPPB} - 479$	0.993	3.4	121	0.11
PPB	Water	$S_{PPB} = 97.5 \times C_{PPB} - 669$	0.995	4.4	164	0.08
	Human plasma	$S_{PPB} = 59.1 \times C_{PPB} - 429$	0.993	3.1	116	0.10
iBPB	Water	$S_{iBPB} = 98.1 \times C_{iBPB} - 789$	0.998	2.8	99	0.08
	Human plasma	$S_{iBPB} = 61.0 \times C_{iBPB} - 557$	0.995	2.6	86	0.09
BPB	Water	$S_{BPB} = 97.5 \times C_{BPB} - 907$	0.9991	2.2	72	0.05
	Human plasma	$S_{BPB} = 60.7 \times C_{BPB} - 566$	0.995	2.5	82	0.09
BzPB	Water	$S_{BzPB} = 92.5 \times C_{BzPB} - 682$	0.998	2.4	87	0.05
	Human plasma	$S_{BzPB} = 46.7 \times C_{BzPB} - 318$	0.998	1.2	45	0.05

[1] MPB for methylparaben, EPB for ethylparaben, iPPB for isopropylparaben, PPB for propylparaben, BPB for butylparaben, iBPB for isobutylparaben, and BzPB for benzylparaben; [2] Peak area signal of each paraben, S, versus the corresponding concentration, C, and over the concentration range 20 to 500 ng mL^{-1}; [3] Correlation coefficient; [4] Standard error of the estimate.

A series of dilute solutions of known concentration spiked in blank human plasma have been analyzed and limits of detection and quantitation (LOD and LOQ) for the targeted analytes were estimated based on signal-to-noise ratios of 3:1 and 10:1. The limits of detection, LOD, and the limits of quantitation, LOQ, were found to be at the level of 7 and 20 ng mL^{-1} for each analyte, respectively.

One-way ANOVA was used for precision and accuracy evaluation, and the results are shown in Table 3. The precision and accuracy tests met the predefined requirements since the repeatability (intraday percentage CVs) ranged from 1.33 to 9.05% and the total accuracy ranged from 96.95 to 105.45%.

Table 3. Accuracy and precision of the seven parabens in human plasma samples at three concentration levels as assessed by the FPSE-HPLC-UV method (*n* = three runs; four replicates per run).

Compound [1]	Concentration (ng mL^{-1})		
Added Concentration	20	100	500
MPB			
Overall mean	20.88 ± 0.45	98.4 ± 3.5	494.2 ± 9.6
Intraday CV(%) [2]	1.93	3.60	1.95
Total precision CV (%) [2]	2.29	3.54	1.93
Total accuracy Er% [3]	104.4	98.4	98.8
EPB			
Overall mean	20.94 ± 0.82	100.5 ± 4.9	504.6 ± 9.1
Intraday CV(%) [2]	4.30	4.33	1.98
Total precision CV (%) [2]	3.76	5.15	1.73
Total accuracy Er% [3]	104.3	100.5	100.9
iPPB			
Overall mean	20.84 ± 0.46	100.2 ± 2.8	499.2 ± 8.9
Intraday CV(%) [2]	6.54	3.05	1.60
Total precision CV (%) [2]	5.90	2.66	1.86
Total accuracy Er%[3]	104.2	100.2	98.8
PPB			
Overall mean	20.4 ± 1.7	101.5 ± 3.2	500.2 ± 7.9
Intraday CV(%) [2]	8.61	3.34	1.56
Total precision CV (%) [2]	8.29	2.97	1.61
Total accuracy Er%[3]	102.0	101.5	100.1
iBPB			
Overall mean	21.1 ± 1.8	98.2 ± 3.8	501.5 ± 9.6
Intraday CV(%) [2]	9.05	4.17	2.83
Total precision CV (%) [2]	8.73	3.76	2.58
Total accuracy Er% [3]	105.1	98.2	100.3
BBP			
Overall mean	19.39 ± 0.55	101.2 ± 1.6	498.8 ± 6.3
Intraday CV(%) [2]	2.71	1.67	1.27
Total precision CV (%) [2]	2.87	1.52	1.26
Total accuracy Er% [3]	96.9	101.2	99.8
BzBP			
Overall mean	21.09 ± 0.55	103.9 ± 1.6	502 ± 11
Intraday CV(%) [2]	2.82	1.56	2.50
Total precision CV (%) [2]	2.53	1.50	2.33
Total accuracy Er% [3]	105.4	103.9	100.4

[1] MPB for methylparaben, EPB for ethylparaben, iPPB for isopropylparaben, PPB for propylparaben, BPB for butylparaben, iBPB for isobutylparaben, and BzPB for benzylparaben; [2] Coefficient of variation; intra- and inter-assay CVs were calculated by ANOVA; [3] Relative recovery percentage.

2.2.3. Recovery

The percentage relative recovery of the method for the seven parabens under study was also calculated as the percentage of the ratio of the slope of the regression equation of spiked human plasma samples to the slope of the regression equation of calibration samples spiked in water at equivalent concentrations (Table 2). All the samples have been analyzed in acetonitrile: 5 mM ammonium formate aqueous solution 40/60, v/v and the calibration spiked water samples have not been processed by the FPSE procedure (Table 2). Based on these data, percentage relative recoveries of 53.7, 57.7, 65.4, 60.6, 62.2, 62.1, and 50.5% were attained for MPB, EPB, iPPB, PPB, iBPB, BPB, and BzPB, respectively.

The percentage absolute recovery was determined by the percentage of the ratio of the peak area measured from human plasma samples spiked at 200 ng mL^{-1} for each analyte to the peak area of blank human plasma samples spiked after the FPSE procedure with analytes at equivalent concentrations. Based on this test, the percentage of absolute recoveries of 50.1, 60.2, 55.4, 65.8, 61.2, 56.9, and 53.3% were attained for MPB, EPB, iPPB, PPB, iBPB, BPB, and BzPB, respectively.

2.2.4. Stability

Human plasma samples spiked with the analytes at 200 ng mL^{-1} and stored at ambient temperature for six hours remained constant. Percentage recoveries of the analytes ranged from 98.2 to 102.5%. Freeze and thaw stability of the targeted analytes was assessed by four consecutive freeze and thaw cycles applied to human plasma samples spiked with the analytes at 200 ng mL^{-1}. The samples were frozen for 7 days at −18 °C and thawed at room temperature (one cycle); the procedure was repeated for three consecutive cycles. To calculate the stability, the data of the stored samples were compared to the data of freshly prepared human plasma samples spiked with the targeted analytes at 200 ng mL^{-1}. Percentage recoveries of the analytes ranged from 97.5 to 102.6%, indicating that the analytes are stable after four complete freeze and thaw cycles.

2.3. Application to Real Human Plasma Samples

A total of 20 human plasma samples were analyzed, and the results are presented in Tables 4 and 5. Half of the samples (n = 10 in the age range between 34 and 83) were collected from women subjected to malignant plastic surgery of breasts, and the rest (n = 10 ranging in age between 33 and 59) were collected from healthy women subjected to non-malignant benign reconstructive and aesthetic surgery of breasts.

Table 4. Results on the concentration levels (in ng mL^{-1}) of the seven parabens in human plasma samples collected from 20 women in Croatia.

Woman No.	Surgery [1]	Age	BMI [2]	Compound [3,4]						
				MPB	EPB	iPPB	PPB	iBPB	BPB	BzPB
1	BC	43	25.0	110	40	50	<LOD	<LOD	<LOD	<LOD
2	BC	59	34.3	40	<LOD	30	<LOD	40	<LOD	<LOD
3	BC	34	28	150	<LOD	detected	<LOD	80	<LOD	70
4	BC	56	25.5	60	<LOD	40	<LOD	<LOD	<LOD	<LOD
5	BC	76	26	170	<LOD	detected	<LOD	<LOD	<LOD	<LOD
6	BC	45	20.9	30	detected	<LOD	detected	<LOD	<LOD	<LOD
7	BC	45	20.4	20	30	<LOD	30	30	<LOD	<LOD
8	BC	83	27.5	140	<LOD	<LOD	10	300	<LOD	90
9	BC	65	23	20	<LOD	<LOD	detected	<LOD	<LOD	<LOD
10	BC	63	26.2	30	detected	<LOD	<LOD	<LOD	<LOD	<LOD
11	AE	46	20.3	150	80	30	<LOD	50	<LOD	<LOD
12	AE	51	28.3	80	detected	70	<LOD	<LOD	<LOD	50
13	AE	48	18.4	detected	60	<LOD	<LOD	<LOD	<LOD	<LOD
14	AE	33	25.1	detected	20	<LOD	<LOD	<LOD	<LOD	<LOD
15	AE	47	27.6	detected	70	<LOD	<LOD	<LOD	<LOD	<LOD
16	AE	38	25.9	30	detected	<LOD	<LOD	<LOD	<LOD	<LOD
17	AE	59	24.9	60	40	<LOD	<LOD	<LOD	<LOD	<LOD
18	AE	55	35.4	detected	30	<LOD	<LOD	<LOD	<LOD	<LOD
19	AE	48	26.6	20	detected	<LOD	<LOD	<LOD	<LOD	<LOD
20	AE	49	22.6	22	50	<LOD	<LOD	<LOD	<LOD	<LOD

[1] Type of surgery: BC for breast cancer; AE for aesthetic reconstructive surgery; [2] BMI for body mass index; [3] MPB for methylparaben, EPB for ethylparaben, iPPB for isopropylparaben, PPB for propylparaben, BPB for butylparaben, iBPB for isobutylparaben, and BzPB for benzylparaben; [4] Samples where parabens were detected but not quantified are denoted as "detected", while the samples where parabens levels were below the limit of detection are denoted as "<LOD".

Table 5. Results of the FPSE-HPLC-UV analysis of the seven parabens in human plasma samples collected from 20 women in Croatia.

Compound [1]	MPB	EPB	iPPB	PPB	iBPB	BPB	BzPB
Percentage of the human plasma samples in which quantified and detected	100	60	30	30	25	Not detected	15
Percentage of the human plasma samples in which quantified	80	45	25	10	25	Not detected	15
Percentage of the human plasma samples from healthy women in which quantified	60	70	20	0	10	Not detected	10
Percentage of the human plasma samples from cancerous cases in which quantified	100	20	30	20	40	Not detected	20
Mean plasma concentration in heathy women (ng mL^{-1})	60.3	50.0	50.0	-	50.0	Not detected	50.0
Mean plasma concentration in cancerous cases (ng mL^{-1})	77.0	35.0	40.0	20.0	112.5	Not detected	80.0
Mean plasma concentration in all the samples (ng mL^{-1})	70.8	46.7	44.0	20.0	100.0	Not detected	70.0

[1] MPB for methylparaben, EPB for ethylparaben, iPPB for isopropylparaben, PPB for propylparaben, BPB for butylparaben, iBPB for isobutylparaben, and BzPB for benzylparaben.

The results indicate that MPB was quantified in 100% of the human plasma samples from cancerous cases with average plasma concentration at 77.0 ng mL^{-1}. The highest quantitation rates in human plasma samples from cancerous cases were found for MPB (100%) and iBPB (40%) with average plasma concentrations at 77.0 and 112.5 ng mL^{-1}, respectively. In healthy women, the highest quantitation rates were observed for both MPB (60%) and EPB (70%) with average plasma concentration at 60.3 and 50.0 ng mL^{-1}, respectively. PPB was quantified only in cancerous cases at a rate of 20% with mean plasma concentration 20 ng mL^{-1}, while iPPB was quantified both in healthy and cancerous cases with mean plasma concentration in all the samples at 44 ng mL^{-1}. BPB was not quantified in any of the analyzed samples, while iBPB was quantified at a higher rate in cancerous cases (40%) than in healthy women (10%) with average plasma concentrations at 112.5 and 50 ng mL^{-1}, respectively. The more lipophilic BzPB was quantified in both cancerous and healthy cases with mean plasma concentration in all the samples at 70 ng mL^{-1}. As this was the pilot study that included a very small population size, statistical analysis was not performed, but the results are still indicative. As can be seen from Table 4, the mean age of woman in the group subjected to breast cancer surgery was higher by 10 years than for the group subjected to reconstructive and aesthetic surgery, but both groups were characterized by similar body mass index (BMI). In the human plasma samples obtained from the healthy women subjected to aesthetic reconstructive surgery, only MBP, EPB, and iPPB were detected, while iBPB and BzPBP were found in only single cases. On the other hand, PPB, iBPB, and BzPB were found in many more samples obtained from woman subjected to the breast cancer surgery. To undoubtedly determine the main factor for such differences, a much larger population group should be studied. Indeed, we will use the method presented here in future epidemiologic studies by evaluating the role of multiple factors on the accumulation of parabens in humans and the possible consequences of such accumulation. The high paraben concentrations detected in this study agree with the previous maximum concentration levels for MPB (142.9 ng mL^{-1}), EPB (45.9 ng mL^{-1}), and PPB (43.9 ng mL^{-1}) reported by Sandanger et al. [25] in plasma samples of postmenopausal women. However, the latter study did not consider the analysis of the isomers, iPPB and iBPB. Furthermore, some studies have shown that conjugated parabens were stable in human serum over 30 days when stored at 37 °C [51], and thus, the contribution of conjugate hydrolysis is considered negligible to the values reported in the current study. Therefore, we agree with Sandager et al. [25] that the high concentration of intact parabens identified in our study is not caused by the hydrolysis of conjugated parabens.

2.4. Comparison with Other Analytical Methods

The proposed FPSE-HPLC-UV method has been compared with other methods dedicated to the analysis of parabens in human plasma as reported in the literature. The results of this literature survey are presented in Table 6. Among the reported methods, only the current method and the FPSE-LC-DAD method [35] allows for the simultaneous quantitation of all seven parabens, including the isomers of PPB and BPB in human plasma samples. In this work, we have optimized both the FPSE protocol to reduce the analysis time and increase the percentage recovery and the chromatographic procedure to achieve adequate separation of all seven parabens within 22 min at a flow rate of 0.25 mL min^{-1}. The method allows for the quantitation of the seven parabens at adequately low LOQ and LOD values with a linearity range that allows the quantitation of parabens in clinical samples using a small sample volume of 50 μL.

Table 6. Comparison of the proposed method with methods published in the literature for the quantitation of parabens in human plasma samples.

Matrix	Analytes	Analytical Method; Column; Flow Rate	Run Time	Sample Preparation/ Extraction Time	Sample Volume	%Recovery	Repeatability (%CV)	Linearity Range	LOQ; LOD	Application to Real Samples	Reference
Human plasma	MPB, EPB, iPPB, PPB, iBPB, BPB, BzPB	RPHPLC-UV; Spherisorb ODS1 (150 × 2.0 mm, 3 μm); 0.25 mL/min isocratic elution	22 min	FPSE/ 40 min	50 μL	50.1–65.8%	1.3 to 9.0% (7 parabens)	20–500 ng/mL	LOQ: 20 ng/mL, LOD: 7 ng/mL	20 samples from healthy and cancerous patients (women)	Current method
Human plasma, whole blood, human urine	MPB, EPB, iPPB, PPB, iBPB, BPB, BzPB	RPHPLC-DAD; Spherisorb C18 (150 × 4.6 mm, 5 μm); 1.0 mL/min Isocratic elution	25 min	FPSE/ 60 min	450 μL	—	1.2 to 10.1% (7 parabens)	0.1–10 μg/mL	LOQ: 100 ng/mL, LOD: 30 ng/mL	6 samples	[38]
Human plasma	MPB, EPB, PPB, BPB, BzPB Bisphenols Estrogens	LC-MS/MS; Kinetex C18 (150 × 3.0 mm, 1.7 μm) 0.4 mL/min Gradient elution	11 min	LLE derivatization with dansyl chloride	500 μL	103.6–112.7%	1.3 to 6% (5 parabens)	MPB: 0.25–32 ng/mL EPB, PPB, BPB, BzPB: 0.094–12 ng/mL	LOQ: 0.134 to 0.202 ng/mL	58 samples from men of reproductive age; 27 maternal and cord plasma samples	[21,31]
Human urine, serum, seminal plasma	MPB, EPB, PPB, BPB, BzPB	LC–MS/MS; Synergi™ Fusion-RP 80 Å (75 × 2.0 mm; 4 μm); 0.3 mL/min Gradient elution	17 min	Automated SPE/ time not specified	500 μL	98.3–101.5%	2.8 to 29.2% (5 parabens)	0.5–500 ng/mL	LODs < 0.41 ng/mL	60 samples from young Danish men	[15]
Human plasma	MPB, EPB, PPB, BPB, BzPB	LC-TOF/MS; Waters®Acquity BEH Phenyl (100 mm × 2.1 mm, 1.7 μm) 0.45 mL/min Gradient elution	10 min	SPE on OASIS HLB cartridges	500 μL	—	< 12% (5 parabens)	50 to 600 pg injected on column	LODs: MPB 7 ng/mL EPB 3 ng/mL PPB 2 ng/mL	332 samples from postmenopausal women	[28]

15

3. Materials and Methods

3.1. Chemical and Reagents

The following 4 parabens—MPB, EPB, PPB, and BPB—were purchased from Acros Organics part of Thermo Fisher Scientific (Geel, Belgium), the two isomers iPPB and iBPB were obtained from Alfa-Aesar part of Thermo Fisher Scientific (Ward Hill, MA, USA), BzPB was acquired from TCI America (Portland, OR, USA), and *p*-hydroxybenzoic acid was purchased from Sigma-Aldrich (Saint Louis, MO, USA). The HPLC grade solvents used in the current study were bought from E. Merck (Darmstadt, Germany). HPLC grade purified water was produced by using a Synergy® UV water purification system (Merck Millipore, Burlington, MA, USA). FPSE membranes were synthesized by Kabir and Furton based on a procedure described elsewhere [38,52]. Hydrophobic polytetrafluorethylene membrane syringe filters (PTFE phobic, 13mm, pore size 0.22 μm) were obtained from RephiLe Bioscience Ltd Europe, Novalab representative (Athens, Greece).

3.2. Human Plasma Samples

Human plasma samples were obtained from Dubrava's University Hospital in Zagreb, Croatia. Particularly, human plasma samples were collected from 10 women subjected to malignant and 10 woman subjected benign plastic, reconstructive, and aesthetic surgery of breasts to evaluate the presence of parabens. The group subjected breast cancer surgery was characterized by the mean age of 56.9 ± 15.5 (in the range of 34 to 83 years old) and the mean body mass index (BMI) of 25.7 ± 3.9. The group subjected to plastic, reconstructive, and aesthetic surgery of breasts has the mean age of 47.4 ± 7.5 (in the range of 33 to 59 years old) and mean BMI of 25.5 ± 4.7. From each study participant, whole blood was collected from decubital vein in BD vacutainer with K_2EDTA. Cells were removed by centrifugation for 15 min at 2000 × *g*, and the resulting plasma samples were immediately dispensed into 0.5 mL aliquots, stored, and transported at −20 °C, to avoid freeze–thaw cycles. Samples that were hemolyzed, icteric, or lipemic were excluded from the study. The survey was accepted and approved by the Ethical Committee of Dubrava's University Hospital and University of Zagreb (380-59-10106-14-4290/82), School of Medicine, Croatia. Informed consents were acquired from all participants before any other action. Blank human plasma samples were taken from National Red Cross General Hospital in Athens, Greece.

3.3. Instrumentation

The analytical instrument used was a Waters® HPLC system (Waters, Milford, MA, USA), consisting of a Waters® 717 plus auto sampler, a temperature oven, a Waters® 1515 isocratic pump, and a Waters® 486 UV detector operated at λ 257 nm, whereas Empower™ Chromatography Data System (Waters Chromatography Europe BV, Etten-Leur, The Netherlands) was used for data acquisition and analysis. The selected parabens were separated on a Spherisorb ODS1 C18 analytical column (150.0 × 2.0 mm i.d., particle size 3 μm) (Waters, Ireland). A mobile phase of 66% 49 mM ammonium formate water solution in acetonitrile containing 0.1% (*v/v*) formic acid was used at a flow rate of 0.25 mL min^{-1}. The mobile phase was always degassed under vacuum while filtering through a 0.45 μm nylon membrane filter prior to use. Chromatographic experiments were performed at ambient temperature with a chromatographic run time of less than 24 min; each sample was injected into a 10 μL loop. A vertical rotating mixer model RS 2O-VS, Heto Lab Equipment, Heto-Holten A/S (Allerød, Denmark) was used for gentle mixing of the samples during FPSE procedure with reciprocal rotation speed set at 10 rpm and reciprocal rotation tilt angle range set at 20° and a Techne Sample concentrator Dri-block DB3 model FDB03OD (Techne Duxford, Cambridge, UK) was used for solvent evaporation.

3.4. Stock and Working Standard Solutions

Stock standard solutions of the analytes at 500 μg mL^{-1} were prepared in methanol. Mixed working standard solutions of the targeted analytes were prepared at the concentration range 25 to 2500 ng mL^{-1} by further dilutions of the stock solutions in water. The stock standard solutions were found stable when stored at -20 °C for 4 months, while the working standard solutions stored at 4 °C were prepared freshly every month.

3.5. Calibration Standards and Quality Control Samples

Calibration standards were prepared in human plasma at 20, 40, 60, 100, 200, 400, and 500 ng mL^{-1} for each analyte. Quality control (QC) was prepared at 20, 100, and 500 ng mL^{-1} in human plasma using separate stock solutions.

3.6. Sample Preparation Procedure

After thawing at room temperature, the samples are vortex mixed to ensure homogeneity prior to the sample preparation procedure, which is performed by an optimized FPSE procedure. The sol–gel Carbowax 20M membrane is cut at a size of 2.25 cm^2 (1.5 × 1.5 cm) and immersed in 2 mL acetonitrile–methanol mixture 50:50, *v/v* for 5 min and then in 2 mL of deionized water for an additional 5 min. Consequently, the FPSE membrane is transferred into a 5 mL screw-capped glass vial, followed by the addition of 50 µL aliquot of each human plasma sample, 550 µL of deionized water, and 200 µL of 0.1% aqueous formic acid. Then, the vial is placed in a vertical rotating mixer with reciprocal rotation speed set at 12 rpm and a reciprocal rotation tilt angle range set at 20° for 20 min. Accordingly, the FPSE membrane is transferred into a clean 5 mL screw-capped glass vial containing 800 µL methanol and then placed in the vertical rotating mixer for 20 min for the elution of the analytes. The methanolic eluent is transferred into an Eppendorf tube, and the content is evaporated within 10 min under a gentle stream of nitrogen at 30 °C. The residue is reconstituted with 150 µL acetonitrile/5mM ammonium formate 40/60, *v/v* and then filtered using a 13 mm PTFE hydrophilic membrane syringe filter with a pore size of 0.22 µm prior to the chromatographic analysis.

3.7. Method Validation

The proposed FPSE-HPLC-UV method was validated for selectivity, specificity, linearity, limit of detection (LOD), lower limit of quantitation (LLOQ), repeatability (intraday precision), interday precision, overall accuracy, absolute recovery, matrix effect, and stability [50]. To evaluate the selectivity over any matrix interference, five blank human plasma samples (negative controls) obtained from different batches have been analyzed following the optimum conditions of the method. Matrix interference was investigated at the retention time window of each paraben set at 5% of the corresponding retention time. The absence of matrix interference is confirmed when the area of any interference is less than 5% of the area determined at the LLOQ level for each analyte. To evaluate the linearity, weighted least-squares linear regressions were used to construct the calibration graphs after the analysis human plasma calibration standards spiked at seven different concentration levels. The quantitation was performed measuring the peak area of each targeted analyte. The analyte response at the LLOQ level should be at least 5 times the signal of a blank sample. Precision and accuracy were established by analyzing four replicates of QC samples at three concentration levels and on three different days. Repeatability (intraday precision) and interday precision were evaluated by calculating the percentage coefficient of variations (% CVs), which based on the acceptance criteria should be less than 15%, and at the LOQ levels should be less than 20%. Overall accuracy was assessed by measuring the percentage relative recovery, and according to predefined criteria, the mean concentration should be within 15% of the nominal values for the QC samples, except for the LLOQ, which should be within 20% of the nominal value.

4. Conclusions

In the current study, the advantages of FPSE on the sample preparation of analytes with diverse lipophilicity and on the enhancement of the sensitivity of the HPLC-UV technique in bioanalysis were demonstrated through the development of an FPSE-HPLC-UV method for the quantitation of seven parabens in human plasma samples. The new and optimized FPSE protocol requires only 50 μL of biological sample for the extraction, allows for the quantitation of the seven parabens at an LOQ of 20 ng mL^{-1} and combined with a semi-micro reversed phase analytical column is in accordance with the philosophy of the Green Analytical Chemistry. The efficiency of the method has been proven with the analysis of the seven parabens in 20 real plasma samples obtained from healthy and cancerous cases. The results for the shorter alkyl chain parabens have shown that MPB was quantified in highest rates in both cancerous and non-cancerous cases, while EPB was present at the highest rates in healthy women, and PPB was present only in cancerous cases. The results for the longer alkyl chain parabens have shown that BPB was not quantified in any of the analyzed samples, while BzPB was quantified at 15% of all the samples. As for the isomers, iPPB was quantified in both cancerous and non-cancerous cases, and iBPB was quantified at the highest quantitation rates in cancerous cases. The high concentration levels detected agree with previous findings for some of the parabens and emphasize the need for further epidemiological research on the possible health effects of the use of these compounds.

Author Contributions: Conceptualization, I.P., V.S., I.V.V., K.G.F. and A.K.; methodology, I.P., A.P., E.Z., N.P. and Ž.R.; validation, I.P. and A.P.; formal analysis, A.P., E.Z. and N.P.; investigation, A.P., E.Z., Z.R and N.P.; resources, I.P., I.V.V., Ž.R., A.K. and K.G.F.; writing—original draft preparation, I.P. and A.P.; writing—review and editing, I.P., V.S., I.V.V., Ž.R. and A.K.; supervision I.P. All authors have read and agreed to the published version of the manuscript.

Funding: This research received no external funding.

Institutional Review Board Statement: The study was conducted according to the guidelines of the Declaration of Helsinki. It was approved by the Ethical Committee of Dubrava's University Hospital (protocol title: "Utjecaj parabena na razvoj rada dojke", date of approval: 11 September 2012.) and by the Ethical Committee of University of Zagreb, School of Medicine, Croatia (protocol code: 380-59-10106-14-4290/82), date of approval 14 September 2014.

Informed Consent Statement: The authors state that they have obtained an appropriate institutional review board for the human experimental investigations. In addition, informed consent has been obtained from the participants involved.

Data Availability Statement: The data presented in this study are available on request from the corresponding author.

Acknowledgments: The authors would like to thank the National Red Cross General Hospital in Athens, Greece for providing blank human plasma samples.

Conflicts of Interest: The authors declare no conflict of interest.

Sample Availability: Samples of the compounds are not available from the authors.

References

1. Liao, C.; Chen, L.; Kannan, K. Occurrence of parabens in foodstuffs from China and its implications for human dietary exposure. *Environ. Int.* **2013**, *57–58*, 68–74. [CrossRef] [PubMed]
2. Abbas, S.; Greige-Gerges, H.; Karam, N.; Piet, M.H.; Netter, P.; Magdalou, J. Metabolism of parabens (4-hydroxybenzoic acid esters) by hepatic esterases and UDP-glucuronosyltransferases in Man. *Drug Metabol. Pharmacokinet.* **2010**, *25*, 568–577. [CrossRef]
3. Soni, M.G.; Carabin, I.G.; Burdock, G.A. Safety assessment of esters of *p*-hydroxybenzoic acid (parabens). *Food Chem. Toxicol.* **2005**, *43*, 985–1015. [CrossRef] [PubMed]
4. Hagedornleweke, U.; Lippold, B.C. Absorption of sunscreens and other compounds through human skin in-vivo—Derivation of a method to predict maximum fluxes. *Pharm. Res.* **1995**, *12*, 1354–1360. [CrossRef]
5. Darbre, P.D.; Harvey, P.W. Parabens can enable hallmarks and characteristics of cancer in human breast epithelial cells: A review of the literature with reference to new exposure data and regulatory status. *J. Appl. Toxicol.* **2014**, *34*, 925–938. [CrossRef] [PubMed]
6. Elder, R.L. Final report on the safety assessment of methylparaben, ethylparaben, propylparaben, and butylparaben. *J. Am. Coll. Toxicol.* **1984**, *3*, 147–209.
7. Fransway, A.F.; Fransway, P.J.; Belsito, D.V.; Yiannias, J.A. Paraben Toxicology. *Dermatitis* **2019**, *30*, 32–45. [CrossRef] [PubMed]
8. Byford, J.R.; Shaw, L.E.; Drew, M.G.B.; Pope, G.S.; Sauer, M.J.; Darbre, P.D. Oestrogenic activity of parabens in MCF7 human breast cancer cells. *J. Steroid Biochem. Mol. Biol.* **2002**, *80*, 49–60. [CrossRef]

9. Darbre, P.D.; Harvey, P.W. Paraben esters: Review of recent studies of endocrine toxicity, absorption, esterase and human exposure, and discussion of potential human health risks. *J. Appl. Toxicol.* **2008**, *28*, 561–578. [CrossRef] [PubMed]
10. Boberg, J.; Taxvig, C.; Christiansen, S.; Hass, U. Possible endocrine disrupting effects of parabens and their metabolites. *Reprod. Toxicol.* **2010**, *30*, 301–312. [CrossRef]
11. Sun, L.; Yu, T.; Guo, J.; Zhang, Z.; Hu, Y.; Xiao, X.; Sun, Y.; Xiao, H.; Li, J.; Zhu, D.; et al. The estrogenicity of methylparaben and ethylparaben at doses close to the acceptable daily intake in immature Sprague-Dawley rats. *Sci. Rep.* **2016**, *6*, 25173. [CrossRef]
12. Darbre, P.D.; Byford, J.R.; Shaw, L.E.; Hall, S.; Coldham, N.G.; Pope, G.S.; Sauer, M.J. Oestrogenic activity of benzylparaben. *J. Appl. Toxicol.* **2003**, *23*, 43–51. [CrossRef]
13. Okubo, T.; Yokoyama, Y.; Kano, K.; Kano, I. ER-dependent estrogenic activity of parabens assessed by proliferation of human breast cancer MCF-7 cells and expression of ER alpha and PR. *Food Chem. Toxicol.* **2001**, *39*, 1225–1232. [CrossRef]
14. Janjua, N.R.; Mortensen, G.K.; Andersson, A.M.; Kongshoj, B.; Skakkebaek, N.E.; Wulf, H.C. Systemic uptake of diethyl phthalate, dibutyl phthalate, and butyl paraben following whole-body topical application and reproductive and thyroid hormone levels in humans. *Environ. Sci. Technol.* **2007**, *41*, 5564–5570. [CrossRef]
15. Frederiksen, H.; Jørgensen, N.; Andersson, A.M. Parabens in urine, serum and seminal plasma from healthy Danish men determined by liquid chromatography–tandem mass spectrometry (LC–MS/MS). *J. Expo. Sci. Environ. Epidemiol.* **2011**, *21*, 262–271. [CrossRef]
16. Barr, L.; Metaxas, G.; Harbach, C.A.J.; Savoy, L.A.; Darbre, P.D. Measurement of paraben concentrations in human breast tissue at serial locations across the breast from axilla to sternum. *J. Appl. Toxicol.* **2012**, *32*, 219–232. [CrossRef] [PubMed]
17. JECFA. Evaluation of certain food additives and contaminants. In *67th Report of the Joint FAO/WHO Expert Committee on Food Additives*; WHO Technical Report Series 940; Rome, Italy, 2006; pp. 1–104. Available online: https://apps.who.int/iris/bitstream/handle/10665/43592/WHO_TRS_940_eng.pdf?sequence=1&isAllowed=y (accessed on 17 January 2021).
18. SCCS/1514/13 Scientific Committee on Consumer Safety (SCCS). Opinion on parabens; Updated request for a scientific opinion on propyl-and butylparaben, COLIPA No 23P82. In *European Commission 2013, Report No SCCS/1514/13*; Luxembourg, 2013; pp. 1–50. Available online: https://ec.europa.eu/health/scientific_committees/consumer_safety/docs/sccs_o_132.pdf (accessed on 17 January 2021).
19. COMMISSION REGULATION (EU) No 358/2014 of 9 April 2014 Amending Annexes II and V to Regulation (EC) No 1223/2009 of the European Parliament and of the Council on Cosmetic Products; Annex II the Following Entries 1374 to 1378. Available online: https://eur-lex.europa.eu/legal-content/EN/TXT/HTML/?uri=CELEX:32014R0358&from=EN#d1e32-7-1 (accessed on 17 January 2021).
20. SCCS/1623/20 Scientific Committee on Consumer Safety (SCCS). Preliminary Opinion on Propylparaben. In *European Commission 2020, Report No. SCCS/1623/20*; Luxembourg, 2020; pp. 1–57. Available online: https://ec.europa.eu/health/sites/health/files/scientific_committees/consumer_safety/docs/sccs_o_243.pdf (accessed on 17 January 2021).
21. Kolatorova, L.; Vitku, J.; Hampl, R.; Adamcova, K.; Skodova, T.; Simkova, M.; Parizek, A.; Starka, P.; Duskova, M. Exposure to bisphenols and parabens during pregnancy and relations to steroid changes. *Environ. Res.* **2018**, *163*, 115–122. [CrossRef] [PubMed]
22. Dogan, S.; Tongur, T.; Erkaymaz, T.; Erdogan, G.; Unal, B.; Sik, B.; Simsek, T. Traces of intact paraben molecules in endometrial carcinoma. *Environ. Sci. Poll. Res.* **2019**, *26*, 31158–31165. [CrossRef]
23. Shirai, S.; Suzuki, Y.; Yoshinaga, J.; Shiraishi, H.; Mizumoto, Y. Urinary excretion of parabens in pregnant Japanese women. *Reprod. Toxicol.* **2013**, *35*, 96–101. [CrossRef] [PubMed]
24. Ünüvar, T.; Büyükgebiz, A. Fetal and Neonatal Endocrine Disruptors. *J. Clin. Res. Pediatr. Endocrinol.* **2012**, *4*, 51–60. [CrossRef] [PubMed]
25. Jiménez-Díaz, I.; Vela-Soria, F.; Rodríguez-Gómez, R.; Zafra-Gómez, A.; Ballesterosa, O.; Navalón, A. Analytical methods for the assessment of endocrine disrupting chemical exposure during human fetal and lactation stages: A review. *Anal. Chim. Acta* **2015**, *892*, 27–48. [CrossRef] [PubMed]
26. Raza, N.; Kim, K.H.; Abdullah, M.; Raza, W.; Brown, R.J.W. Recent developments in analytical quantitation approaches for parabens in human-associated samples. *TrAC Trends Anal. Chem.* **2018**, *98*, 161–173. [CrossRef]
27. Grecco, C.F.; Souza, I.D.; Queiroz, M.E.C. Recent development of chromatographic methods to determine parabens in breast milk samples: A review. *J. Chromatogr. B Analyt. Technol. Biomed. Life Sci.* **2018**, *1093–1094*, 82–90. [CrossRef] [PubMed]
28. Sandanger, T.M.; Huber, S.; Moe, M.K.; Braathen, T.; Leknes, H.; Lund, E. Plasma concentrations of parabens in postmenopausal women and self-reported use of personal care products: The NOWAC postgenome study. *J. Exp. Sci. Environ. Epidemiol.* **2011**, *21*, 595–600. [CrossRef] [PubMed]
29. Grzeskowiak, T.; Czarczynska-Goslinska, B.; Zgoła-Grzeskowiak, A. Current approaches in sample preparation for trace analysis of selected endocrine disrupting compounds: Focus on polychlorinated biphenyls, alkylphenols, and parabens. *TrAC Trends Anal. Chem.* **2016**, *75*, 209–226. [CrossRef]
30. Asimakopoulos, A.G.; Wang, L.; Thomaidis, N.S.; Kannan, K. A multi-class bioanalytical methodology for the determination of bisphenol A diglycidyl ethers, *p*-hydroxybenzoic acid esters, benzophenone-type ultraviolet filters, triclosan, and triclocarban in human urine by liquid chromatography–tandem mass spectrometry. *J. Chromatogr. A* **2014**, *1324*, 141–148.

31. Kolatorova Sosvorova, L.; Chlupacova, T.; Vitku, J.; Vlk, M.; Heracek, J.; Starka, L.; Saman, D.; Simkova, M.; Hampl, R. Determination of selected bisphenols, parabens and estrogens in human plasma using LC-MS/MS. *Talanta* **2017**, *174*, 21–28. [CrossRef]

32. Azzouz, A.; Rascón, A.J.; Ballesteros, E. Simultaneous determination of parabens, alkylphenols, phenylphenols, bisphenol A and triclosan in human urine, blood and breast milk by continuous solid-phase extraction and gas chromatography–mass spectrometry. *J. Pharm. Biomed. Anal.* **2016**, *119*, 16–26. [CrossRef]

33. Melo, L.P.; Queiroz, M.E.C. A molecularly imprinted polymer for microdisc solid phase extraction of parabens from human milk samples. *Anal. Methods* **2013**, *5*, 3538. [CrossRef]

34. Rodríguez-Gómez, R.; Zafra-Gómez, A.; Camino-Sánchez, F.J.; Ballesteros, O.; Navalón, A. Gas chromatography and ultra-high performance liquid chromatography tandem mass spectrometry methods for the determination of selected endocrine disrupting chemicals in human breast milk after stir-bar sorptive extraction. *J. Chromatogr. A* **2014**, *1349*, 69–79. [CrossRef]

35. Fotouhi, M.; Seidi, S.; Shanehsaz, M.; Naseri, M.T. Magnetically assisted matrix solid phase dispersion for extraction of parabens from M. breast milks. *J. Chromatogr. A* **2017**, *1504*, 17–26. [CrossRef] [PubMed]

36. Vela-Soria, F.; Iribarne-Durán, L.M.; Mustielesa, V.; Jiménez-Díaz, I.; Fernández, M.F.; Olea, N. QuEChERS and ultra-high performance liquid chromatography–tandem mass spectrometry method for the determination of parabens and ultraviolet filters in human milk samples. *J. Chromatogr. A* **2018**, *1546*, 1–9. [CrossRef] [PubMed]

37. Vela-Soria, F.; Gallardo-Torres, M.E.; Ballesteros, O.; Díaz, C.; Pérez, J.; Navalón, A.; Fernández, M.F.; Olea, N. Assessment of parabens and ultraviolet filters in human placenta tissue by ultrasound-assisted extraction and ultra-high performance liquid chromatography-tandem mass spectrometry. *J. Chromatogr. A* **2017**, *1487*, 153–161. [CrossRef]

38. Tartaglia, A.; Kabir, A.; Ulusoy, S.; Sperandio, E.; Piccolantonio, S.; Ulusoy, H.I.; Furton, K.G.; Locatelli, M. FPSE-HPLC-PDA analysis of seven paraben residues in human whole blood, plasma, and urine. *J. Chromatogr. B* **2019**, *1125*, 121707. [CrossRef]

39. Rigkos, G.; Alampanos, V.; Kabir, A.; Furton, K.G.; Roje, Ž.; Vrček, I.V.; Panderi, I.; Samanidou, V. An improved fabric-phase sorptive extraction protocol for the determination of seven parabens in human urine by HPLC-DAD. *Biomed. Chromatogr.* **2020**, e4974. [CrossRef]

40. Alampanos, V.; Kabir, A.; Furton, K.G.; Roje, Ž.; Vrček, I.V.; Samanidou, V. Fabric phase sorptive extraction combined with high-performance-liquid chromatography-photodiode array analysis for the determination of seven parabens in human breast tissues: Application to cancerous and non-cancerous samples. *J. Chromatogr. A* **2020**, *1630*, 461530. [CrossRef]

41. Kabir, A.; Furton, K.G. Fabric Phase Sorptive Extractors (FPSE). United States Patent Application Publication US 2014/0274660A1, 18 September 2014. Available online: https://patentimages.storage.googleapis.com/c2/36/06/2787b6030e4008/US20140274660A1.pdf (accessed on 17 January 2021).

42. Locatelli, M.; Tinari, N.; Grassadonia, A.; Tartaglia, A.; Macerola, D.; Piccolantonio, S.; Sperandio, E.; D'Ovidio, C.; Carradori, S.; Ulusoy, H.I.; et al. FPSE-HPLC-DAD method for the quantitation of anticancer drugs in human whole blood, plasma, and urine. *J. Chromatogr. B* **2018**, *1095*, 204–213. [CrossRef]

43. Locatelli, M.; Kabir, A.; Innosa, D.; Lopatriello, T.; Furton, K.G. A fabric phase sorptive extraction-high performance liquid chromatography-photo diode array detection method for the determination of twelve azole antimicrobial drug residues in human plasma and urine. *J. Chromatogr. B* **2017**, *1040*, 192–198. [CrossRef] [PubMed]

44. Lioupi, A.; Kabir, A.; Furton, K.G.; Samanidou, V. Fabric phase sorptive extraction for the isolation of five common antidepressants from human urine prior to HPLC-DAD analysis. *J. Chromatogr. B* **2019**, *1118–1119*, 171–179. [CrossRef] [PubMed]

45. Lakade, S.S.; Borrull, F.; Furton, K.F.; Kabir, A.; Marcé, R.M.; Fontanals, N. Dynamic fabric phase sorptive extraction for a group of pharmaceuticals and personal care products from environmental waters. *J. Chromatogr. A* **2016**, *1456*, 19–26. [CrossRef]

46. Valkova, N.; Lépine, F.; Valeanu, L.; Dupont, M.; Labrie, L.; Bisaillon, J.G.; Beaudet, R.; Shareck, F.; Richard, V. Hydrolysis of 4-Hydroxybenzoic Acid Esters (Parabens) and Their Aerobic Transformation into Phenol by the Resistant Enterobacter cloacae Strain EM. *Appl. Environ. Microbiol.* **2001**, *67*, 2404–2409. [CrossRef]

47. Kabir, A.; Mesa, R.; Jurmain, J.; Furton, K. Fabric Phase Sorptive Extraction Explained. *Separations* **2017**, *4*, 21. [CrossRef]

48. Seethapathy, S.; Gorecki, T. Applications of polydimethylsiloxane in analytical chemistry: A review. *Anal. Chim. Acta* **2012**, *750*, 48–62. [CrossRef] [PubMed]

49. Galuszka, A.; Migaszewski, Z.; Namiesnik, J. The 12 principles of green analytical chemistry and the SIGNIFICANCE mnemonic of green analytical practices. *TrAC Trends Anal. Chem.* **2013**, *50*, 78–84. [CrossRef]

50. European Medicines Agency. *Guideline on Bioanalytical Method Validation, Committee for Medicinal Products for Human Use (CHMP)*; EMEA/CHMP/EWP/192217/2009 Rev. 1; European Medical Agency: London, UK, 2015.

51. Ye, X.Y.; Wong, L.Y.; Jia, L.T.; Needham, L.L.; Calafat, A.M. Stability of the conjugated species of environmental phenols and parabens in human serum. *Environ. Int.* **2009**, *35*, 1160–1163. [CrossRef] [PubMed]

52. Tartaglia, A.; Locatelli, M.; Kabir, A.; Furton, K.W.; Macerola, D.; Sperandio, E.; Piccolantonio, S.; Ulusoy, H.I.; Maroni, F.; Bruni, P.; et al. Comparison between Exhaustive and Equilibrium Extraction Using Different SPE Sorbents and Sol-Gel Carbowax 20M Coated FPSE Media. *Molecules* **2019**, *24*, 382. [CrossRef]

Article

Fourier Transform Infrared (FTIR) Spectroscopic Analyses of Microbiological Samples and Biogenic Selenium Nanoparticles of Microbial Origin: Sample Preparation Effects

Alexander A. Kamnev *, Yulia A. Dyatlova, Odissey A. Kenzhegulov, Anastasiya A. Vladimirova, Polina V. Mamchenkova and Anna V. Tugarova

Laboratory of Biochemistry, Institute of Biochemistry and Physiology of Plants and Microorganisms, Russian Academy of Sciences, 410049 Saratov, Russia; jdyatlowa2013@yandex.ru (Y.A.D.); odissey94.sid@mail.ru (O.A.K.); vladimirova-nastyusha@bk.ru (A.A.V.); norgeadress@gmail.com (P.V.M.); tugarova_anna@mail.ru (A.V.T.)
* Correspondence: aakamnev@ibppm.ru or a.a.kamnev@mail.ru

Citation: Kamnev, A.A.; Dyatlova, Y.A.; Kenzhegulov, O.A.; Vladimirova, A.A.; Mamchenkova, P.V.; Tugarova, A.V. Fourier Transform Infrared (FTIR) Spectroscopic Analyses of Microbiological Samples and Biogenic Selenium Nanoparticles of Microbial Origin: Sample Preparation Effects. *Molecules* **2021**, 26, 1146. https://doi.org/10.3390/molecules 26041146

Academic Editors: Victoria Samanidou, Irene Panderi and Alberto Pettignano

Received: 13 January 2021
Accepted: 18 February 2021
Published: 21 February 2021

Abstract: To demonstrate the importance of sample preparation used in Fourier transform infrared (FTIR) spectroscopy of microbiological materials, bacterial biomass samples with and without grinding and after different drying periods (1.5–23 h at 45 °C), as well as biogenic selenium nanoparticles (SeNPs; without washing and after one to three washing steps) were comparatively studied by transmission FTIR spectroscopy. For preparing bacterial biomass samples, *Azospirillum brasilense* Sp7 and *A. baldaniorum* Sp245 (earlier known as *A. brasilense* Sp245) were used. The SeNPs were obtained using *A. brasilense* Sp7 incubated with selenite. Grinding of the biomass samples was shown to result in slight downshifting of the bands related to cellular poly-3-hydroxybutyrate (PHB) present in the samples in small amounts (under ~10%), reflecting its partial crystallisation. Drying for 23 h was shown to give more reproducible FTIR spectra of bacterial samples. SeNPs were shown to contain capping layers of proteins, polysaccharides and lipids. The as-prepared SeNPs contained significant amounts of carboxylated components in their bioorganic capping, which appeared to be weakly bound and were largely removed after washing. Spectroscopic characteristics and changes induced by various sample preparation steps are discussed with regard to optimising sample treatment procedures for FTIR spectroscopic analyses of microbiological specimens.

Keywords: sample preparation; FTIR spectroscopy; bacterial biomass; biogenic selenium nanoparticles; *Azospirillum brasilense*; *Azospirillum baldaniorum*

1. Introduction

The Fourier transform infrared (FTIR) spectroscopic technique is versatile and sensitive to the molecular composition and fine structural features, as well as intra- and intermolecular interactions, of functional groups in samples virtually in all aggregation states. This has made it indispensable for both theoretical and experimental structural and spectrochemical analytical studies of diverse materials ranging from small molecules (see, e.g., [1–6]) to more complicated materials, macromolecules and supramolecular structures [7–11], up to prokaryotic or eukaryotic cells and tissues [12–18]. Over recent decades, FTIR spectroscopy has been increasingly used in microbiological studies for the identification and classification of microorganisms, as well as for solving various bioanalytical problems related to microbiology [12,13,15,19–25]. Nevertheless, standardised sample preparation for biological objects, including microbial cells, as well as mathematical methods for analysing the resulting complicated spectra, are still under development. To date, a few topical articles have been published in which the preparation of microbiological samples for analysis by using FTIR spectroscopy and some specific features of the technique are discussed (see, e.g., [19,20,22–24]). Nevertheless, further development and optimisation of methodologies for preparing various microbiological samples for FTIR spectroscopic analysis are still of significance to ensure obtaining reliable spectroscopic data. The latter are indispensable for the most meaningful interpretation adequately reflecting the objects under study.

In this report, we consider some methodological approaches in sample preparation and their effects when using FTIR spectroscopy as applied to bacterial cultures (dried biomass) and biogenic selenium (Se^0) nanoparticles (SeNPs) of bacterial origin. In our work, two widely studied strains were used which belong to the genus *Azospirillum*, Gram-negative alphaproteobacteria, among which there are many ubiquitous rhizobacteria with phytostimulating capabilities and a number of other biotechnologically attractive traits (for reviews, see, e.g., [26–28], as well as some of our earlier experimental reports [21,24,25] and references cited therein). FTIR spectroscopy in its various variants is a useful technique providing a wealth of information on their ecology and physiological behaviour, particularly under stress conditions [21,24,25]. The strains under study in this work, *A. brasilense* Sp7 [29] and *A. baldaniorum* Sp245 (earlier known as *A. brasilense* Sp245 [30] and reclassified only recently [31]), have also been documented to be capable of reducing selenite ($Se^{IV}O_3^{2-}$) with the formation of SeNPs [32,33]. This trait, which is common for a number of microorganisms [34,35], is of importance for agrobiotechnology (e.g., bioremediation of seleniferous soils and aquifers) and nanobiotechnology (green synthesis of biogenic SeNPs and other Se-containing nanostructures) [35]. Thus, analysing such biogenic SeNPs using instrumental techniques, including FTIR spectroscopy, particularly with regard to the bioorganic capping layer of such nanostructures [34,36], is of primary importance.

2. Results and Discussion

2.1. Bacterial Biomass: Sample Treatment Effects in FTIR Spectroscopic Analysis

To date, large amounts of results and data obtained using FTIR spectroscopy have made it possible to form an extensive database for the analysis and interpretation of FTIR spectra of various microbiological objects. Nevertheless, as noted above, many aspects of sample preparation of such samples, which can commonly be structurally and compositionally complicated and non-uniform, are still not fully standardised. As has been mentioned earlier (see, e.g., [19,20,24,25] and references therein), FTIR spectra of microbial biomass can be obtained in various ways. Each way of sample preparation has its peculiarities which, if even slightly altered, may result in some changes (sometimes directly visible or resolvable using special approaches) in spectroscopic images, reflecting some structural changes in the sample. Therefore, it is of importance to have information on how various sample preparation steps can influence the resulting spectra and ultimately to standardise and develop a valid methodology for preparing bacterial samples for a reliable FTIR spectroscopic analysis. In this work, our attention was directed to some specific and important processing steps (grinding and drying) and to studying their effects on the measured FTIR spectra in the case of bacterial samples.

2.1.1. Effects of Grinding

Prior to measurement, the dried bacterial culture was pretreated in two variants: (1) the sample was thoroughly powdered (ground in a mortar), and the resulting powder was resuspended in Milli-Q water and processed as described in Section 3.3; (2) the grinding stage was excluded from sample preparation, so that the dry biomass was directly processed as described in Section 3.3. Traditionally, the grinding stage is an important part of sample preparation in the FTIR spectroscopy of materials (especially non-uniform or heterogeneous materials), since it allows for obtaining a more homogeneous aqueous or oil suspension. In the case of an aqueous suspension (as used by us previously [25] and in this work), when dried, it forms a uniform thin film on ZnSe glasses. Such films make it possible to obtain high-quality transmission FTIR spectra with a high signal-to-noise ratio and, therefore, to greatly facilitate further analysis of the data obtained. (Note that, while grinding, part of material may be lost, which has to be taken into account when the amounts of samples are limited.)

Figure 1 shows FTIR spectra of *A. baldaniorum* Sp245 and *A. brasilense* Sp7 biomass samples (dried for 23 h) with and without grinding. As can be seen, the spectra contain all the bands typical of bacterial FTIR spectra [19,20] (see Table 1 for band assignments of typical bands for *A. baldaniorum* Sp245), and all the spectra generally look very similar.

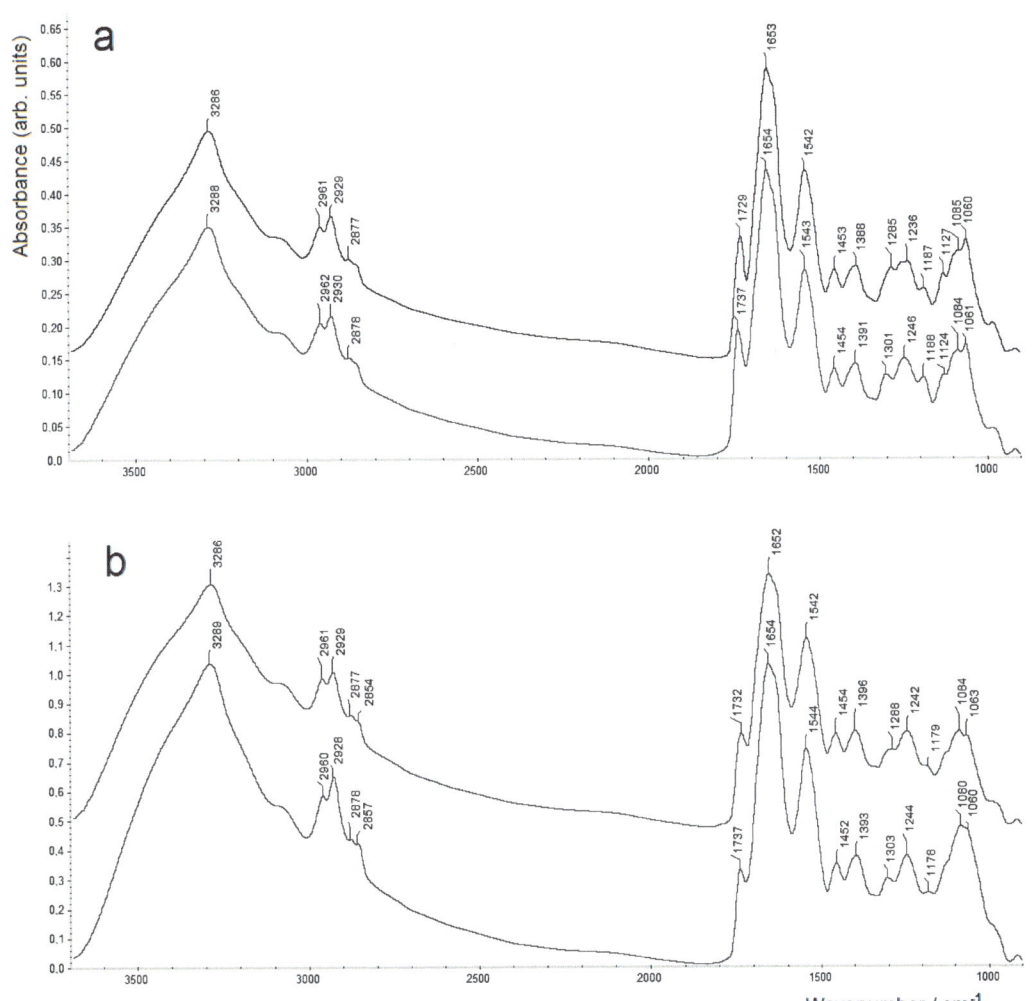

Figure 1. Transmission FTIR spectra of dry bacterial biomass of *Azospirillum baldaniorum* Sp245 (**a**) and *A. brasilense* Sp7 (**b**) measured with grinding (upper spectra) or without grinding (lower spectra). The upper spectra are vertically shifted from the baseline (zero absorbance) for clarity.

Table 1. Band maxima of typical vibration bands in FTIR spectra of dry biomass samples of *A. baldaniorum* Sp245 (see Figure 1a) and their assignments [1] [19–25].

Samples of *A. baldaniorum* Sp245		Assignment (Functional Groups)
Without Grinding	**With Grinding**	
3288	3286	O–H; N–H (amide A in proteins), ν
2962	2961	C–H in –CH3, ν_{as}
2930	2929	C–H in >CH2, ν_{as}
2878	2877	C–H in –CH3, ν_s
2854	2854	C–H in >CH2, ν_s
1737	1729	Ester C=O, ν (PHB; phospholipids)
1654	1653	Amide I (proteins)
1543	1542	Amide II (proteins)
1454	1453	–CH$_3$, δ (in proteins, lipids, polyesters, etc.)
1391	1388	COO$^-$, ν_s (in amino acid side chains and carboxylated polysaccharides) [2]
1301	1285	C–O–C/C–C–O, ν (in esters; PHB)
1246	1236	C–O–C (esters)/amide III/O–P=O, ν_{as}
1188 1124	1187 1127	C–O, C–C, C–OH, ν; C–O–H, C–O–C, δ (polysaccharides)
1084	1085	O–P=O, ν_s

[1] Notations: ν—stretching vibrations; ν_s—symmetric stretching vibrations; ν_{as}—antisymmetric stretching vibrations; δ—bending vibrations.
[2] The corresponding antisymmetric stretching vibrations (ν_{as} of COO$^-$, usually of somewhat higher intensities than ν_s) may vary in wavenumbers (observed commonly around ~1650–1580 cm^{-1}) and in microbial biomass are commonly masked by significantly more intensive amide I/II bands of cellular proteins.

In the FTIR spectra of the samples in Figure 1, some differences (exceeding the spectral resolution of 4 cm^{-1}) are observed only in the region around ~1730 cm^{-1}, as well as ~1300 cm^{-1}. For the samples subjected to grinding, the values of the maxima of these bands are noticeably lower. Thus, for *A. baldaniorum* Sp245, the bands observed at 1737 and 1301 cm^{-1} shifted to 1729 and 1285 cm^{-1}, respectively, after grinding (see Figure 1a). A similar shift is observed for *A. brasilense* Sp7 (the bands observed at 1737 and 1303 cm^{-1} shifted to 1732 and 1288 cm^{-1}, respectively; see Figure 1b).

It is common knowledge that the aforementioned bands correspond to the functional groups of the intracellular reserve biopolyesters of the polyhydroxyalkanoate (PHA) series [19,37], which in azospirilla are represented by the homopolymer poly-3-hydroxy-butyrate (PHB) (see [21,24,25] and references cited therein). In PHB, polyester chains are interconnected by weak C–H⋯O hydrogen bonds [37]. Changes in the intensity of inter- and intramolecular interactions formed by these hydrogen bonds between the ester carbonyl group (showing a band at ca. 1720–1750 cm^{-1} due to C=O stretching vibrations, which are sensitive to H-bonding) and the –CH$_3$ group in the polymer chains cause some variability in the degree of ordering (crystallinity), which is one of the most important properties of native PHB. As the degree of ordering decreases, the aforementioned bands in FTIR spectra shift to higher frequencies, and vice versa [37–39]. (In our case, a shift is also observed in the region of ~1240 cm^{-1} related in part to C–O–C vibrations of ester moieties.)

It has to be mentioned that the capability of PHA biosynthesis and accumulation as intracellular granules is of primary importance for bacterial survival and stress endurance [40–44]. Bacteria of the genus *Azospirillum* are known to be capable of accumulating relatively large amounts of PHB which, under appropriate conditions (e.g., lack of bound nitrogen, i.e., a high C:N ratio in the medium), may exceed 60–70% of dry cell weight (d.c.w.) [21,25,45]. Thus, as follows from the spectra (see Figure 1), the relative amounts of intracellular PHB in the samples studied in this work are low, around or below

~10% d.c.w. (cf., e.g., [25]), because of the presence of a minimal normal concentration [21] of bound nitrogen (as NH_4^+) in the culture medium (see Section 3.1). Nevertheless, the observed downshifting of several PHB-related bands in samples after grinding (vide supra) to slightly but statistically significantly lower wavenumbers unambiguously show that the grinding step induces partial transition from the metastable and more amorphous state of the intracellular PHB to its more ordered state (i.e., its partial crystallisation). Note that a very similar but even more strongly expressed downshifting of the main PHB-related bands, ν(C=O) around ~1740 cm^{-1} and ν(C–O–C/C–C–O) at ~1300 cm^{-1}, was shown to be induced by the sample preparation procedure that involves grinding and pressing the bacterial biomass with KBr [24] often used in FTIR spectroscopy.

2.1.2. Effects of Drying

Another variation in sample preparation conditions tested in this work was associated with the duration of drying. The presence of water (which is featured by strong vibration bands, particularly the "scissoring" mode of δ(H–O–H) vibrations at ~1640–1650 cm^{-1} which falls within the amide I region of proteins [19,46]), even in traces, in a sample can alter the measured FTIR spectrum.

The drying periods for the bacterial samples adopted in this study were 1.5 h and 23 h (at 45 °C). Photographs of the samples just applied to the ZnSe glass and dried for 1.5 h and 23 h are shown in Figure 2.

Figure 2. Photographs of aqueous suspension of *A. brasilense* Sp7 bacterial biomass freshly applied to a ZnSe glass (**A**), dried for 1.5 h (**B**) and for 23 h (**C**).

From the same sample of the dried bacterial culture of *A. brasilense* Sp7, three separate ZnSe glasses were prepared for FTIR spectroscopy of these parallel measurements. The amide I band (~1645–1655 cm^{-1}, peptide bonds in proteins) was used as a normalisation standard for the FTIR spectra. As can be seen from Figure 3, some differences were observed in the intensities for the three parallel samples (each dried for 1.5 h) in the region of the broad band at 3700–2700 cm^{-1} (the region of stretching vibrations of O–H and N–H groups), as well as in the region of 1485–1000 cm^{-1} (C–O, C–C, C–O–H, C–O–C in polysaccharides and polyesters).

However, the intensities of all the bands for three similar replicate samples of *A. brasilense* Sp7 biomass dried for 23 h were virtually the same (Figure 4). It is important to emphasise that, despite some differences in the intensities of the bands in the FTIR spectra between shorter (1.5 h) and longer (23 h) drying periods, the positions of the maxima of all bands in the FTIR spectra of all samples remained unchanged. Thus, with a longer drying time (23 h in our case), greater reproducibility was observed in measuring transmission FTIR spectra with regard to absorption band intensities. Accordingly, it may be recommended to dry bacterial samples being prepared for FTIR spectroscopic analysis at moderate temperatures up to 40–45 °C (to avoid denaturation of proteins) overnight to ensure a good reproducibility of both the intensities and band positions.

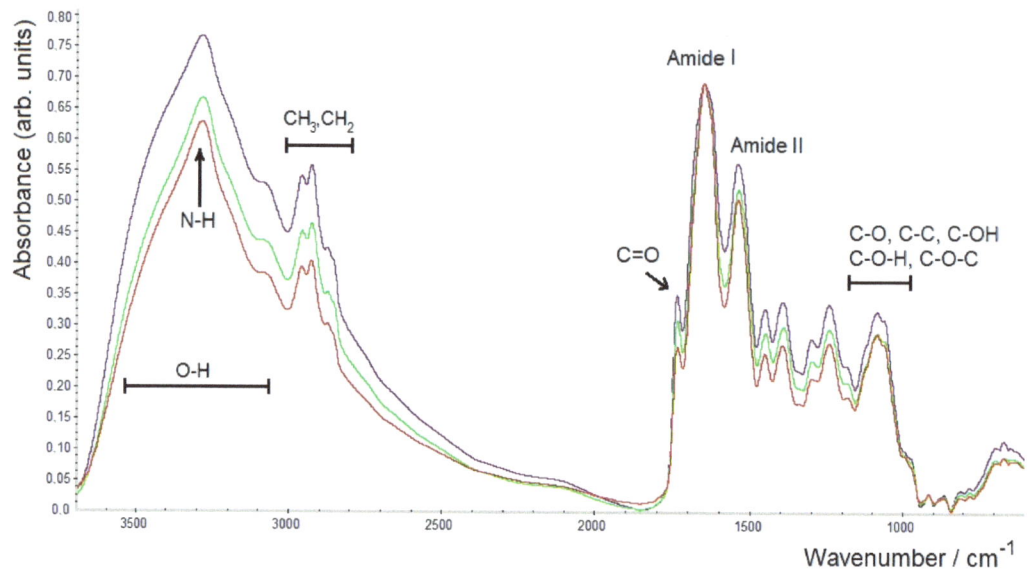

Figure 3. FTIR spectra of three samples of *A. brasilense* Sp7 biomass dried for 1.5 h (3 replicates) normalised by the amide I band.

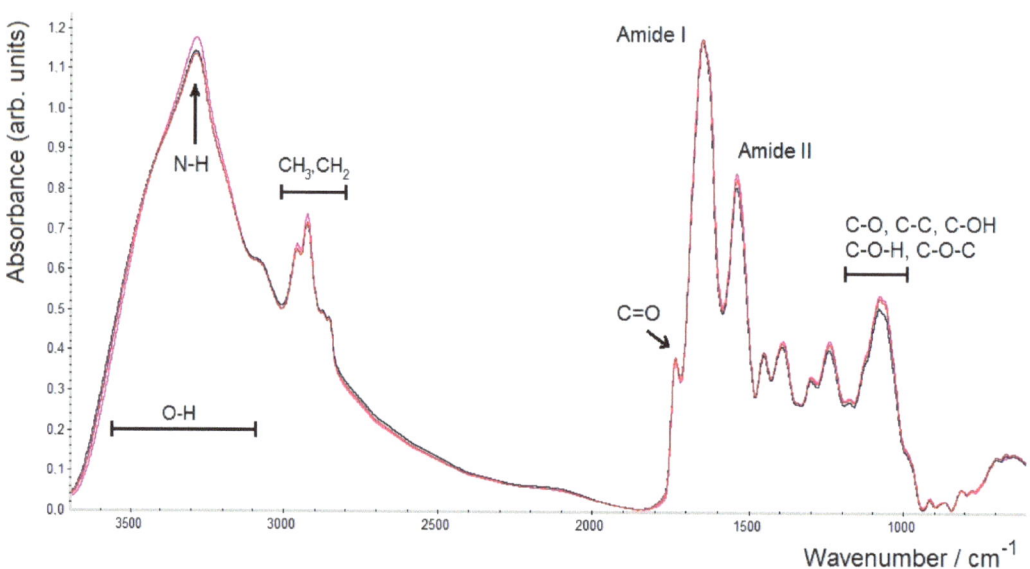

Figure 4. FTIR spectra of three samples of *A. brasilense* Sp7 biomass dried for 23 h (3 replicates) normalised by the amide I band.

It is also worth noting a specific difficulty that we encountered in our work. In some cases, sample preparation of the bacterial biomass for transmission FTIR spectroscopic measurements (for obtaining aqueous suspensions to be applied to a ZnSe glass) may be hampered both in the case of grinding and without it. This is observed when bacteria have accumulated significant amounts of PHB (e.g., over ~40% d.c.w.). Since this polyester has hydrophobic properties, and its physical properties are similar to those of some commercial plastics, large PHB amounts in bacterial culture lead to difficulties in the process of sample preparation; the culture becomes difficult to grind and is also practically not resuspendable in Milli-Q water [25]. Such difficulties in sample preparation require a special approach to bacterial biomass with a high PHB content and the development of additional steps that would make it possible to obtain the most homogeneous sample appropriate for FTIR measurements.

Thus, in this part, it has been shown that varying the conditions in some stages of sample processing, such as drying time, as well as the use of the grinding of bacterial biomass samples, can lead to changes in the obtained FTIR spectra when analysing microbiological objects. Consequently, the application and details of such steps can be optimised with regard to the expected composition of the bacterial specimens, which has been attempted in this study.

2.2. Analysis of Bacterially Synthesised Selenium Nanoparticles by FTIR Spectroscopy

In this part of the work, using the example of biogenic SeNPs, we discuss the influence of the sample preparation process on the state of the samples reflected in their FTIR spectra. The isolated SeNPs of bacterial origin obtained using *A. brasilense* Sp7, with different numbers of washing steps, were studied by FTIR spectroscopy.

Figure 5 shows the region 2000–700 cm^{-1}, which reflects the greatest changes in the samples under study and is most informative in FTIR spectra when studying biological samples. Note that the FTIR spectra of samples with two and three washing steps, owing to a partial loss of material occurring during the purification of SeNPs performed by washing, are characterised by a lower signal-to-noise ratio, which has led to some increase in noise in the FTIR spectra, as can be seen in spectra C and D under 1000 cm^{-1} (however, not impairing the analysis).

First of all, from Figure 5, it is clearly seen that the FTIR spectrum of isolated biogenic SeNPs without additional washing steps (spectrum **A**) is noticeably different from FTIR spectra of those after one to three washing steps (see spectra **B–D** which have much fewer differences between them). As has been well documented, microbially synthesised SeNPs always contain specific capping layers of biomacromolecules originating from the biological system in which they were synthesised [34–36]. We performed a comparative analysis of the spectra in Figure 5 (as the SeNPs were obtained using *A. brasilense* Sp7, they are expected to contain bioorganic components from this bacterium; hence part of the assignments listed in Table 1, which are typical of bacterial cell biomass, may be used).

For the FTIR spectrum of nanoparticles that were not washed (spectrum **A**), as compared to the other spectra, the most significant difference is the presence of a strong band at 1564 cm^{-1}. This band may be assigned to antisymmetric stretching vibrations of ionised carboxylate residues (salts of carboxylic acids), $\nu_{as}(COO^-)$, in the biomacromolecular shell of SeNPs. This assignment is also confirmed by the presence of the accompanying band related to the symmetric stretching vibrations $\nu_s(COO^-)$ at 1412 cm^{-1}, as well as of the bands related to its bending vibrations $\delta(COO^-)$ (at 821 and 772 cm^{-1}). The carboxylates may evidently represent various amino acid residues and be contained in carboxylated polysaccharides (the typical polysaccharide region within 1200–950 cm^{-1} is also seen in spectrum **A**). Note that the positions of carboxylate-related vibration bands are known to vary depending on the interactions with the surrounding biomolecules. The band at 1654 cm^{-1} in spectrum **A** represents the amide I region of proteins (see below); the accompanying amide II band around 1540 cm^{-1} is definitely overlapped by the strong and broad $\nu_{as}(COO^-)$ absorption. Very similar results were reported earlier for biogenic SeNPs isolated from *A. brasilense* Sp7 biomass without the additional washing steps [47].

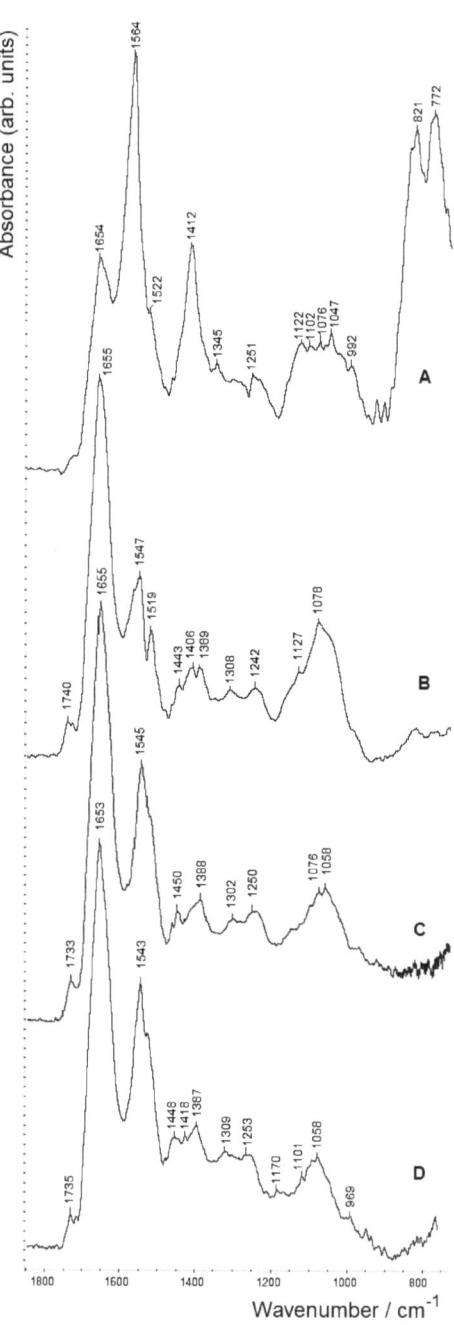

Figure 5. FTIR spectra (transmission mode, on ZnSe discs) of isolated biogenic selenium nanoparticles (SeNPs) obtained using *A. brasilense* Sp7, measured without washing (**A**) and after 1 (**B**), 2 (**C**) and 3 (**D**) washing steps. Spectra **A–C** are vertically shifted from the baseline (which corresponds to spectrum **D**) for clarity.

As can be seen in spectrum **B**, the amount of carboxylic residues significantly decreased after the first washing step; further washing brings about only minor changes (spectra **C,D**). Thus, the carboxylate-containing components are most likely rather weakly bound to the surface of SeNPs, in contrast to the rest of the biomacromolecular shell, which evidently remains stable. This is in line with the recently reported comparative data on SeNPs of bacterial origin, where the bioorganic capping layers (showing differences when obtained using different bacteria) are postulated to contain an outer, more weakly bound shell and an inner part more strongly bound to the Se core [36].

In spectra **B–D**, both amide I (1655–1653 cm^{-1}) and amide II (1547–1543 cm^{-1}) bands related to proteins are more pronounced. The polysaccharide region (1200–950 cm^{-1}) is slightly diminishing with each additional washing step (cf. spectra **B–D**), indicating that part of carboxylic groups may indeed be associated with weakly bound carboxypolysaccharides removed upon washing. Besides proteins and polysaccharides, the presence of lipids at all steps is corroborated by the typical ester ν(C=O) band around 1740 cm^{-1} (which in spectrum **A** is seen as a weaker shoulder and appears to be somewhat more pronounced after even the first washing; cf. spectra **B–D**).

In order to reveal unresolved (closely overlapping) bands, second derivatives of the spectra can be informative, especially for complicated spectra of microbiological samples [19,48]. Using OMNIC software, the second derivatives of the FTIR spectra shown in Figure 5 were calculated and presented in the most informative spectroscopic region (~1800–1400 cm^{-1}; Figure 6, Table 2). Minima on the second derivatives (below zero point) correspond to both well-resolved spectral bands and inflection points (poorly resolved bands that may be seen as shoulders, overlapping with stronger adjacent bands) in the original spectrum [19].

Table 2. Peak assignments [1] in second derivatives of FTIR spectra of biogenic SeNPs obtained using *A. brasilense* Sp7 measured without washing and after 1–3 washing steps (see Figure 6, curves **A–D**, respectively) [19–25,46–48].

Functional Groups	SeNPs without Washing	SeNPs after 1 Washing Step	SeNPs after 2 Washing Steps	SeNPs after 3 Washing Steps
C=O (ester), ν	1737 (v.w.) 1728 (v.w.)	1742 1726 (w.)	1737	1734
Amide I (in proteins)	1696 (v.w.) 1679 (v.w.) 1657 1635 (v.w.)	1690 1676 (w.) 1658 (s.) 1644	1690 1675 (w.) 1655 (s.) 1627	1684 1668 (w.) 1654 (s.) 1633
Carboxylate (COO$^-$, ν_{as})	1561 (v.s.)	1566	1583 (w.)	1556 (w.)
Amide II (in proteins)	1539 (v.w.)	1546 (s.)	1549 (s.)	1542 (s.)
"Tyrosine" band	1518 (w.)	1516	1518	1520
Carboxylate (COO$^-$, ν_s)	1410 (s.)	1409	1413 (w.)	1418 (w.)

[1] Notations: ν—stretching vibrations; ν_s—symmetric stretching vibrations; ν_{as}—antisymmetric stretching vibrations; δ—bending vibrations; v.s.—very strong, s.—strong, w.—weak, v.w.—very weak.

As can be seen from Figure 6, the two typical protein bands, amide I and amide II, in curve **A** (SeNPs separated without additional washing) at 1657 and 1539 cm^{-1}, respectively, are weaker than the two dominating peaks assigned to ν_{as} and ν_s of carboxylate (at 1561 and 1410 cm^{-1}). The ester ν(C=O) band in curve **A** (split into two components, at 1737 and 1728 cm^{-1}) is also very weak. However, after the first washing, when a significant part of the carboxylate-containing components have evidently been removed, and after the next two to three washing steps, the protein bands at ~1650 and ~1540 cm^{-1}, as well as the ν(C=O) bands related to lipids (at ~1740 cm^{-1}), are much more clearly seen (see curves **B–D**). It is necessary to add that, owing to the presence of carboxylic groups, especially in the form of ionised carboxylates (salts which dissociate in solution), in the surface capping layer of biogenic SeNPs, the latter are most often characterised by negative zeta potentials which stabilise their aqueous suspensions [33,34,36].

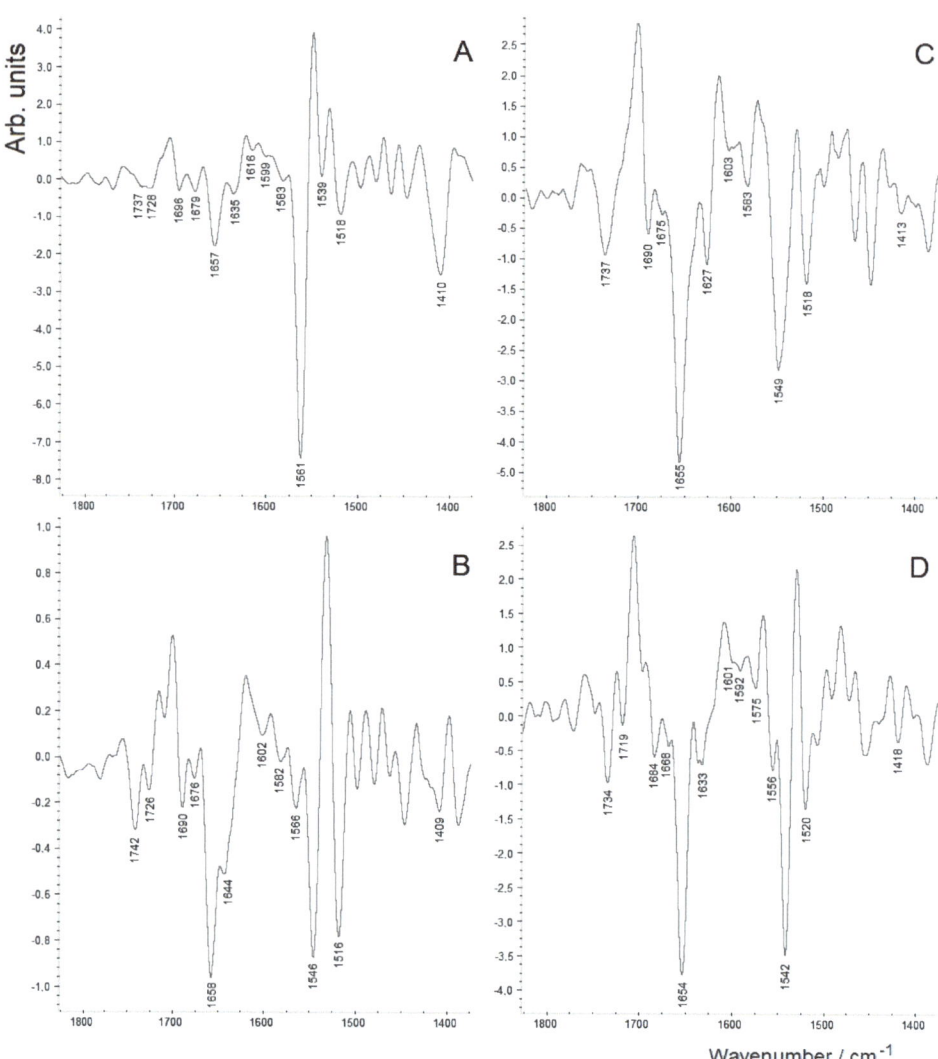

Figure 6. Second derivatives of FTIR spectra of isolated biogenic SeNPs obtained using *A. brasilense* Sp7 (see Figure 5) measured without washing (**A**) and after 1 (**B**), 2 (**C**) and 3 (**D**) washing steps.

It may also be noted that for the amide I band, which is sensitive to the secondary structure of protein and is known to contain several closely overlapping bands within the region ~1690–1620 cm^{-1} [19,21], besides the main band within 1658–1654 cm^{-1} (the region typical of the dominating α-helix), there are several weaker component bands within the "full" amide I region clearly seen in curves **B–D** in Figure 6, which correspond to several β-structured protein components [19,21,48]. (Their detailed discussion is, however, out of the scope of this paper.)

Thus, it has been shown that, in the case of biogenic SeNPs, their sample preparation for FTIR spectroscopic analysis is an important step. As has been found, additional washing (even one step) decreases the content of weakly bound carboxylic components in the sample, which is reflected in the FTIR spectra. On the one hand, generally during sample preparation, it is necessary to take into account that the components of the buffer or medium can contribute to the measured spectrum. On the other hand, the procedures for removing these components, particularly the widely used method of washing a biological sample in combination with centrifugation, can concomitantly lead to a change in its state and/or composition, which is reflected in FTIR spectra, as shown in this work.

3. Materials and Methods

3.1. Bacterial Strains and Growth Conditions

Wild-type strains *Azospirillum brasilense* Sp7 [29] (ATCC 29145) and *Azospirillum baldaniorum* Sp245 [31] (previously known as *Azospirillum brasilense* Sp245 [30]) were taken from the Collection of Rhizosphere Microorganisms [WDCM 1021] maintained at the Institute of Biochemistry and Physiology of Plants and Microorganisms, Russian Academy of Sciences, Saratov, Russia [49] (URL: http://collection.ibppm.ru/catalogue/azospirillum/azospirillum-brasilense/ accessed on 13 January 2021). The bacteria were cultivated in a liquid modified malate salt medium (MSM) as reported earlier [24,25] which contained the following salts (g·L^{-1}): K$_2$HPO$_4$, 3.0; KH$_2$PO$_4$, 2.0; NH$_4$Cl, 0.5; NaCl, 0.1; FeSO$_4$·7H$_2$O, 0.02 (added as chelate with nitrilotriacetic acid); CaCl$_2$, 0.02; MgSO$_4$·7H$_2$O, 0.2; Na$_2$MoO$_4$·2H$_2$O, 0.002; sodium malate, 5.0 (obtained by mixing 3.76 g of malic acid with 2.24 g NaOH per litre), yeast extract, 0.1, pH 6.8–7.0. The cultures (100 mL in 250 mL Erlenmeyer flasks) were grown under aerobic conditions on a shaker (180 rpm) for up to 19 h. Cell growth was monitored at λ = 595 nm (Spekol 221, Germany); the optical density (A$_{595}$) values of the resulting culture suspensions were about 1.0.

3.2. Bacterial Synthesis of SeNPs and Their Purification

The SeNPs were obtained according to the procedure reported elsewhere [47] with minor modifications. Briefly, bacterial cells of *A. brasilense* Sp7 (grown as described in Section 3.1) were harvested by centrifugation in 2 mL Eppendorf tubes (Minispin centrifuge; 15 min, 7000*g* ×) and washed three times with sterile saline solution (0.85% NaCl aqueous solution) to remove the culture medium components. All the next steps were performed under sterile conditions. The resulting wet biomass pellet was resuspended in half of the initial volume of sterile saline solution. Sodium selenite (Na$_2$SeO$_3$·5H$_2$O, "Merck") as 0.5 M stock aqueous solution was added to the suspensions up to 10 mM. Suspensions containing the cells (washed as above) and selenite were placed in a thermostat (at 31–32 °C). The SeNPs thereby formed were monitored by transmission electron microscopy (TEM; data not shown; see, e.g., [33,47]). After 24 h, the bacterial cells were removed from the suspension by "soft" centrifugation (1400*g* ×, 5 min); the supernatant with SeNPs was collected and filtered through a 0.22 or 0.44 mm PVDF filter to remove occasional bacterial cells. The suspensions of SeNPs were further centrifuged at 12,000*g* × for 30 min, and the collected precipitate pellet was resuspended in a minimum volume of Milli-Q directly for FTIR spectroscopic analysis (on a ZnSe disc) or after 1 to 3 additional washing steps in Milli-Q water and centrifugation at 12,000*g* × for 30 min.

3.3. Sample Preparation for FTIR Spectroscopic Analyses

The bacterial cells of *A. brasilense* Sp7 and *A. baldaniorum* Sp245 (see Section 3.1) were collected by centrifugation (10,000*g* ×, 10 min, 4 °C), washed 3 times with physiological solution and dried (on open Petri dishes, ø 3.5 cm) in a thermostatted desiccator at 45 °C up to a constant weight. For infrared spectroscopic measurements, the samples of dried biomass were prepared in several ways: with/without grinding and with different drying periods (1.5 or 23 h). As for grinding, dry bacterial biomass was powdered in a small agate mortar (for about 5 min). The bacterial samples were resuspended in a small volume of

Milli-Q water. SeNPs were prepared as described above (Section 3.2). Then the resulting aqueous suspensions (about 30–70 µL) were placed as thin films on clean flat ZnSe discs (CVD-ZnSe, "R'AIN Optics", Dzerzhinsk, Russia; ø 1.0 cm, thickness 0.2 cm) and dried at 45 °C again as described above.

3.4. FTIR Spectroscopic Measurements

Transmission FTIR spectroscopic measurements were performed as described elsewhere [47] on a Nicolet 6700 FTIR spectrometer (Thermo Electron Corporation, Waltham, MA, USA; DTGS detector; KBr beam splitter). Spectra were collected with a total of 64 scans (resolution 2 cm^{-1} for spectra of SeNPs and 4 cm^{-1} for bacterial biomass samples) against the ZnSe disc background and manipulated using OMNIC software (version 8.2.0.387, Thermo Electron Corporation, Waltham, MA, USA) supplied by the manufacturer of the spectrometer. For each spectrum, the baseline was corrected using the "automatic baseline correct" function, and then each spectrum was smoothed using the standard "automatic smooth" function of the software which uses the Savitsky–Golay algorithm (95-point moving second-degree polynomial). All the FTIR spectroscopic measurements were repeated two (for SeNPs) or three times (for bacteria) for each sample and were well reproducible.

4. Conclusions

It has been shown that preliminary sample preparation steps (such as grinding, washing and drying) of microbiological specimens for transmission FTIR spectroscopic measurements may in some cases alter the composition and other properties of samples, which is reflected in their FTIR spectra. Thus, special care should be taken to ensure that samples are analysed by FTIR spectroscopy in their stable state which would ensure obtaining reproducible spectra. However, if any sample preparation step is expected to alter the properties of the samples under study, this can be checked by comparing their FTIR spectra before and after a sample treatment step. For complicated microbiological objects (such as bacterial biomass or nanoparticles of microbial origin, as studied in this work), the described examples could allow a most adequate protocol for sample preparation to be chosen to ensure obtaining reliable and reproducible spectroscopic data.

Author Contributions: Conceptualisation, A.A.K. and A.V.T.; methodology, Y.A.D., O.A.K., A.A.V. and P.V.M.; software, A.V.T., Y.A.D. and O.A.K..; validation and formal analysis, A.A.K. and A.V.T.; investigation, A.A.K., A.V.T., Y.A.D., O.A.K., A.A.V. and P.V.M.; resources and data curation, A.A.K. and A.V.T.; writing—original draft preparation, Y.A.D., O.A.K. and A.A.V.; writing—review and editing, A.A.K. and A.V.T.; visualisation, A.A.K. and A.V.T.; supervision, A.A.K. and A.V.T.; project administration, A.A.K. and A.V.T.; funding acquisition, A.A.K. and A.V.T. All authors have read and agreed to the published version of the manuscript.

Funding: Part of this study was performed with support from the Russian Foundation for Basic Research (Grants 17-08-01696-a, 16-08-01302-a).

Institutional Review Board Statement: Not applicable.

Informed Consent Statement: Not applicable.

Data Availability Statement: The data presented in this study are available in this article.

Acknowledgments: In this work, some experiments were performed on the equipment of the "Simbioz" Centre for the Collective Use of Research Equipment in the field of physical–chemical biology and nanobiotechnology at IBPPM RAS, Saratov, Russia (FTIR spectrometer Nicolet 6700, USA).

Conflicts of Interest: The authors declare no conflict of interest. The funders had no role in the design of the study; in the collection, analyses, or interpretation of data; in the writing of the manuscript, or in the decision to publish the results.

Sample Availability: Samples of the bacterial biomass and SeNPs are available from the authors.

References

1. Xu, J.-L.; Gowen, A.A. Time series Fourier transform infrared spectroscopy for characterization of water vapor sorption in hydrophilic and hydrophobic polymeric films. *Spectrochim. Acta Part A Mol. Biomol. Spectrosc.* **2021**, *250*, 119371. [CrossRef] [PubMed]

2. Frandsen, B.N.; Deal, A.M.; Lane, J.R.; Vaida, V. Lactic acid spectroscopy: Intra- and intermolecular interactions. *J. Phys. Chem. A* **2021**, *125*, 218–229. [CrossRef]

3. Kannan, P.P.; Karthick, N.K.; Arivazhagan, G. Hydrogen bond interactions in the binary solutions of formamide with methanol: FTIR spectroscopic and theoretical studies. *Spectrochim. Acta Part A Mol. Biomol. Spectrosc.* **2020**, *229*, 117892. [CrossRef]

4. Camisasca, G.; Schlesinger, D.; Zhovtobriukh, I.; Pitsevich, G.; Pettersson, L.G.M. A proposal for the structure of high- and low-density fluctuations in liquid water. *J. Chem. Phys.* **2019**, *151*, 034508. [CrossRef]

5. Duarte, L.J.; Bruns, R.E. FTIR and dispersive gas phase absolute infrared intensities of hydrocarbon fundamental bands. *Spectrochim. Acta Part A Mol. Biomol. Spectrosc.* **2019**, *214*, 1–6. [CrossRef]

6. Gao, X.; Leng, C.; Zeng, G.; Fu, D.; Zhang, Y.; Liu, Y. Ozone initiated heterogeneous oxidation of unsaturated carboxylic acids by ATR-FTIR spectroscopy. *Spectrochim. Acta Part A Mol. Biomol. Spectrosc.* **2019**, *214*, 177–183. [CrossRef]

7. Sato, E.T.; Machado, N.; Araújo, D.R.; Paulino, L.C.; Martinho, H. Fourier transform infrared absorption (FTIR) on dry *stratum corneum*, corneocyte-lipid interfaces: Experimental and vibrational spectroscopy calculations. *Spectrochim. Acta Part A Mol. Biomol. Spectrosc.* **2021**, *249*, 119218. [CrossRef]

8. Chrisikou, I.; Orkoula, M.; Kontoyannis, C. FT-IR/ATR solid film formation: Qualitative and quantitative analysis of a piperacillin-tazobactam formulation. *Molecules* **2020**, *25*, 6051. [CrossRef]

9. Gorbikova, E.; Samsonov, S.A.; Kalendar, R. Probing the proton-loading site of cytochrome *c* oxidase using time-resolved Fourier transform infrared spectroscopy. *Molecules* **2020**, *25*, 3393. [CrossRef] [PubMed]

10. Bekiaris, G.; Koutrotsios, G.; Tarantilis, P.A.; Pappas, C.S.; Zervakis, G.I. FTIR assessment of compositional changes in ligno-cellulosic wastes during cultivation of *Cyclocybe cylindracea* mushrooms and use of chemometric models to predict production performance. *J. Mater. Cycles Waste Manag.* **2020**, *22*, 1027–1035. [CrossRef]

11. Andrushchenko, V.; Pohle, W. Influence of the hydrophobic domain on the self-assembly and hydrogen bonding of hydroxy-amphiphiles. *Phys. Chem. Chem. Phys.* **2019**, *21*, 11242–11258. [CrossRef]

12. Procacci, B.; Rutherford, S.H.; Greetham, G.M.; Towrie, M.; Parker, A.W.; Robinson, C.V.; Howle, C.R.; Hunt, N.T. Differentiation of bacterial spores via 2D-IR spectroscopy. *Spectrochim. Acta Part A Mol. Biomol. Spectrosc.* **2021**, *249*, 119319. [CrossRef]

13. Elkadi, O.A.; Hassan, R.; Elanany, M.; Byrne, H.J.; Ramadan, M.A. Identification of *Aspergillus* species in human blood plasma by infrared spectroscopy and machine learning. *Spectrochim. Acta Part A Mol. Biomol. Spectrosc.* **2021**, *248*, 119259. [CrossRef]

14. Yang, J.; Yin, C.; Miao, X.; Meng, X.; Liu, Z.; Hu, L. Rapid discrimination of adulteration in *Radix astragali* combining diffuse reflectance mid-infrared Fourier transform spectroscopy with chemometrics. *Spectrochim. Acta Part A Mol. Biomol. Spectrosc.* **2021**, *248*, 119251. [CrossRef]

15. Grace, C.E.E.; Lakshmi, P.K.; Meenakshi, S.; Vaidyanathan, S.; Srisudha, S.; Mary, M.B. Biomolecular transitions and lipid accumulation in green microalgae monitored by FTIR and Raman analysis. *Spectrochim. Acta Part A Mol. Biomol. Spectrosc.* **2020**, *224*, 117382. [CrossRef] [PubMed]

16. Alugoju, P.; Narsimulu, D.; Bhanu, J.U.; Satyanarayana, N.; Periyasamy, L. Role of quercetin and caloric restriction on the biomolecular composition of aged rat cerebral cortex: An FTIR study. *Spectrochim. Acta Part A Mol. Biomol. Spectrosc.* **2019**, *220*, 117128. [CrossRef]

17. Marques, V.; Cunha, B.; Couto, A.; Sampaio, P.; Fonseca, L.P.; Aleixo, S.; Calado, C.R.C. Characterization of gastric cells infection by diverse *Helicobacter pylori* strains through Fourier-transform infrared spectroscopy. *Spectrochim. Acta Part A Mol. Biomol. Spectrosc.* **2019**, *210*, 193–202. [CrossRef] [PubMed]

18. Kar, S.; Katti, D.R.; Katti, K.S. Fourier transform infrared spectroscopy based spectral biomarkers of metastasized breast cancer progression. *Spectrochim. Acta Part A Mol. Biomol. Spectrosc.* **2019**, *208*, 85–96. [CrossRef]

19. Naumann, D. Infrared spectroscopy in microbiology. In *Encyclopedia of Analytical Chemistry*; Meyers, R.A., Ed.; Wiley: Chichester, UK, 2000; pp. 102–131. [CrossRef]

20. Ojeda, J.J.; Dittrich, M. Fourier transform infrared spectroscopy for molecular analysis of microbial cells. In *Microbial Systems Biology: Methods and Protocols. Methods in Molecular Biology*; Navid, A., Ed.; Humana Press: Totowa, NJ, USA, 2012; Volume 881, Chapter 8; pp. 187–211. [CrossRef]

21. Kamnev, A.A.; Sadovnikova, J.N.; Tarantilis, P.A.; Polissiou, M.G.; Antonyuk, L.P. Responses of *Azospirillum brasilense* to nitrogen deficiency and to wheat lectin: A diffuse reflectance infrared Fourier transform (DRIFT) spectroscopic study. *Microb. Ecol.* **2008**, *56*, 615–624. [CrossRef] [PubMed]

22. Pistorius, A.M.A.; DeGrip, W.J.; Egorova-Zachernyuk, T.A. Monitoring of biomass composition from microbiological sources by means of FT-IR spectroscopy. *Biotechnol. Bioeng.* **2009**, *103*, 123–129. [CrossRef]

23. Maity, J.P.; Kar, S.; Lin, C.-M.; Chen, C.-Y.; Chang, Y.-F.; Jean, J.-S.; Kulp, T.R. Identification and discrimination of bacteria using Fourier transform infrared spectroscopy. *Spectrochim. Acta Part A Mol. Biomol. Spectrosc.* **2013**, *116*, 478–484. [CrossRef]

24. Kamnev, A.A.; Tugarova, A.V.; Dyatlova, Y.A.; Tarantilis, P.A.; Grigoryeva, O.P.; Fainleib, A.M.; De Luca, S. Methodological effects in Fourier transform infrared (FTIR) spectroscopy: Implications for structural analyses of biomacromolecular samples. *Spectrochim. Acta Part A Mol. Biomol. Spectrosc.* **2018**, *193*, 558–564. [CrossRef]

25. Tugarova, A.V.; Dyatlova, Y.A.; Kenzhegulov, O.A.; Kamnev, A.A. Poly-3-hydroxybutyrate synthesis by different *Azospirillum brasilense* strains under varying nitrogen deficiency: A comparative in-situ FTIR spectroscopic analysis. *Spectrochim. Acta Part A Mol. Biomol. Spectrosc.* **2021**, *252*, 119458. [CrossRef] [PubMed]
26. Bashan, Y.; de-Bashan, L.E. How the plant growth-promoting bacterium *Azospirillum* promotes plant growth—A critical assessment. *Adv. Agron.* **2010**, *108*, 77–136. [CrossRef]
27. Cassán, F.; Okon, Y.; Creus, C. (Eds.) *Handbook for Azospirillum. Technical Issues and Protocols*; Springer International Publishing: Cham, Switzerland, 2015. [CrossRef]
28. Cassán, F.; Coniglio, A.; López, G.; Molina, R.; Nievas, S.; de Carlan, C.L.N.; Donadio, F.; Torres, D.; Rosas, S.; Pedrosa, F.O.; et al. Everything you must know about *Azospirillum* and its impact on agriculture and beyond. *Biol. Fertil. Soils* **2020**, *56*, 461–479. [CrossRef]
29. Tarrand, J.J.; Krieg, N.R.; Döbereiner, J. A taxonomic study of the *Spirillum lipoferum* group, with description of a new genus, *Azospirillum* gen. nov. and two species, *Azospirillum lipoferum* (Beijerink) comb. nov. and *Azospirillum brasilense* sp. nov. *Can. J. Microbiol.* **1978**, *24*, 967–980. [CrossRef]
30. Baldani, V.L.D.; Baldani, J.I.; Döbereiner, J. Effects of *Azospirillum* inoculation on root infection and nitrogen incorporation in wheat. *Can. J. Microbiol.* **1983**, *29*, 924–929. [CrossRef]
31. Ferreira, N.d.S.; Sant'Anna, F.H.; Reis, V.M.; Ambrosini, A.; Volpiano, C.G.; Rothballer, M.; Schwab, S.; Baura, V.A.; Balsanelli, E.; Pedrosa, F.d.O.; et al. Genome-based reclassification of *Azospirillum brasilense* Sp245 as the type strain of *Azospirillum baldaniorum* sp. nov. *Int. J. Syst. Evol. Microbiol.* **2020**, *70*. [CrossRef]
32. Tugarova, A.V.; Vetchinkina, E.P.; Loshchinina, E.A.; Burov, A.M.; Nikitina, V.E.; Kamnev, A.A. Reduction of selenite by *Azospirillum brasilense* with the formation of selenium nanoparticles. *Microb. Ecol.* **2014**, *68*, 495–503. [CrossRef]
33. Tugarova, A.V.; Mamchenkova, P.V.; Khanadeev, V.A.; Kamnev, A.A. Selenite reduction by the rhizobacterium *Azospirillum brasilense*, synthesis of extracellular selenium nanoparticles and their characterisation. *New Biotechnol.* **2020**, *58*, 17–24. [CrossRef]
34. Tugarova, A.V.; Kamnev, A.A. Proteins in microbial synthesis of selenium nanoparticles. *Talanta* **2017**, *174*, 539–547. [CrossRef]
35. Ojeda, J.J.; Merroun, M.L.; Tugarova, A.V.; Lampis, S.; Kamnev, A.A.; Gardiner, P.H.E. Developments in the study and applications of bacterial transformations of selenium species. *Crit. Rev. Biotechnol.* **2020**, *40*, 1250–1264. [CrossRef]
36. Bulgarini, A.; Lampis, S.; Turner, R.J.; Vallini, J. Biomolecular composition of capping layer and stability of biogenic selenium nanoparticles synthesized by five bacterial species. *Microb. Biotechnol.* **2021**, *14*. [CrossRef]
37. Sato, H.; Dybal, J.; Murakami, R.; Noda, I.; Ozaki, Y. Infrared and Raman spectroscopy and quantum chemistry calculation studies of C–H···O hydrogen bondings and thermal behavior of biodegradable polyhydroxyalkanoate. *J. Mol. Struct.* **2005**, *744–747*, 35–46. [CrossRef]
38. Padermshoke, A.; Katsumoto, Y.; Sato, H.; Ekgasit, S.; Noda, I.; Ozaki, Y. Melting behavior of poly(3-hydroxybutyrate) investigated by two-dimensional infrared correlation spectroscopy. *Spectrochim. Acta Part A Mol. Biomol. Spectrosc.* **2005**, *61*, 541–550. [CrossRef]
39. Kansiz, M.; Domínguez-Vidal, A.; McNaughton, D.; Lendl, B. Fourier-transform infrared (FTIR) spectroscopy for monitoring and determining the degree of crystallisation of polyhydroxyalkanoates (PHAs). *Anal. Bioanal. Chem.* **2007**, *388*, 1207–1213. [CrossRef]
40. Müller-Santos, M.; Koskimäki, J.J.; Alves, L.P.S.; de Souza, E.M.; Jendrossek, D.; Pirttilä, A.M. The protective role of PHB and its degradation products against stress situations in bacteria. *FEMS Microbiol. Rev.* **2021**, *45*. [CrossRef]
41. Obruca, S.; Sedlacek, P.; Slaninova, E.; Fritz, I.; Daffert, C.; Meixner, K.; Sedrlova, Z.; Koller, M. Novel unexpected functions of PHA granules. *Appl. Microbiol. Biotechnol.* **2020**, *104*, 4795–4810. [CrossRef] [PubMed]
42. Sedlacek, P.; Slaninova, E.; Koller, M.; Nebesarova, J.; Marova, I.; Krzyzanek, V.; Obruca, S. PHA granules help bacterial cells to preserve cell integrity when exposed to sudden osmotic imbalances. *New Biotechnol.* **2019**, *49*, 129–136. [CrossRef] [PubMed]
43. Sedlacek, P.; Slaninova, E.; Enev, V.; Koller, M.; Nebesarova, J.; Marova, I.; Hrubanova, K.; Krzyzanek, V.; Samek, O.; Obruca, S. What keeps polyhydroxyalkanoates in bacterial cells amorphous? A derivation from stress exposure experiments. *Appl. Microbiol. Biotechnol.* **2019**, *103*, 1905–1917. [CrossRef]
44. Obruca, S.; Sedlacek, P.; Koller, M.; Kucera, D.; Pernicova, I. Involvement of polyhydroxyalkanoates in stress resistance of microbial cells: Biotechnological consequences and applications. *Biotechnol. Adv.* **2018**, *36*, 856–870. [CrossRef] [PubMed]
45. Müller-Santos, M.; de Souza, E.M.; Pedrosa, F.d.O.; Chubatsu, L.S. Polyhydroxybutyrate in *Azospirillum brasilense*. In *Handbook for Azospirillum*; Cassán, F., Okon, Y., Creus, C., Eds.; Springer International Publishing: Cham, Switzerland, 2015; Chapter 13; pp. 241–250. [CrossRef]
46. Talari, A.C.S.; Martinez, M.A.G.; Movasaghi, Z.; Rehman, S.; Rehman, I.U. Advances in Fourier transform infrared (FTIR) spectroscopy of biological tissues. *Appl. Spectrosc. Rev.* **2017**, *52*, 456–506. [CrossRef]
47. Kamnev, A.A.; Mamchenkova, P.V.; Dyatlova, Y.A.; Tugarova, A.V. FTIR spectroscopic studies of selenite reduction by cells of the rhizobacterium *Azospirillum brasilense* Sp7 and the formation of selenium nanoparticles. *J. Mol. Struct.* **2017**, *1140*, 106–112. [CrossRef]
48. Tugarova, A.V.; Scheludko, A.V.; Dyatlova, Y.A.; Filip'echeva, Y.A.; Kamnev, A.A. FTIR spectroscopic study of biofilms formed by the rhizobacterium *Azospirillum brasilense* Sp245 and its mutant *Azospirillum brasilense* Sp245.1610. *J. Mol. Struct.* **2017**, *1140*, 142–147. [CrossRef]
49. Turkovskaya, O.V.; Golubev, S.N. The Collection of Rhizosphere Microorganisms: Its importance for the study of associative plant-bacterium interactions. *Vavilov J. Genet. Breed.* **2020**, *24*, 315–324. [CrossRef]

Review

Fabric Phase Sorptive Extraction: A Paradigm Shift Approach in Analytical and Bioanalytical Sample Preparation

Abuzar Kabir [1] and Victoria Samanidou [2,*]

[1] Department of Chemistry and Biochemistry, International Forensic Research Institute,
    Florida International University, 11200 SW 8th St, Miami, FL 33199, USA; akabir@fiu.edu
[2] Laboratory of Analytical Chemistry, Department of Chemistry, Aristotle University of Thessaloniki,
    54124 Thessaloniki, Greece
*   Correspondence: samanidu@chem.auth.gr

Abstract: Fabric phase sorptive extraction (FPSE) is an evolutionary sample preparation approach which was introduced in 2014, meeting all green analytical chemistry (GAC) requirements by implementing a natural or synthetic permeable and flexible fabric substrate to host a chemically coated sol–gel organic–inorganic hybrid sorbent in the form of an ultra-thin coating. This construction results in a versatile, fast, and sensitive micro-extraction device. The user-friendly FPSE membrane allows direct extraction of analytes with no sample modification, thus eliminating/minimizing the sample pre-treatment steps, which are not only time consuming, but are also considered the primary source of major analyte loss. Sol–gel sorbent-coated FPSE membranes possess high chemical, solvent, and thermal stability due to the strong covalent bonding between the fabric substrate and the sol–gel sorbent coating. Subsequent to the extraction on FPSE membrane, a wide range of organic solvents can be used in a small volume to exhaustively back-extract the analytes after FPSE process, leading to a high preconcentration factor. In most cases, no solvent evaporation and sample reconstitution are necessary. In addition to the extensive simplification of the sample preparation workflow, FPSE has also innovatively combined the extraction principle of two major, yet competing sample preparation techniques: solid phase extraction (SPE) with its characteristic exhaustive extraction, and solid phase microextraction (SPME) with its characteristic equilibrium driven extraction mechanism. Furthermore, FPSE has offered the most comprehensive cache of sorbent chemistry by successfully combining almost all of the sorbents traditionally used exclusively in either SPE or in SPME. FPSE is the first sample preparation technique to exploit the substrate surface chemistry that complements the overall selectivity and the extraction efficiency of the device. As such, FPSE indeed represents a paradigm shift approach in analytical/bioanalytical sample preparation. Furthermore, an FPSE membrane can be used as an SPME fiber or as an SPE disk for sample preparation, owing to its special geometric advantage. So far, FPSE has overwhelmingly attracted the interest of the separation scientist community, and many analytical scientists have been developing new methodologies by implementing this cutting-edge technique for the extraction and determination of many analytes at their trace and ultra-trace level concentrations in environmental samples as well as in food, pharmaceutical, and biological samples. FPSE offers a total sample preparation solution by providing neutral, cation exchanger, anion exchanger, mixed mode cation exchanger, mixed mode anion exchanger, zwitterionic, and mixed mode zwitterionic sorbents to deal with any analyte regardless of its polarity, ionic state, or the sample matrix where it resides. Herein we present the theoretical background, synthesis, mechanisms of extraction and desorption, the types of sorbents, and the main applications of FPSE so far according to different sample categories, and to briefly show the progress, advantages, and the main principles of the proposed technique.

Keywords: extraction; sample preparation; green analytical chemistry; fabric phase sorptive extraction

check for
updates

Citation: Kabir, A.; Samanidou, V. Fabric Phase Sorptive Extraction: A Paradigm Shift Approach in Analytical and Bioanalytical Sample Preparation. *Molecules* **2021**, *26*, 865. https://doi.org/10.3390/ molecules26040865

Academic Editor: Nuno Neng
Received: 31 December 2020
Accepted: 4 February 2021
Published: 6 February 2021

Publisher's Note: MDPI stays neutral with regard to jurisdictional claims in published maps and institutional affiliations.

1. Introduction

When an analytical or bioanalytical chemist is presented with a sample for analysis, regardless of the nature of the sample, a number of important decisions must be made such as which chromatographic/electrophoretic instrument will be used and what the sample preparation strategy would be, among others. Unless the analyst is imposed with some regulatory restrictions, the analyst may independently decide as to whether a solvent based extraction technique (e.g., liquid–liquid extraction, liquid phase microextraction) or a sorbent based extraction technique (e.g., solid phase extraction, solid phase microextraction, stir bar sorptive extraction) [1–3] will be deployed. If the goals of the sample preparation are to achieve highly selective extraction of the target analytes as well as to minimize the matrix interference, the obvious choice would be sorbent-based extraction techniques. Subsequently, another major decision point would be whether the sample preparation technique is an exhaustive one as used in solid phase extraction (SPE), or an equilibrium driven one as used in solid phase microextraction (SPME). Both the techniques have some advantages and shortcomings. In addition to the differences in the extraction mechanism, they use almost exclusively two different sets of sorbents (with a few exception). SPME and its different modifications can be deployed in the field, whereas SPE is not generally field deployable. What if an analyst wants to exploit all the advantageous features of both the techniques while minimizing the inherent shortcomings? Keeping this dilemma in mind, Kabir and Furton [4] developed fabric phase sorptive extraction in 2014 as a new generation sample preparation technique that innovatively combines both SPE and SPME in a single sample preparation technology platform. Fabric phase sorptive extraction (FPSE) simultaneously exerts exhaustive extraction mechanism as well as equilibrium driven extraction during the sample preparation process and consequently accomplishes exhaustive or near exhaustive extraction even when the extraction is carried out under equilibrium extraction conditions (e.g., direct immersion extraction). As such, FPSE is neither a new format of SPME nor a new format of SPE, but a true combination of both the techniques.

FPSE has not only combined the extraction mechanisms of SPE and SPME, it has also successfully made available all the sorbents which are exclusively used in either SPE or in SPME. For example, poly(dimethylsiloxane), PDMS, is a popular sorbent coating used in SPME. On the other hand, the C18 phase is predominantly used in SPE. Now, an analyst may use both the sorbents in FPSE.

FPSE is the first sample preparation technology that exploits the surface chemistry of the substrate. In fact, the selectivity and the extraction efficiency of the FPSE membrane originate from the organic polymer, one or more organically modified inorganic precursor and the surface chemistry of the fabric substrate. As such, the selectivity and extraction efficiency of sol–gel PDMS coating on cellulose fabric is substantially different from that of sol–gel PDMS coating on polyester or fiberglass fabric.

FPSE also enjoys the enormous advantages of sol–gel synthesis process that chemically binds the organic polymer/ligand to the substrate using an inorganic/organically modified linker. The chemical bonding between the substrate and the polymer assures very high thermal, solvent, and chemical stability of the FPSE membrane. As a result, the FPSE membranes can be exposed to pH 1–13, as well as to any organic solvent without compromising structural and chemical integrity of the extracting polymer. Sol–gel based chemical coating process provides unprecedented batch-to-batch reproducibility. It is worth mentioning that classical extraction and microextraction techniques often use physical coating processes to immobilize the polymer on the substrate surface, resulting in poor reproducibility, limited range of pH stability, and the tendency to swell when exposed to organic sorbents. Sol–gel-derived sorbents are inherently porous with their characteristic sponge-like porous architecture [5]. As such, the sample matrix can easily permeate through the micro and mesopores of the sol–gel sorbents for rapid analyte–sorbent interaction leading to fast extraction equilibrium.

Due to the open bed, planar geometry, the FPSE membrane can be used in an equilibrium-based extraction mode (as in direct immersion extraction in SPME) or in an exhaustive extraction mode (as an SPE disk). Although, the application potential of FPSE membrane as an SPE disk has not fully explored, Lakade et al. [6] has demonstrated that the FPSE membrane can be used as an SPE disk without compromising the quality of the analytical data.

2. Theoretical Background

In its classical operational mode (direct immersion extraction), FPSE mimics the extraction principle of solid phase microextraction (SPME) and similar microextraction techniques, including stir bar sorptive extraction (SBSE) and thin film microextraction (TFME). Extraction of the analyte(s) on the FPSE membrane is primarily governed by the difference in partition coefficient of an analyte between the sample matrix and the FPSE membrane, and the mass transfer of the analyte(s) from the bulk of the sample matrix towards the FPSE membrane continues until an equilibrium is established between the two phases. The mass of the analyte(s) extracted by the FPSE membrane (n), under equilibrium extraction conditions, is proportional to the partition coefficient between the FPSE membrane (which varies with different fabric substrates and the sorbent coatings on the substrate surface) and the sample matrix (K_{es}), volume of the extracting phase (V_e), volume of the sample (V_s), and the initial concentration of the analyte (C_o).

The mass of the analyte extracted by the FPSE membrane at equilibrium (n) can be expressed as:

$$n = \frac{K_{es}V_eV_sC_o}{K_{es}V_e + V_s} \tag{1}$$

When the sample volume is too large compared to the volume of the extracting sorbent ($V_e \ll V_s$),

Equation (1) can be simplified as:

$$n = K_{es}V_eC_o \tag{2}$$

As can be inferred from Equation (2), the mass of the analyte extracted by the FPSE membrane (n) is directly proportional to the volume of the extracting sorbent and is independent of the sample volume. As such, the value of n can be increased by increasing the volume of the extracting sorbent if the initial concentration of the analyte(s) (C_o) is kept constant [7].

The extraction efficiency of an FPSE membrane depends on: (a) thermodynamic factors, and (b) kinetic factors [8]. The partition coefficient for an analyte between the FPSE membrane and the sample matrix is a thermodynamic criteria that depends on the material properties of FPSE substrate, coated sorbent, mass of the extracting sorbent, temperature at which the extraction is carried out, and chemical state of analyte in the sample matrix, as well as other factors. However, the most simplistic way to increase the mass of the extracted analyte by the FPSE membrane is to increase the volume of the extracting sorbent.

The kinetic criteria determine the rate at which the equilibrium is reached and can be dramatically improved by maximizing the contact surface between the sample matrix and the extracting phase as well as by applying external energetics (e.g., stirring, sonication, orbital shaking) to diffuse the analyte(s) through the boundary layer between the bulk solution and the extracting phase.

According to the kinetic theory of extraction, increasing the volume of the extracting phase by increasing the thickness of the coated sorbent may lead to unsustainably long extraction equilibrium time as demonstrated in Equation (3).

As the extracting phase in a microextraction technique is generally immobilized on the substrate surface in the form of a thin film, the diffusion of the analyte(s) through the boundary layer regulates the rate of extraction (extraction kinetics). The time required to extract 95% of the equilibrium extraction amount of the analyte $t_{e,95\%}$ can be calculated as:

$$te, 95\% = \frac{B\delta bKes}{Ds} \tag{3}$$

where, b = the thickness of the extracting sorbent; δ = thickness of the boundary layer; K_{es} = distribution constant for the analyte between the extracting sorbent and the sample matrix; Ds = diffusion coefficient of the analyte in the sample matrix; B = geometric factors related to the geometry on which the extracting sorbent is immobilized.

It is evident from Equation (3) that the equilibrium extraction time can be reduced by (i) reducing the coating thickness of the sorbent, (ii) increasing the primary contact surface area of the extracting media (smaller B value), and (iii) increasing the analyte diffusion in the sample matrix by applying external energetics such as magnetic stirring, sonication, orbital shaking, etc.

The rate of extraction in the FPSE membrane, like other microextraction techniques, is not linear. The extraction proceeds very fast in the beginning of the extraction process, and the extraction rate steadily decreases as the extraction progresses towards the equilibrium. The initial rate of extraction ($\frac{dn}{dt}$) is directly proportional to the surface area of the extracting phase A, as shown in Equation (4).

$$\frac{dn}{dt} = \left(\frac{Ds\ A}{\delta}\right) \cdot Co \tag{4}$$

As such, if one is to increase the sensitivity of the microextraction technique, the volume of the extracting sorbent must be increased. In the same time, in order to reduce the extraction equilibrium time, the primary contact surface area of the extraction device must be augmented.

Fabric phase sorptive extraction has eloquently exploited both the thermodynamic and the kinetic criteria of the microextraction process. It utilizes sorbent loading approximately $40,000\times$ times higher than SPME fiber and $150\times$ times higher than stir bar. Regarding primary contact surface area, the FPSE membrane (in its typical 2.5 cm $\times$ 2.0 membrane size) is 50–100 times higher than SPME fiber and 10 times higher than stir bar.

In addition, the sponge-like porous architecture of sol–gel-derived hybrid inorganic–organic sorbent is highly favorable for achieving a fast extraction equilibrium compared to their classical counterparts: highly viscous pristine organic/inorganic polymers traditionally used in most of the microextraction techniques.

3. Preparation of Sol–Gel Sorbent Coated FPSE Membranes

Preparation of sol–gel sorbent coated FPSE membrane involves a number of decision points, including:

(1) Selection and pretreatment of the fabric substrate.
(2) Design and preparation of the sol solution for creating the sol–gel sorbent coating on the treated fabric substrate.
(3) Sol–gel sorbent coating process using dip coating technology.
(4) Aging, thermal conditioning, and cleaning of the sol–gel sorbent coated FPSE membrane.
(5) Based on the sample volume, cutting the FPSE membrane into the appropriate size.

It is worth mentioning that steps 1–3 primarily depend on the physicochemical properties of analytes, especially the polarity and molecular state of the analytes.

3.1. Selection and Pretreatment of Fabric Substrate

Among all the microextraction techniques, FPSE is the only sample preparation technique that exploits the surface chemistry of the fabric substrate. In general, if the analytes are nonpolar, a hydrophobic substrate such as polyester is the rational choice.

When the analytes are polar or medium-polar, a hydrophilic fabric substrate such as 100% cotton cellulose is the judicious selection. Sol–gel sorbents are chemically bonded to the fabric substrate. To ensure this chemical bonding, the fabric substrate should possess abundant surface hydroxyl functional groups. Another important selection criterion for the fabric substrate is its permeability so that the aqueous sample containing the analyte(s) of interest can permeate through the FPSE membrane easily even after creating the sol–gel sorbent coating on the substrate surface. The through pores of the FPSE membrane can extract the analyte(s) almost exhaustively at a short period. The selection of the fabric substrate is followed by the pretreatment of substrate to remove any residual finishing chemicals from the fabric surface. To clean the fabric substrate and to activate surface hydroxyl groups, a fabric treatment protocol has been developed. The protocol can be found elsewhere [7,9].

3.2. Design and Preparation of the Sol Solution for Creating the Sol–Gel Sorbent Coating on the Treated Fabric Substrate

The most important step in preparing the FPSE membrane is the design of sol solution. The sol solution for creating sol–gel sorbent coating on the substrate surface consists of (a) one or more inorganic/organically modified sol–gel precursors, (b) a sol–gel active inorganic/organic polymer, (c) a compatible solvent system, (d) an acid catalyst, and water for hydrolysis. Among numerous available organically modified silane precursors, methyl trimethoxysilane (MTMS) is the most commonly used sol–gel precursor. Other popular sol–gel precursors include phenyl trimethoxysilane (PTMS) and 3-aminopropyl trimethoxysilane (3-APTMS).

Commercially available sol–gel active inorganic/organic polymers are abundant in number and many of them are yet to be explored as viable candidates for an FPSE sorbent. Popular polymers used in FPSE include poly(dimethyl siloxane) (PDMS), poly(ethylene glycol) (PEG), poly(tetrahydrofuran) (PTHF), and poly(dimethyl diphenyl siloxane) (PDMDPS).

Among many commercially available acid catalysts (HCl, acetic acid, hydrofluoric acid, trifluoroacetic acid, oxalic acid), trifluoroacetic acid (TFA) is the most commonly used acid catalyst in sol–gel synthesis.

The sol solution for sol–gel sorbent coating on a fabric substrate is generally prepared in an amber reaction vessel (2 oz.) by sequential addition and subsequent vortexing of the sol–gel precursor, solvent, organic/inorganic polymer, acid catalyst, and water.

A detail account on the potential chemical reactions involved in the sol–gel sorbent coating process can be found elsewhere [5].

The primary criteria for selecting the sol–gel precursor and the inorganic/organic polymer are based on the polarity and functional makeup of the target analytes. Generally speaking, the higher the number of the intermolecular interactions between the FPSE membrane and the target analytes, the higher the extraction efficiency of an FPSE membrane. The overall selectivity and extraction efficiency of a sol–gel sorbent coated FPSE membrane depend combinedly on the surface chemistry of the fabric substrate, the sol–gel precursor, and the inorganic/organic polymer. As such, the selectivity of the pristine polymers such as PDMS and PEG used in SPME and similar microextraction devices are substantially different than that of sol–gel PDMS and sol–gel PEG coated FPSE membranes. Sol–gel sorbents are highly porous and easily accessible for the aqueous/gaseous sample matrices due to their sponge-like porous 3D polymeric network.

3.3. Sol–Gel Sorbent Coating Process Using Dip Coating Technology

The sol solution prepared in step 2 is employed in the sol–gel dip coating process. To initiate the coating process, a segment of the pretreated fabric is carefully submerged into the sol solution. The coating process begins as soon as the fabric substrate is introduced into the sol solution. Typically, the sol–gel coating process continues for 12 h at room temperature. Once the predetermined residence in the sol solution is over, the sol solution is discarded from the reaction vessel, and the sol–gel sorbent coated FPSE membrane is air dried for 1 h.

3.4. Aging, Thermal Conditioning, and Cleaning of Sol–Gel Sorbent Coated FPSE Membrane

The air-dried sol–gel sorbent coated FPSE membrane is thermally conditioned in a special conditioning device built inside a gas chromatograph (GC) oven under continuous helium gas flow for 24 h. The temperature of the GC oven is set at 50 °C. After conditioning at 50 °C for 24 h, the FPSE membrane undergoes a cleaning protocol established to remove unbonded sol solution ingredients and reaction byproducts. The FPSE membrane cleaning protocol involves rinsing the membrane in methylene chloride: a methanol mixture (50:50 v/v) under sonication for 1 h. The rinsing solvent mixture is then drained from the rinsing vessel, the FPSE membrane is air dried for 1 h and thermally condition at 50 °C for 24 h under a helium environment. This step completes the sequence of steps involved in creating the sol–gel sorbent coated FPSE membrane. The FPSE membrane is stored in an air-tight container until it is used in fabric phase sorptive extraction.

3.5. Cutting the FPSE Membrane into Appropriate Size

Unlike classical microextraction techniques such as SPME, SBSE, and TFME, the membrane size in FPSE is not fixed and can be adjusted based on the analytical need. For a small volume of sample (for example, blood, plasma, saliva), a small FPSE membrane disc (1 cm diameter) can be used. For a larger sample volume (5–20 mL), a larger membrane size (e.g., 2.5 cm × 2.0 cm) is recommended. Although a larger size of an FPSE membrane favors a faster extraction equilibrium due to higher contact surface area, it requires a relatively larger volume of organic solvent for quantitative back-extraction, which may unnecessarily dilute the analytes prior to injection into the chromatographic system. It is important to note that FPSE eliminates the solvent evaporation and sample reconstitution from the sample preparation workflow, an inevitable step in the SPE workflow. As such, the volume of solvent usage in FPSE back-extraction must be kept at its lowest level as possible.

4. Mechanism of Extraction in FPSE

Classical microextraction techniques such as SPME, SBSE, and TFME preferentially employ highly viscous pristine polymeric sorbents including PDMS, PEG, and PA, etc., as the extracting phase. During extraction, the analytes are solvated by the extracting polymeric phase. The diffusion coefficient in the highly viscous polymeric coating enables the analytes to penetrate the whole volume of the coating if enough time is allowed. As such, the mass transfer rate as well as the extraction kinetic is relatively slow in the viscous polymeric sorbent coating. When the analytes are heavier (high molar mass), the diffusion into the polymeric extracting phase is even slower. The extraction kinetic can be enhanced by impregnating the viscous polymeric phases with high surface area carbonaceous particulates material such as divinyl benzene (DVB) and Carboxen. These particles act as a bridge inside the liquid polymeric phases and facilitate faster extraction kinetics.

Unlike pristine polymers used in classical microextraction techniques, FPSE utilizes sol–gel sorbent coating technology that chemically binds the organic/inorganic polymer to the fabric substrate via sol–gel precursor as a cross-linker. The resulting sol–gel sorbent is a 3D polymeric network possessing random linkage between the sol–gel precursor and the inorganic/organic polymer. Sol–gel sorbents are inherently porous with sponge-like porous architecture containing numerous mesopores and micropores. During the analyte extraction, the fabric substrate attracts aqueous sample/analytes via its hydrophilic/hydrophobic surface property. As the analytes approach towards the FPSE membrane, multiple intermolecular interactions between the sol–gel sorbent and the analytes come into play, resulting in successful extraction of the analytes into the sol–gel sorbent. The sorbent loading in the FPSE membrane is very high compared to SPME/SBSE/TFME. Due to the high sorbent loading, only a fraction of analyte retention capacity is utilized during the extraction process (even after the exhaustive extraction) as demonstrated by Mesa et al. [10].

5. Types of Sorbents in FPSE

Fabric phase sorptive extraction is the only microextraction technique that offers a complete range of sorbent chemistries including polar, medium polar, nonpolar, cation exchanger, anion exchanger, mixed mode, zwitterionic, as well as zwitterionic mixed mode sorbents. Table 1 provides a list of major FPSE sorbent chemistries. It should be noted that all of these sorbents can be coated either on 100% cotton cellulose (hydrophilic) or on fiber glass (neutral) or on polyester (hydrophobic) substrates.

Table 1. List of major fabric phase sorptive extraction (FPSE) sorbents.

	Name of the SorbeSorbent Coating	Polarity of the Sorbent
	Neutral Sorbents	
1.	Sol–gel poly(dimethylsiloxane)	Nonpolar
2.	Sol–gel poly(dimethyldiphenylsiloxane)	Nonpolar
3.	Sol–gel methyl	Nonpolar
4.	Sol–gel C4	Nonpolar
5.	Sol–gel C8	Nonpolar
6.	Sol–gel C12	Nonpolar
7.	Sol–gel C18	Nonpolar
8.	Sol–gel Graphene	Nonpolar
9.	Sol–gel Multi Wall Carbon Nanotubes	Nonpolar
10.	Sol–gel Single Wall Carbon Nanotubes	Nonpolar
11.	Sol–gel Activated Carbon	Nonpolar
12.	Sol–gel poly(tetrahydrofuran)	Medium polar
13.	Sol–gel poly(ethylene glycol)-poly(propylene glycol)-poly(ethylene glycol)	Medium polar
14.	Sol–gel poly(propylene glycol)-poly(ethylene glycol)-poly(propylene glycol)	Medium polar
15.	Sol–gel propyl methacrylate	Medium polar
16.	Sol–gel poly(caprolactone)-poly(dimethylsiloxane)-poly(caprolactone	Medium polar
17.	Sol–gel poly(caprolactone)-poy(tetrahydrofuran)-poly(caprolactone)	Medium polar
18.	Sol–gel poly(caprolactone diol)	Medium polar
19.	Sol–gel poly(caprolactone triol)	Medium polar
20.	Sol–gel Silica	Polar
21.	Sol–gel Sucrose	Polar
22.	Sol–gel Sucralose	Polar
23.	Sol–gel Chitosan	Polar
24.	Sol–gel Carbowax 20M	Polar
25.	Sol–gel poly(ethylene glycol), 300	Polar
26.	Sol–gel poly(ethylene glycol), 10,000	Polar
	Mixed Mode Sorbents	
27.	Sol–gel cation exchanger, C18	Ion exchanger/nonpolar
28.	Sol–gel anion exchanger, C18	Ion exchanger/nonpolar
29.	Sol–gel Zwitterionic cation exchanger, anion exchanger, C18	Dual ion exchanger/nonpolar
	Ion Exchanger Sorbents	
30.	Sol–gel cation exchanger	Ion exchanger
31.	Sol–gel anion exchanger	Ion exchanger
32.	Sol–gel Zwitterionic anion and cation exchanger	Ion exchanger

6. FPSE Method Development

Unlike solid phase microextraction and similar sorbent based microextraction techniques, method development in fabric phase sorptive extraction is simple and straight forward. Figure 1 presents a graphical schematic of a typical FPSE workflow.

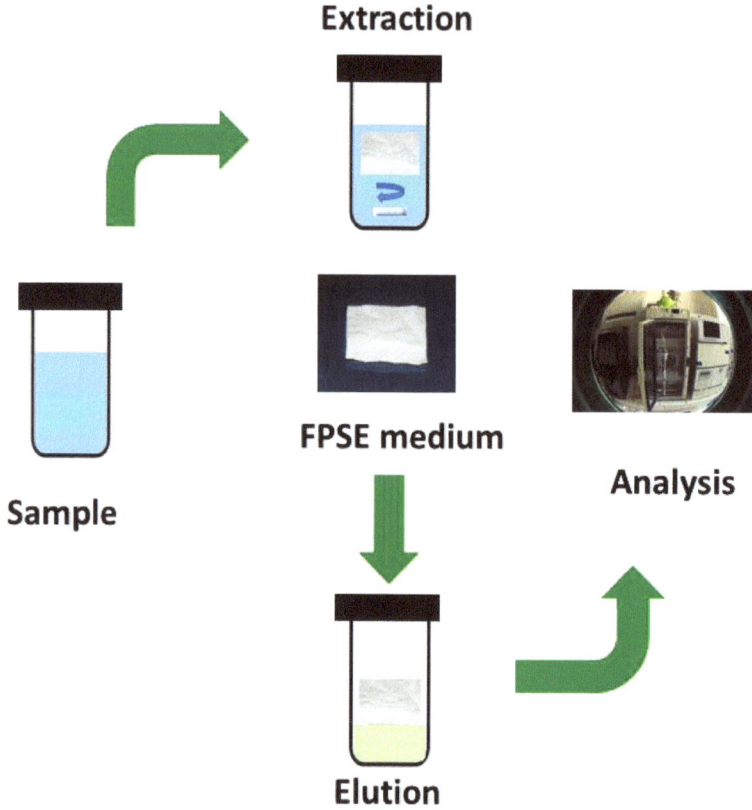

Figure 1. Typical FPSE workflow.

FPSE does not require any sample pre-treatment process to reduce/minimize matrix interferents such as filtration, protein precipitation, or centrifugation, and the FPSE membrane can be introduced directly into the sample, regardless of the complexity of the sample. However, the extraction efficiency can be substantially improved when a systematic method development strategy is followed to optimize a number of factors that directly impact on the overall extraction efficiency of the FPSE membrane. The factors are presented in Figure 2 with their relative significance. As such, an analyst may decide which factor(s) should be given more attention during the method development exercises. The factors include:

(i)   Sorbent chemistry;
(ii)   Substrate surface chemistry;
(iii)   Extraction equilibrium time;
(iv)   Sample volume;
(v)   Desorption time;
(vi)   Desorption solvent;
(vii) Ionic strength;
(viii) Sample pH;
(ix)   Agitation mode;
(x)   FPSE membrane dimension.

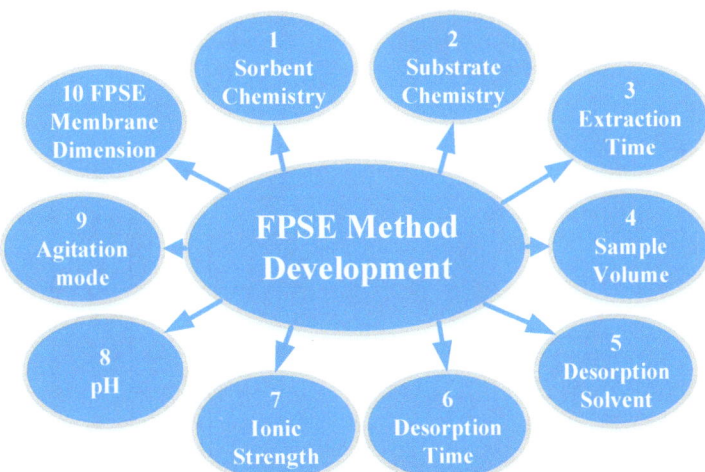

Figure 2. Factors and their relative importance on fabric phase sorptive extraction method development.

FPSE method development exercises can be carried out using a conventional one-factor-at-a-time (One FAT) approach or using a chemometric design of experiment approach. The later approach is the more scientific and green approach, as it provides deep insight about the overall extraction process and sheds light as to whether different factors interact with each other or not. A screening design can be carried out to select factors with the most influence on the overall extraction efficiency. Subsequently, a response surface model (RSM) design can be employed to find the optimum levels of the most influential factors.

6.1. Selection of FPSE Sorbent Chemistry

As can be seen in Table 1, FPSE offers a broad range of sorbents spanning from nonpolar, to medium polar, to polar, to ionized, to mixed mode, and to zwitterionic. As such, it is practically impossible for one to determine the most efficient sorbent by real experimentation. As such, for the first time, a new approach for selecting FPSE sorbent chemistry has been developed based on an absolute recovery percentage calculator that utilizes the logKow of an analyte to predict an estimated absolute recovery of an analyte for a given FPSE sorbent chemistry. For example, the absolute recovery on sol–gel Carbowax 20M (sol-gel CW 20M) sorbent coated on 100% cotton cellulose fabric can be expressed as:

Absolute Recovery % = $4.2977487 + 22.823041 \times \text{Log Kow} - 3.1343544 \times (\text{Log Kow} - 2.737)^2$

This equation is valid for any analyte possessing a logKow value between 0.3 and 5.07, and majority of the analytes we generally encounter fall in this range. During the FPSE method development exercises, it is recommended that an analyst select the 3 best FPSE membranes, and subsequently determine the best FPSE membrane by exposing them under identical FPSE conditions. A good starting time can be:

FPSE membrane size: 2.5 cm × 2.0 cm;
Sample volume: 10 mL;
Analyte concentration: 1 µg/mL;
Extraction time: 1 h;
Stirring speed: 800 rpm;
Desorption solvent: methanol;
Desorption solvent volume: 500 µL;
Desorption time: 10 min.

The prepared sample can be injected into a gas chromatograph or high-performance liquid chromatograph to obtain the chromatographic signal area for an analyte or group of analytes.

Absolute recovery calculations for major FPSE sorbent chemistries are presented in Table 2.

Table 2. Absolute recovery calculator for selected FPSE membranes.

Sorbent (Substrate)	Equation for Recovery% Calculation
Si-CW20M (Cellulose)	$4.2977487 + 22.823041 \log Kow - 3.1343544 (\log Kow - 2.737)^2$
Si-PEG1000 (Cellulose)	$-11.53483 + 20.950137 \log Kow - 0.4017218 (\log Kow - 2.737)^2$
Si-PEG300 (Cellulose)	$14.758805 + 16.309632 \log Kow - 5.5504622 (\log Kow - 2.737)^2$
Si-CN-CW20M (Cellulose)	$-24.39275 + 23.940499 \log Kow + 1.247171 (\log Kow - 2.737)^2$
Si-PPG-PEG-PPG (Cellulose)	$-3.648816 + 21.546191 \log Kow - 2.878525 (\log Kow - 2.737)^2$
Si-PEG-PPG-PEG (Cellulose)	$-7.680093 + 23.069108 \log Kow - 1.7262745 (\log Kow - 2.737)^2$
Si-PTHF (Cellulose)	$12.40054 + 17.848979 \log Kow + 17.848979 (\log Kow - 2.737)^2$
Si-PTHF (Fiber Glass)	$-28.44237 + 20.9507 \log Kow + 3.3273496 (\log Kow - 2.737)^2$
Si-C18 (Cellulose)	$-2.274875 + 20.816015 \log Kow - 4.1478973 (\log Kow - 2.737)^2$
Si-C8 (Cellulose)	$-3.392783 + 21.261305 \log Kow - 3.7724155 (\log Kow - 2.737)^2$
Si-PDPS (Cellulose)	$-10.30009 + 17.450029 \log Kow - 0.2880039 (\log Kow - 2.737)^2$
Si-PDMDPS (Polyester)	$-9.185327 + 17.815515 \log Kow - 1.9655752 (\log Kow - 2.737)^2$
Si-PDMDPS (Cellulose)	$-19.60225 + 15.453851 \log Kow - 1.62186 (\log Kow - 2.737)^2$

The predicted absolute recovery values often corroborate with the actual recovery values obtained from real experimentation, as demonstrated by several researchers [11,12]. It is important to note that this model was developed using analyte solution in deionized water. When the sample matrix contains too many matrix interferents, substantial deviation from the expected recovery of the analyte may be observed [13].

6.2. Selection of FPSE Substrate

FPSE is the only microextraction technique that exploits the substrate surface chemistry to compliment to the overall selectivity and the extraction efficiency of an FPSE membrane. The surface property of the fabric substrate substantially impacts on the selectivity and extraction efficiency of an FPSE membrane. The dependence of the analytes' polarity (logKow) on the nature of different fabric substrates has been estimated using a compound mixture consisting of furfural alcohol (FA, logKow 0.3), piperonal (PIP, logKow 1.05), phenol (PHE, logKow 1.5), benzodioxole (BDO, logKow 2.08), 4-nitrotoluene (4NT, logKow 2.45), 9-anthracene methanol (9AM, logKow 3.04), 1,2,45-tetramethyl benzene (TMB, logKow 4.0), triclosan (TCL, logKow 4.53), and diethylstilbestrol (DES, logKow 5.07). As can be seen from Table 3, for the sol–gel PTHF sorbent, cellulose fabric is favored for polar analytes extraction, whereas fiber glass fabric is suitable for nonpolar analyte extraction. Between sol–gel PDMDPS sorbents coated on polyester fabric and cellulose fabric, polyester is better for nonpolar analyte extraction. Since PDMDPS is a nonpolar polymer, extraction of polar analytes is not favored in either polyester fabric or in cellulose fabric. As expected, the organic/inorganic polymer plays the most significant role in the overall selectivity and extraction efficiency of FPSE membrane. However, the role of the fabric substrate cannot be ignored.

Table 3. Comparison of extraction recovery between different fabric substrates.

Sorbent	FA (%)	PIP (%)	PHE (%)	BDO (%)	4NT (%)	9AM (%)	NAP (%)	TMB (%)	TCL (%)	DES (%)
Sol–gel PTHF (Cellulose)	0	9.8	4.0	45.6	46.7	66.1	86.8	93.4	82.1	49.5
Sol–gel PTHF (Fiber Glass)	0	1.2	4.1	21.8	25.4	25.4	38.4	77.8	74.9	91.7
Sol–gel PDMDPS (Polyester)	0	1.7	0	13.4	13.7	67.7	51.8	83.5	74.9	46.7
Sol–gel PDMDPS (Cellulose)	0	0.1	1.8	9.7	10.0	11.0	68.7	50.4	42.1	68.1

6.3. Optimization of Extraction Equilibrium Time

Extraction efficiency is one of the most important factors that influence the extraction efficiency of an FPSE membrane. Generally, extraction efficiency is verified between 0 and 60 min, when most of the analytes reach the plateau of the extraction kinetic curve, and exposing the FPSE membrane longer than this time period does not yield any improvement in the extraction efficiency of an analyte in a given FPSE membrane. In some cases, when a high mass of matrix interferents are present in the sample matrix, as in the case of an environmental or biological sample, longer extraction equilibrium time may be observed.

6.4. Optimization of Sample Volume

Sample volume requirement in FPSE is flexible and depends on the availability and nature of the sample. For a smaller sample volume, a smaller FPSE membrane size can be used. If the sample is freely available, a larger FPSE membrane size (e.g., 2.5 cm × 2.0 cm) can be used, and a sample volume from 10 mL to 30 mL may be systematically investigated to determine the optimum sample volume.

6.5. Optimization of Desorption Solvent

Due to the strong chemical bonding between the fabric substrate and the sol–gel sorbent coating, an FPSE membrane can be exposed to any organic solvent for quantitative back-extraction of the analytes after the extraction process. As such, a single solvent or a mixture of solvents can be used to efficiently back-extract the analytes. The solvent or solvent system (mixture of multiple solvents) should be optimized to ensure quantitative back-extraction of the extracted analytes.

6.6. Optimization of Desorption Time

Since the sol–gel sorbents are inherently porous with sponge-like morphology, the diffusion of the solvent during solvent mediated back-extraction does not need any external energetic stimulus such as magnetic stirring. However, it is imperative to allow adequate time for the solvent to exhaustively scavenge the extracted analytes from the sol–gel sorbents. Most researchers have reported 5 min as the optimum desorption time, although in some cases 7.5 min or 10 min as the optimum desorption times are not unusual. For method development, a time range between 0 and 10 min can be investigated.

6.7. Optimization of Ionic Strength of the Sample Matrix

Ionic strength of the sample matrix can be increased by the addition of NaCl or another suitable salt to the sample to compel polar analytes out of the aqueous solution and become available for being extracted into the FPSE membrane. The optimum salt concentration can be determined by monitoring the increase in the extraction efficiency with the concentration of salt in the solution.

6.8. Optimization of pH of the Sample Matrix

When acidic or basic analytes are extracted on a neutral FPSE membrane, pH adjustment of the sample matrix may be used to force the analytes to remain in their neutral state so that the neutral FPSE membrane can maximize its extraction efficiency under the given extraction conditions. It is a cumbersome process and requires obtaining an optimum matrix pH value via a series of experiments. In order to eliminate this cumbersome drill, a mixed mode sorbent coated FPSE membrane can be used.

6.9. Optimization of Sample Matrix Agitation

Extraction kinetics can be expedited by applying external stimuli such as magnetic stirring, ultra-sonication, or orbital shaking during the FPSE process. The optimum stirring speed should be established experimentally during the FPSE method development.

6.10. Selection of FPSE Membrane Size

FPSE is the only microextraction technique that allows the analyst to determine the size of the FPSE membrane. Although the typical size for a small volume of sample is a 1 cm diameter disc, or a 2.5 cm × 2.0 cm rectangular block for a larger sample volume, the analyst may use any size of the FPSE membrane depending on the analytical need.

7. Applications

All developed methodologies reported in the literature since 2014 are briefly described below, showing the wide range of applicability of fabric phase sorptive extraction in terms of sample matrix and analyte diversity.

Many research groups all over the world have adopted this innovative sample preparation approach and have developed new analytical strategies to deal with significant analytical problems encountered in virtually all facets of analytical fields.

Kumar et al. in 2014 were the first group to implement fabric phase sorptive extraction (FPSE) in the development of a simple, fast, and sensitive analytical method using a sol–gel poly(tetrahydrofuran) (sol–gel PTHF) coated FPSE membrane for the quantification of endocrine disrupting chemicals (EDCs), including 17α-ethinyl estradiol (EE2), β-estradiol (E2) and bisphenol A (BPA). Analysis was performed by high performance liquid chromatography with fluorescence detection (HPLC-FLD). In their work, the authors have investigated and optimized various factors that influence the efficiency of FPSE technique. The developed method was applied successfully for the analysis of the examined estrogen molecules in urine and various kinds of aqueous samples with good reported recoveries, i.e., 96–98% for drinking water, 94–95% for ground water, 92–94% for river water, and 90–91% for urine samples, while lower detection limits of BPA, E2, and EE2 over previously reported methods were achieved within the range of 20 to 42 pg/mL. Linearity, precision, and accuracy results proved that the developed method is rapid, precise, reproducible, and sensitive for the determination of estrogens in urine and aqueous samples [9].

A year later, in 2015, Roldán-Pijuán et al. [14] presented for the first time a novel technique: the approach of stir fabric phase sorptive extraction (SFPSE), which integrates sol–gel hybrid organic–inorganic coated fabric phase sorptive extraction media with a magnetic stirring mechanism. Two flexible fabric substrates, namely cellulose and polyester, were utilized as the host matrix for three different sorbents, e.g., sol–gel poly(tetrahydrofuran) (sol–gel PTHF), sol–gel poly(ethylene glycol) (sol–gel PEG), and sol–gel poly(dimethyldiphenylsiloxane) (sol–gel PDMDPS). Triazine herbicides were selected as model compounds to evaluate the operational performance of this unique microextraction device. The factors affecting the extraction efficiency of SFPSE have been investigated, and the optimal extraction conditions using sol–gel PEG coated SFPSE device in combination with UPLC-DAD yielded limits of quantification (LOQs) for the seven triazine herbicides in the range of 0.26–1.50 µg/L, while the hyphenation with LC-MS/MS allowed the improvement of the method sensitivity to the range of 0.015 µg/L to 0.026 µg/L. Enrichment factors between 444 and 1411 were achieved. The developed method was finally applied for the determination of selected triazine herbicides from three river water samples. Relative recoveries of the target analytes, in the range of 75% to 126%, were found to be satisfactory, while absolute extraction recoveries were in the range of 22.2–70.5%.

In the same year, Montesdeoca-Esponda et al. [15] developed a fast and sensitive sample preparation methodology using fabric phase sorptive extraction followed by ultra-high-performance liquid chromatography and tandem mass spectrometry detection for the determination of a group of compounds added to sunscreens and other personal care products which may present detrimental effects to aquatic ecosystems, i.e., benzotriazole UV stabilizer compounds in aqueous samples. In their work, the authors optimized the extraction of seven benzotriazole UV filters in terms of several parameters influencing the extraction process, such as sorbent chemistry selection, extraction time, back-extraction solvent, back-extraction time, and the impact of ionic strength. Under the optimized conditions, which included polyester fabric that was used as the substrate for sol–gel PDMDPS coating, FPSE provided enrichment factors of 10 times with detection limits ranging from 6.01 to 60.7 ng L^{-1}. Ultra-high-performance liquid chromatography and tandem mass spectrometry detection were used for the determination of target analytes in sewage samples from wastewater treatment plants with different purification processes of Gran Canaria Island (Spain).

No expensive commercial supplies or instrument were needed. Thus, FPSE was proved to be a cost-effective alternative to other expensive extraction and microextraction methods for the target analytes.

Kumar et al. in 2015 [12] developed and validated a novel analytical method for the quantification of endocrine-disruptor alkyl phenols, namely: 4-tert-butylphenol, 4-sec-butylphenol, 4-tert-amylphenol, and 4-cumylphenol, in aqueous and soil samples. Analysis was subsequently performed by high-performance liquid chromatography with ultraviolet detection. Various parameters influencing the fabric phase sorptive extraction performance, such as extraction time, eluting solvent, elution time, and pH of the sample matrix, were optimized. Sol–gel PTHF coated FPSE media on cellulose substrate was proved to show the best extraction efficiency with methanol as the extraction solvent, yielding recovery rates of 74.0, 75.6, 78.0, and 78.3 for 4-tert-butylphenol, 4-sec-butylphenol, 4-tert-amylphenol, and 4-cumylphenol, respectively. Optimum conditions offer high preconcentration and enrichment factors and significantly reduced sample preparation time. The limits of detection ranged from 0.161 to 0.192 ng/mL, and the method was successfully applied for the recovery of alkyl phenols from spiked ground water, river water, and treated water obtained from a sewage treatment plant, and a soil and sludge sample. The method reduces the use of organic solvents substantially, meeting the criteria of the green analytical chemistry principle.

Racamonde et al. in 2015 [16] investigated the use of FPSE for the determination of four nonsteroidal anti-inflammatory drugs (ibuprofen, naproxen, ketoprofen and diclofenac) in environmental water samples prior to their determination by gas chromatography mass spectrometry. Various factors affecting FPSE, namely: sorbent chemistry, matrix pH, and ionic strength, were investigated using a mixed level factorial design (31×22), while other important parameters, e.g., sample volume, extraction kinetics, desorption time, and volume, were also optimized. Three different FPSE sorbent chemistries, sol–gel PDMDPS, sol–gel PTHF, and sol–gel PEG, were investigated. Sol–gel PEG coatings on the cellulose substrate with ethyl acetate as the eluent proved to provide optimal operational conditions, leading to the limits of detection (S/N = 3) in the range of 0.8–5 ng L^{-1}. The enrichment factors ranged from 162 to 418, while absolute extraction efficiencies varied from 27% to 70%. Satisfactory relative recoveries within the range 82–116% demonstrated that the proposed method can be readily applied to routine environmental pollution monitoring. Actually, the proposed method was successfully applied to the analysis of examined analytes in two influent and effluent samples from a wastewater treatment plant and two river water samples in Spain.

Compared to other sorptive microextraction techniques, FPSE showed many benefits such as simplicity in device fabrication, low cost, high enrichment factors and faster extraction equilibrium.

Samanidou et al. in 2015 [17] applied FPSE to develop a simple, sensitive, reliable, and fast analytical methods for the simultaneous determination of residual highly polar amphenicol antibiotics (amphenicols) in raw milk, followed by high-performance liquid chromatography–diode array analysis. A highly polar polymer coated FPSE membrane using short-chain poly (ethylene glycol) (sol–gel PEG) was used. The intense affinity of amphenicols towards the strongly polar sol–gel PEG-coated FPSE device yielded absolute recovery of the selected antibiotics residues in the range of 44% for thiamphenicol, 66.4% for florfenicol, and 81.4% for chloramphenicol. The developed method was validated in terms of sensitivity, linearity, accuracy, precision, and selectivity according to European Decision 657/2002/EC. The decision limit (CCα) values obtained were 52.49 µg kg^{-1} for thiamphenicol, 55.23 µg kg^{-1} for florfenicol, and 53.8 µg kg^{-1} for chloramphenicol, while the corresponding results for detection capability (CCβ) were 56.8 µg kg^{-1}, 58.99 µg kg^{-1}, and 55.9 µg kg^{-1}, respectively.

Lakade et al. in 2015 [18] proposed the use of FPSE applying different coating chemistries, namely: nonpolar sol–gel PDMDPS, medium polar sol–gel PTHF, and polar sol–gel poly(ethylene glycol)-block-poly(propylene glycol)-block-poly(ethylene glycol) (sol–gel PEG-PPG-PEG), and sol–gel Carbowax 20M (sol–gel CW 20M) to the extraction of a group of pharmaceuticals and personal care products (PPCPs) with a wide range of polarity from environmental aqueous samples. Several factors influencing FPSE, such as sample pH, stirring speed, addition of salt, extraction time, sample volume, elution solvent, and desorption time, were investigated and optimized for each sorbent coated FPSE membrane. Optimum conditions included the FPSE membrane coated with sol–gel CW 20M that provided the highest absolute recoveries (77–85%) for most of the analytes, except for the most polar ones. All examined sorbents offered better recovery compared to the commercially available coating for stir bar sorptive extraction based on ethylene glycol/silicone (EG/Silicone). The method based on FPSE with sol–gel CW 20M membrane and liquid chromatography-(electrospray ionization) tandem mass spectrometry (LC-(ESI) MS/MS) was applied to environmental water samples. Good apparent recoveries (41–80%) and detection limits (1–50 ng L^{-1}) were achieved.

One year later, in 2016, Anthemidis et al. [19] developed a novel flow injection fabric disk sorptive extraction (FI-FDSE) system for the automated determination of trace metals. The platform was based on a mini-column packed with sol–gel coated fabric membrane in the form of disks, incorporated into an on-line solid-phase extraction system, coupled with flame atomic absorption spectrometry (FAAS). This configuration resulted in high loading flow rates and shorter analytical cycles due to the minor observed backpressure. The potentials of this technique were demonstrated for trace lead and cadmium determination in environmental water samples. Various sol–gel coated FPSE media were investigated. The on-line formed complex of metal with ammonium pyrrolidine dithiocarbamate (APDC) was retained onto the fabric surface. Among the examined sol–gel coated FPSE membranes, sol–gel PDMDPS coated membrane provided the best extraction sensitivity and excellent reproducibility due to its hydrophobic nature similar to that of metal-APDC complex. The analytes were subsequently eluted by methyl isobutyl ketone (MIBK) prior to atomization. Optimum parameters included 90 s preconcentration time, with a sampling frequency of 30 h^{-1}, and thus enrichment factors of 140 and 38 and detection limits of 1.8 and 0.4 µg L^{-1} were achieved for lead and cadmium.

Huang et al. in 2016 [20] proposed the use of the cellulose fabric as the host matrix, for three extraction sorbents, namely: sol–gel PTHF, sol–gel PEG, and sol–gel PDMDPS, which were prepared on the surface of the cellulose fabric. Two extraction techniques have been proposed. The first one included stir bar fabric phase sorptive extraction (stir bar-FPSE), and the second one magnetic stir fabric phase sorptive extraction (magnetic stir-FPSE), both allowing stirring of fabric phase sorbent during every step of the extraction process. Three brominated flame retardants (BFRs) [tetrabromobisphenol A (TBBPA), tetrabromobisphenol A bisallylether (TBBPA-BAE), tetrabromobisphenol A bis(2,3-dibromopropyl)ether (TBBPA-BDBPE)] were selected as model analytes for the practical evaluation of the two proposed techniques using high-performance liquid chromatography (HPLC). Several experimental conditions which mainly affect the extraction process such as the type of fabric phase, extraction time, the amount of salt, and elution conditions were studied and optimized. Both techniques possessed high extraction capability and fast extraction equilibrium as a result of the large sorbent loading capacity and unique stirring performance. High recoveries (90–99%) and low limits of detection (LODs) (0.01–0.05 μg.L^{-1}) were achieved using the optimized conditions. The results were promising, and the methods were shown to be practical for monitoring of hazardous pollutants in the water sample, meeting green analytical chemistry requirements, mainly due to low solvent consumption.

Guedes-Alonso et al. in 2016 [21] in their study proposed an extraction method based on sorptive fabric phase coupled to ultra-high-performance liquid chromatography tandem mass spectrometry detection (FPSE-UHPLC-MS/MS) for the determination of four progestogens and six androgens in environmental and biological samples. These analytes consist two important groups of endocrine disrupting compounds (EDCs) which may have severe harmful impact on aquatic biota, even at very low concentrations. All the experimental parameters involved in the extraction, such as sample volume, extraction and desorption times, desorption solvent volume, and sample pH values have been optimized. Sol–gel PTHF coated FPSE and analyte desorption with methanol proved to show the best results. The developed method showed satisfactory limits of detection (between 1.7 and 264 ng L^{-1}), and good recoveries. The applicability of the method was examined by its use of the analysis of tap water, wastewater treated with different processing technologies, and urine samples. The concentrations of the detected hormones ranged from 28.3 to 227.3 ng L^{-1} in water samples and from 1.1 to 3.7 μg L^{-1} in urine samples. The method showed significant benefits such as minimum usage of organic solvents, short extraction times, small sample volumes, and high analyte preconcentration factors.

Karageorgou et al. in 2016 [22] used FPSE for the determination of sulfonamides residues in milk using a highly polar sol–gel PEG coated membrane. The developed HPLC method was validated according to the European Union Decision 2002/657/EC. Due to the low organic solvent consumption, the FPSE-based method meets all green analytical chemistry (GAC) criteria. The decision limit (CC$_\alpha$) values were 116.5 μg kg^{-1} for sulfamethazine, 114.4 μg kg^{-1} for sulfisoxazole, and 94.7 μg kg^{-1} for sulfadimethoxine, whereas the corresponding results for detection capability (CC$_\beta$) were 120.4 μg kg^{-1} for sulfamethazine, 118.5 μg kg^{-1} for sulfisoxazole, and 104.1 μg kg^{-1} for sulfadimethoxine.

Samanidou et al. in 2016 [13] evaluated the application of FPSE for the extraction of benzodiazepines from human blood serum. Benzodiazepines were selected as model analytes because they represent one of the most widely used therapeutic drugs in psychiatry and are also amongst the most frequently encountered drugs in forensic toxicology. FPSE was performed using cellulose fabric extraction media coated with sol–gel PEG. Absolute recovery values in the equilibrium state for the examined benzodiazepines were found to be 27% for bromazepam, 63% for lorazepam, 42% for diazepam, and 39% for alprazolam.

Lakade et al. in 2016 [6] described the use of a new extraction approach based on fabric phase sorptive extraction (FPSE). This new mode proposes the extraction of the analytes in dynamic mode in order to reduce the extraction time. Dynamic fabric phase sorptive extraction (DFPSE) was applied using sol–gel Carbowax 20M material, followed by liquid chromatography–tandem mass spectrometry. This approach was evaluated for the extraction of a group of pharmaceuticals and personal care products (PPCPs) from environmental water samples. Different experimental parameters affecting the extraction were investigated and optimized. Best performance was achieved using ethyl acetate as elution solvent. Recovery rates were higher than 60% for most of the compounds, with the exception of the most polar ones (between 8% and 38%). The analytical method was validated and applied to river water, effluent and influent wastewater, and good performance was obtained. The analysis of samples revealed the presence of some PPCPs at low ng L^{-1} concentrations.

Aznar et al. in 2016 [23] in order to investigate the migration of additives added to food packaging materials to food in contact with them during storage and shelf life, developed a novel simple, fast, and sensitive analyte extraction method based on fabric phase sorptive extraction (FPSE), followed by analysis using ultra-high performance liquid chromatography and mass spectrometry detection (UPLC-MS). The method was applied to the analysis of 18 common non-volatile plastic additives. Three FPSE media coated with different sol–gel sorbents characterized by different polarities, including sol–gel poly(dimethyl siloxane) (sol–gel PDMS), sol–gel PEG, and sol–gel PTHF, were investigated, and all showed very satisfactory results. Analytes with low logP values (polar analytes) showed higher enrichment factors (EFs), especially with sol–gel PTHF and sol–gel PEG membrane. For compounds with high logP values (nonpolar compounds), the use of sol–gel PDMS improved the enrichment capacity.

For compounds with low logP values (logP < 5), sol–gel PEG coated FPSE media showed higher enrichment factors. Sol–gel PTHF coated FPSE media, with an intermediate polarity, showed the best EFs values.

Ten compounds obtained enrichment factors above 3 with sol–gel PTHF coated FPSE membrane, whereas for sol–gel PDMS or sol–gel PEG, only six compounds were above this value.

Acetonitrile showed best desorption efficiency yielding recoveries over 70% for 13 out of 18 selected compounds in all FPSE media.

Alcudia Leon et al. in 2017 [24] presented a novel sampling device that integrates air sampling and preconcentration based on fabric phase sorptive extraction principles. The determination of the main components of the sexual pheromone of *Tuta absoluta* [(3E,8Z,11Z)-tetradecatrien-1-yl acetate and (3E,8Z)-tetradecadien-1-yl acetate] traces in environmental air in tomato crops has been selected as a model system. A laboratory-built unit made up of commercial brass elements as a holder of the sol–gel coated fabric extracting phase was designed and optimized. The unit proved to efficiently work under sampling and analysis modes which eliminated any need for sorptive phase manipulation prior to instrumental analysis. In the sampling mode, the unit is connected to a sampling pump to pass the air through the sorptive phase under controlled flowrate. In the analysis mode, the unit is placed in the gas chromatograph autosampler without any instrumental modification, thus eliminating the risk of cross contamination between sampling and analysis. The limits of detection for both compounds resulted to be 1.6 μg and 0.8 μg.

Three different fabric phases coated with sol–gel PEG, sol–gel PTHF, and sol–gel PDMDPS were evaluated for the extraction of two sexual pheromones components from a gaseous standard. The results indicated that the sol–gel PDMDPS proved to show optimum results. In fact, the results confirmed the expected behavior considering the high hydrophobicity (log $K_{o/w}$ are 5.76 and 6.28 for component A and B, respectively) of the target compounds.

Locatelli et al. in 2017 [25] developed a fabric phase sorptive extraction high-performance liquid chromatography-photodiode array detection (FPSE-HPLC-PDA) method for the simultaneous extraction and analysis of twelve azole antimicrobial drug residues (i.e., ketoconazole, terconazole, voriconazole, bifonazole, clotrimazole, tioconazole, econazole, butoconazole, miconazole, posaconazole, ravuconazole, and itraconazole) in human plasma and urine samples. The limit of quantification of the FPSE-HPLC-PDA method was found as 0.1 μg/mL, and good linearity was observed up to a concentration of 8 μg/mL. The performance of the developed method was investigated on real samples from healthy volunteers after a single dose administration of itraconazole and miconazole. The method proved to be a rapid and robust green analytical tool for clinical and pharmaceutical applications. Three different FPSE membranes were investigated: sol–gel silica Carbowax® 20 M (sol–gel CW20 M) with 8.63 mg/cm^2 sorbent loading, sol–gel polydimethylsiloxane (sol–gel PDMS) with a sorbent loading of 4.56 mg/cm^2, sol–gel caprolactone-dimethylsiloxane-caprolactone (sol–gel

CAP-DMS-CAP) with a sorbent loading of 6.14 mg/cm^2, while optimal effectiveness was observed by sol–gel Carbowax® 20 M and methanol as elution solvent.

Heena et al. in 2017 [26] developed a method for the determination of Co(II), Ni(II), and Pd(II) in aqueous samples using fabric phase sorptive extraction high-performance liquid chromatography-UV detection (FPSE-HPLC-UV). A preconcentration step was necessary due to the trace level concentrations of these elements in aqueous samples. Sol–gel polytetrahydrofuran nanocomposite was selected as the optimum sorbent. The limit of detection for Co(II), Ni(II), and Pd(II) morpholino dithiocarbamate complexes were found at much lower concentration levels as compared to earlier reported data with excellent reproducibility. The new FPSE-HPLC-UV method can be used for routine determination of these metal species in various aqueous environmental samples and in different alloys.

Kazantzi and Anthemidis in 2017 [27] developed a novel flow injection on-line fiber fabric sorptive extraction (FI-FFSE) platform, taking advantage of the benefits of the FPSE technique in automatic mode. A microcolumn packed with a sol–gel coated fiber fabric medium, the poly(dimethylsiloxane) (sol–gel PDMS), incorporated into a FI-SPE system, was presented. The low backpressure in this configuration results in high loading flow rates and shorter analytical cycles. The on-line formed complex of metal with sodium diethyl dithiocarbamate (DDTC) is retained onto the fabric surface, while analytes are eluted by methyl isobutyl ketone (MIBK) prior to atomization. For 90 s preconcentration time, enrichment factors of 165 and 43 and detection limits (3 s) of 1.6 and 0.3 µg L^{-1} were achieved for lead and cadmium determination, respectively, with a sampling frequency of 30 h^{-1}. The developed method has been successfully applied to the on-line lead and cadmium determination by FAAS in energy and refreshment drinks.

In 2017, Samanidou et al. [28] evaluated fabric phase sorptive extraction (FPSE) as a simple and rapid strategy for the extraction of four penicillin antibiotic residues (benzylpenicillin, cloxacillin, dicloxacillin, and oxacillin) from cows' milk, without prior protein precipitation. Time-consuming solvent evaporation and reconstitution steps were eliminated successfully from the sample preparation workflow. Short-chain poly(ethylene glycol) provided optimum extraction sensitivity for the selected penicillins, which were analyzed using a Reversed Phase (RP) HPLC method, validated according to the European Decision 657/2002/EC. The limits of quantitation achieved were a similar order of magnitude with those reported in the literature (with the exception of benzylpenicillin) and less than the maximum residue limits (MRL) set by European legislation.

Saini et al. in 2017 [29] applied FPSE to the trace-level determination of four selected polycyclic aromatic hydrocarbons (PAHs), namely: fluoranthene, phenanthrene, anthracene, and pyrene, in environmental water samples using a nonpolar sol–gel C18 coated FPSE media. Extraction efficiency was optimized. Limits of detection (LODs) and quantification (LOQs) were found to be at pg/mL levels: 0.1–1 pg/mL and 0.3–3 pg/mL, respectively. Average absolute recovery rates were in the range of 88.1–90.5%. The applicability of the developed FPSE-HPLC-FLD protocol was proved by the analysis of eight environmental water samples and proved to be simple, green, fast, and cost effective, with adequate sensitivity, and thus it can be applied for routine monitoring of water quality and safety.

Samanidou et al. in 2017 [30] investigated the synergistic combination of the advanced material properties offered by sol–gel graphene sorbent and the simplicity of the fabric phase sorptive extraction approach in selectively extracting bisphenol A and residual monomers including bisphenol A glycerolatedimethacrylate, urethane dimethacrylate, and triethylene glycol dimethacrylate-derived dental restorative materials from cow and human breast milk samples. After evaluation of the extraction efficiency of different coatings, sol–gel graphene coated media proved to show the best results. The main experimental parameters affecting the analytes' extraction, namely: sorbent chemistry used, sample loading conditions, elution solvent, sorption stirring time, elution time, impact of protein precipitation, amount of sample, and matrix effect, were investigated and optimized. Absolute recovery values from standard solutions were 50% for bisphenol

A, 78% for triethylene glycol dimethacrylate, 110% for urethane dimethacrylate, and 103% for bisphenol A glycerolatedimethacrylate, while respective absolute recovery values from milk were 30%, 52%, 104%, and 42%. The developed method was validated according to European Decision 657/2002/EC.

Aznar et al. in 2017 [31] proposed a simple, fast, and sensitive analyte extraction method based on fabric phase sorptive extraction (FPSE) followed by gas chromatography mass spectrometry (GC-MS) and ultra-performance liquid chromatography-quadrupole time of flight mass spectrometry (UPLC-QTOF-MS) analysis for the analysis of 12 volatile and semi-volatile compounds, namely: furfuryl alcohol, butyric acid, cis-3-hexen-1-ol, ethyl butyrate, vanillin, ethyl isovalerate, linalool, 1-octen-3-one, eugenol, octanal, ethyl octanoate, and limonene, which represent most of the principal chemical families possessing different polarities and volatilities. Five FPSE membranes coated with different sol–gel sorbent chemistries having different polarities and selectivities were evaluated: long chain poly(dimethylsiloxane) (sol–gel PDMS), short chain poly(tetrahydrofuran) (sol–gel PTHF), Carbowax 20M (sol–gel CW20M), short chain poly(dimethyl siloxane) (sol–gel SC PDMS), and polyethylene glycol-polypropylene glycol-polyethylene glycol triblock copolymer (sol–gel PEG-PPG-PEG). Sol–gel CW20M coated FPSE media showed the best extraction performance. The developed methodology was applied to the analysis of orange juice obtained from fresh oranges and oranges after storing at 5 °C for two months in order to identify the best chemical markers, both volatiles and non-volatiles, attributed to the freshness of the orange.

Santana Viera et al. in 2017 [32] developed an FPSE based method for the analysis of seven cytostatic drug compounds that are commonly used in anti-cancer therapies prior to ultra-high-performance liquid chromatography tandem mass spectrometry (UHPLC-MS/MS). The extraction protocol was optimized after investigation of the major parameters that affect the extraction efficiency. The detection limit of the method was within the values at which these compounds are usually found in environmental water (0.20 ng L^{-1} to 80 ng L^{-1}). The applicability of the method was proved by the analysis of real wastewater samples from an effluent obtained from a hospital area and three wastewater treatment plants located in Gran Canaria Island, Spain.

Sol–gel M-CW20M was the fabric media that achieved better results, with a range of absolute recoveries between 25% and 90% using methanol as the eluting solvent.

Yang et al. in 2018 [33] proposed a green, simple, inexpensive, and sensitive ionic liquid immobilized fabric phase sorptive extraction method coupled with high performance liquid chromatography for the rapid screening and simultaneous determination of four fungicides (azoxystrobin, chlorothalonil, cyprodinil, and trifloxystrobin) residues in tea infusions. The optimum conditions were found to be 10% [HIMIM]NTf$_2$ as coating solution, 2 min vortex time, 500 µL acetonitrile as dispersive solvent, and 2 min desorption time. Under these conditions, the proposed technique was applied to detect fungicides from real tea water samples with satisfactory results.

Rekhi et al. in 2018 [34] reported on the determination of trace levels of aluminum by high-performance liquid chromatography (HPLC) with UV detection using quercetin, a natural bioactive flavonol, as a metal complexation agent. The developed method has been successfully applied to the direct determination of aluminum in water samples derived from various sources. Fabric phase sorptive extraction (FPSE) was applied for the preconcentration of aluminum due to its presence in environmental water at trace levels. Efficient extraction of the quercetin-Al(III) complex from aqueous samples has been accomplished by applying FPSE using a cellulose fabric substrate coated with the sol–gel C18 hybrid nanocomposite sorbent.

Kabir et al. in 2018 [35] proposed a novel fabric phase sorptive extraction high-performance liquid chromatography-photodiode array detection (FPSE-HPLC-PDA) method for the simultaneous extraction and analysis of three drug residues (ciprofloxacin, sulfasalazine, and cortisone) in human whole blood, plasma, and urine samples, which are generally administered in human patients to treat inflammatory bowel disease (IBD). The analytical method was optimized and validated in the range 0.05–10 µg/mL for whole blood, 0.25–10 µg/mL for human plasma, and 0.10–10 µg/mL for human urine. The performance of the validated FPSE-HPLC-PDA was proved on real IBD patient samples. The developed method was shown to be a rapid, robust, and green analytical tool for clinical and pharmaceutical applications.

Sol–gel CW 20M media were found to yield the best recoveries using methanol for back extraction. The FPSE membrane can be reused up to approximately 30 times when washed by 2 mL acetonitrile: methanol (50:50, *v:v*) for 5 min and subsequently dried and stored in a hermetically sealed glass manifold, with no appreciable carry-over and no efficiency loss.

In 2018, Kazantzi et al. [36] proposed an automatic sample preparation (preconcentration/separation) based on a novel sol–gel sorbent based on caprolactone-dimethylsiloxane-caprolactone block polymer comprised of a nonpolar dimethylsiloxane and hydrophilic caprolactone as a coating on hydrophobic polyester fabric substrate and investigated its evaluation in an automatic FDSE on-line fabric disk sorptive extraction (FDSE) system coupled with flame atomic absorption spectrometry (FAAS). The proposed flow injection system was evaluated for the analysis of trace Cu(II), Ni(II), Zn(II), Pb(II), and Cd(II) in urine samples. The method was based on the on-line formation of target analytes with ammonium pyrrolidine dithiocarbamate (APDC) and their retention onto the surface of the fabric disk medium. Methyl isobutyl ketone (MIBK) was used to elute metal-APDC complexes directly into the nebulizer–burner system of FAAS. For 90 s of preconcentration time, enhancement factors of 250, 130, 185, and 36 and detection limits (3 s) of 0.15, 0.41, 1.62, and 0.49 µg L^{-1} were obtained for Cu(II), Ni(II), Pb(II), and Cd(II), respectively. For 30 s of preconcentration time, an enhancement factor of 49 and a detection limit of 0.12 µg L^{-1} were achieved for Zn(II) determination. The method was tested by analyzing certified reference materials and biological samples.

Locatelli et al. in 2018 [37] described a fast, sensitive, and selective procedure for the analysis of aromatase inhibitors including anastrozole, letrozole, and exemestane used in the treatment of metastatic breast cancer by high performance liquid chromatography (HPLC) in human whole blood, plasma, and urine samples based on fabric phase sorptive extraction (FPSE). Validation was performed following the demands of international guidelines on bioanalytical methods validation. The analytical performance was proved on real human biological samples. The developed protocol can be readily applied for clinical and pharmaceutical analyses.

Six different FPSE membrane chemistries were primarily evaluated: sol–gel CW 20M, sol–gel PEG-PPG-PEG, sol–gel PCAP-PDMS-PCAP, sol–gel octadecyl (C18), sol–gel polycaprolactone A, and sol–gel sucrose. Three of these membranes performed better for the extraction of the examined analytes: sol–gel CW 20M, sol–gel PEG-PPG-PEG, and sol–gel polycaprolactone using methanol for back extraction.

Kaur et al. in 2019 [11] combined FPSE with gas chromatography mass spectrometry for the rapid extraction and determination of nineteen organochlorine pesticides in various fruit juices and water samples. The extraction approach was optimized in terms of sorbent chemistry, extraction time, stirring speed, type and volume of back-extraction solvent, and back-extraction time. Optimum conditions yielded limits of detection in a range of 0.007–0.032 ng/mL. The relative recoveries obtained by spiking organochlorine pesticides in water and selected juice samples were in the range of 91.56–99.83%. Sol–gel poly(ethylene glycol)-poly(propylene glycol)-poly(ethylene glycol) was proved to be the best sorbent for the extraction and preconcentration of organochlorine pesticides in aqueous and fruit juice samples prior to analysis with gas chromatography mass spectrometry.

Tartaglia et al. in 2019 [38] reported on the performance comparison between the exhaustive and equilibrium extraction using classical Avantor C18 solid phase extraction (SPE) sorbent, hydrophilic-lipophilic balance (HLB) SPE sorbent, Sep-Pak C18 SPE sorbent, novel sol–gel Carbowax 20M (sol–gel CW 20M) SPE sorbent, and sol–gel CW 20M coated fabric phase sorptive extraction (FPSE) media for the extraction of three inflammatory bowel disease (IBD) drugs. Both the commercial SPE phases and in-house synthesized sol–gel CW 20M SPE phases were loaded into SPE cartridges and the extractions were carried out under an exhaustive extraction mode, while FPSE was carried out under an equilibrium extraction mode. The method was validated in compliance with international guidelines for the bioanalytical method validation. Novel in-house synthesized and loaded sol–gel CW 20M SPE sorbent cartridges were characterized in terms of their extraction capability, breakthrough volume, retention volume, hold-up volume, number of the theoretical plate, and the retention factor.

The performance of FPSE and SPE techniques was evaluated by comparing the breakthrough volume and enrichment factors. The authors found that for the examined analytes, SPE showed the highest enrichment factors; consequently, this method is more suitable for samples with low analytes concentration.

Perez Mayan in 2019 [39] investigated the use of FPSE for the extraction and preconcentration of ultra-trace level residues of fungicides (19 compounds) and insecticides (3 species) in wine samples. Subsequently, the preconcentrated analytes were determined using ultra-performance liquid chromatography–tandem mass spectrometry (UPLC-MS/MS). Experimental extraction parameters affecting the efficiency and repeatability of the extraction were optimized. Optimized conditions included cellulose fabric coated with a sol–gel polyethylene glycol sorbent and back extraction using ACN-MeOH (80:20 v/v) mixture. Limits of quantification (LOQs) ranged between 0.03 and 0.3 ng mL^{-1}. Relative recoveries ranged from $77 \pm 6\%$ to $118 \pm 4\%$, and from $87 \pm 4\%$ to $121 \pm 6\%$ for red and white wines, respectively. The applicability of the method was proved for commercial wines.

In 2019, Lioupi et al. [40] developed and validated an innovative fabric phase sorptive extraction high-performance liquid chromatography–diode array detection (FPSE-HPLC-DAD) method for the extraction of five common antidepressants (venlafaxine, paroxetine, fluoxetine, amitriptyline, clomipramine) in human urine samples. The extraction protocol was optimized with regards to the extraction main parameters. Sol–gel graphene sorbent, coated on cellulose FPSE media, were the most efficient among other with different polarities using $CH_3OH:CH_3CN$ (50:50 v/v) for back-extraction. The absolute recovery values were 25.5% for venlafaxine, 33.9% for paroxetine, 67.0% for fluoxetine, 43.0% for amitriptyline, and 29.0% for clomipramine, while relative recoveries were higher than 90%. The developed method provides satisfactory limit of detection 0.15 ng/μL.

Locatelli in 2019 [41] proposed a fabric phase sorptive extraction based method prior to high-performance liquid chromatography–photodiode array detection (FPSE-HPLC-PDA) for the simultaneous extraction and analysis of six benzophenone derivative UV filters, including benzophenone (BZ), 5-benzoyl-4-hydroxy-methoxybenzenesulfonic acid (BP-4), bis(4-hydroxyphenyl)methanone (4-DHB), bis(2,4-dihydroxyphenyl)methanone (BP-2), (2,4-dihydroxybenzophenone) (BP-1), and 2,2′-dihydroxy-4-methoxybenzophenone (DHMB) in human whole blood, plasma, and urine samples. The limit of quantification was found to be 0.1 μg/mL. This new approach shows promising results with high potential for direct adaptation as a rapid, robust, and green analytical tool for several applications, e.g., in the current sample preparation practices used in many bioanalytical fields including pharmacokinetics (PK), pharmacodynamics (PD), therapeutic drug monitoring (TDM), clinical and forensic toxicology, disease diagnosis, and drug discovery. Optimized conditions included the use of sol–gel CW®20M FPSE membrane with a 20:80 (% $v\!:\!v$) mixture of phosphate buffer 40 mM at pH 3 and methanol.

Taraboletti et al. in 2019 [42] reported a metabolomics workflow using a mass spectrometry-compatible fabric phase sorptive extraction (FPSE) technique implementing a matrix coated with sol–gel poly(caprolactone-b-dimethylsiloxane-b-caprolactone) that binds both polar and nonpolar metabolites in whole blood, eliminating serum processing steps. FPSE preparation technique combined with liquid chromatography–mass spectrometry can distinguish radiation exposure markers such as taurine, carnitine, arachidonic acid, α-linolenic acid, and oleic acid found 24 h after 8 Gy irradiation. These findings suggest that the FPSE approach could work in future technology to triage irradiated individuals accurately, via biomarker screening, by providing a novel method to stabilize biofluids between collection and sample analysis.

Alampanos et al. in 2019 [43] proposed an environmentally friendly method by making use of high-performance liquid chromatography and photo-diode array detection (HPLC-PDA) for the determination of four penicillin antibiotics residues (benzylpenicillin, cloxacillin, dicloxacillin, and oxacillin) in human blood serum after FPSE. Solvent evaporation and reconstitution steps, which are considered to be rather time-consuming, were eradicated successfully from the sample preparation workflow, organic solvent consumption was brought to a minimum, while protein precipitation was assessed as impractical. Thus, the proposed method met all green analytical chemistry (GAC) criteria. The microextraction device was characterized by high chemical and solvent stability owing to the strong chemical bonds formed between the sol–gel sorbent and the substrate. Therefore, any organic solvent/solvent mixture can serve as the eluent/back-extraction solvent. The authors, after optimization of FPSE experimental parameters, propose sol–gel poly(tetrahydrofuran) coated FPSE membrane as the optimum extraction sensitivity for the selected penicillin antibiotics, after back-extraction using 90:10 v/v acetonitrile and ammonium acetate (0.01M). For all four penicillin antibiotics, the limit of detection was 0.15 ng/μL.

Zilfidou et al. in 2019 [44] applied FPSE for the simple and rapid simultaneous extraction of five common antidepressant drug residues (venlafaxine, paroxetine, fluoxetine, amitriptyline, and clomipramine) from human blood serum. Elimination of protein precipitation step and minimized solvent consumption led to a sample preparation workflow compliant with the principles of green analytical chemistry (GAC). Among all the membrane examined, sol–gel polycaprolactone-dimethylsiloxane-polycaprolactone coated polyester substrate presented optimum extraction efficiency and was found to be reusable for at least 30 times. Back-extraction was achieved by methanol: acetonitrile (50:50 v/v). The limit of detection was found at 0.15 ng μL^{-1}, while good absolute recoveries (9.4–88.1%) were obtained.

Tartaglia et al. in 2019 [45] described an FPSE based method for the simultaneous determination of seven paraben residues including methyl paraben (MPB), ethyl paraben (EPB), propyl paraben (PPB), isopropyl paraben (iPPB), butyl paraben (BPB), isobutyl paraben (iBPB), and benzyl paraben (BzPB) in human whole blood, plasma and urine, prior to high-performance liquid chromatography (HPLC) coupled with photo diode array detector (PDA) analysis. The analytical method has been validated according to the international guidelines.

The performance of the analytical method was evaluated on real biological samples. The proposed innovative method allows simultaneous analysis of seven paraben residues in three different biological matrices, including whole blood, plasma, and urine, and therefore it is easily applicable to monitor these substances in different biological samples. Furthermore, the extraction technique used in this work is fast, easy to use, and in accordance with the modern green analytical chemistry (GAC) principles. Sol–gel CW 20M FPSE media and back-extraction with methanol provided the best recovery rates.

Kaur et al. in 2019 [11] combined FPSE with gas chromatography–mass spectrometry for the rapid extraction and determination of nineteen organochlorine pesticides in various fruit juices and water samples. FPSE efficiency was optimized in terms of sorbent chemistry, extraction time, stirring speed, type and volume of back-extraction solvent, and back-extraction time. Under optimum conditions, the limits of detection were obtained in a range of 0.007–0.032 ng/mL. The relative recoveries obtained by spiking organochlorine pesticides in water and selected juice samples were in the range of 91.56–99.83%. The sorbent sol–gel poly(ethylene glycol)-poly(propylene glycol)-poly(ethylene glycol) was applied for the extraction and preconcentration of organochlorine pesticides in aqueous and fruit juice samples prior to analysis with gas chromatography–mass spectrometry.

Otoukesh et al. in 2019 [46] proposed a fabric phase sorptive extraction (FPSE) for the enrichment of acrylate compounds coming from acrylic adhesives used commonly for sticking the paper labels on polyethylene terephthalate (PET) bottles, and therefore they may exist in recycled polyethylene terephthalate (rPET) in different food simulants: simulant A (ethanol 10%), simulant B (acetic acid 3%), and simulant C (ethanol 20%), and their respective extracts by ultra-high-performance liquid chromatography with mass spectrometric detection (UPLC-MS). Four acrylates were studied: ethylene glycol dimethacrylate (EGDM), pentaerythritol triacrylate (PETA), triethylene glycol diacrylate (TEGDA), and trimethylolpropane triacrylate (TMPTA). Five different types of FPSE membrane coated with different sol–gel sorbents were studied, and finally sol–gel polyethylene glycol-polypropylene glycol-polyethylene glycol triblock copolymer (PEG-PPG-PEG) coated FPSE membrane was chosen for its satisfactory results combined with methanol for back-extraction since it provided an elution ability slightly higher than acetonitrile. Under the optimized conditions, the method provided limits of detection of the compounds in the range of $(0.1–1.9 \text{ ng g}^{-1}, 0.1–1.2 \text{ ng g}^{-1}, 0.2–2.3 \text{ ng g}^{-1})$ in EtOH 10%, HAc 3%, and EtOH 20%, and the enrichment factor values (EFs) after applying N_2 were in the range of 11.1–25.0, 13.8–26.3, 8.3–21.9, in simulants A, B, and C, respectively. The optimized method was applied successfully to analyze thirteen types of recycled PET samples.

Mesa et al. in 2019 [10] developed a simple and sensitive analytical methodology for rapid screening and quantification of selected estrogenic endocrine disrupting chemicals including α-estradiol, hexestrol, estrone, 17α-ethinyl estradiol, diethylstilbestrol, and bisphenol A from intact milk using fabric phase sorptive extraction in combination with high-performance liquid chromatography coupled to ultraviolet detection/tandem mass spectrometry. The new approach eliminates protein precipitation and defatting step from the sample preparation workflow, while the error prone and time-consuming solvent evaporation and sample reconstitution steps have also been eliminated. Parameters which mostly affect the extraction efficiency of fabric phase sorptive extraction, including sorbent chemistry, sample volume, and extraction time, were optimized. The limit of detection values obtained in fabric phase sorptive extraction with high-performance liquid chromatography with ultraviolet detection ranged from 25.0 to 50.0 ng/mL.

Two sol–gel sorbent coatings were tested to determine the better sorbent coating for the selected EDCs, sol–gel PTHF, and sol–gel PDMS. Sol–gel PTHF was distinctly superior in extraction efficiencies for all compounds, with acetonitrile used for back extraction.

Lastovka et al. in 2019 [47] developed a method for the quantification of highly potent analgesic agent (2*R*,4a*R*,7*R*,8a*R*)-4,7-dimethyl-2-(thiophen-2-yl)octahydro-2*H*-chromen-4-ol in rat whole blood and plasma using dried matrix spots (DMS) and fabric phase sorptive extraction (FPSE) techniques in combination with LC–MS/MS. The linearity was obtained in the range of 20–5000 ng/mL and 50–5000 ng/mL for plasma-FPSE and blood-FPSE experiments, respectively. The mean extraction recovery (%) was 26 for plasma-DMS, 25 for blood DMS, 38 for plasma-FPSE, and 31 for blood-FPSE.

A sol–gel PCAP-PDMS-PCAP sorbent-coated FPSE biofluid sampler FPSE blood sampler was compared to a DBS card and has been used under a different sampling and extraction mode (DBS card with direct spotting, and FPSE biofluid sampler with equilibrium extraction mode); both perform satisfactorily with different sample matrices. However, the FPSE biofluid sampler was found more selective in preparing an interferents-free sample for instrumental analysis. Due to the exploitation of high-performance sol–gel sorbent, the FPSE biofluid sampler has the potential to streamline the current practice of blood analysis.

Gulle et al. in 2019 [48] developed a fabric phase sorptive extraction (FPSE)-based sample preparation method for methyl paraben (MP), propyl paraben (PP), and butyl paraben (BP) in cosmetic and environmental samples, prior to high performance liquid chromatography–photodiode array (HPLC-PDA) detection. In the proposed method, MP, PP, and BP molecules were efficiently retained on a sol–gel Carbowax-20M sorbent-coated FPSE membrane when the matrix pH was adjusted to 5. Subsequently, the extracted analytes were desorbed from the FPSE membrane with methanol. Experimental conditions were studied to optimize variables such as pH, adsorption time, and desorption solvent. Using the optimal conditions, analytical parameters such as linearity ranges, detection limits, and preconcentration factors for each of the selected parabens were calculated from experimental data. The limit of detection (LOD) values for MP, PP, and BP were calculated as 2.85, 2.98, and 2.75 ng mL^{-1}, respectively. Finally, the developed method was applied to cosmetic and environmental samples.

Kaur et al. in 2019 [49] developed a high-efficiency and solvent minimized microextraction technique, fabric phase sorptive extraction followed by gas chromatography and mass spectrometry analysis for the rapid determination of four organophosphorus pesticides (terbufos, malathion, chlorpyrifos, and triazofos) in vegetable samples including beans, tomato, brinjal, and cabbage. The most important fabric phase sorptive extraction parameters were investigated and optimized. Under optimum experimental conditions, the limits of detection were found in the range of 0.033 to 0.136 ng/g. Three different sol–gel sorbent coated FPSE membranes were evaluated, including sol–gel Carbowax 20 M (sol–gel CW 20 M), sol–gel poly(tetrahydrofuran) (sol–gel PTHF), and sol–gel poly(dimethyl siloxane) (sol–gel PDMS). Sol–gel CW 20 M coated FPSE membrane was selected as the suitable FPSE membrane for the selected OPPs.

Sun et al. in 2019 [50] developed a new method which coupled FPSE with ion mobility spectrometry (IMS) for the rapid detection of polycyclic aromatic hydrocarbons (PAHs) in water present in the field. Polydimethylsiloxane (PDMS) was coated on the glass fiber cloth through a sol–gel reaction. After extracting the PAHs in water, the fabric coated PDMS could be directly put into the inlet of IMS instrument for thermal desorption. The PAHs were analyzed by the IMS instrument operated in the positive ion mode with a corona discharge (CD) ionization source. The primary parameters affecting extraction efficiency such as extraction time, extraction temperature, and ionic strength were investigated and optimized by using phenanthrene (Phe), benzo[a]anthracene (BaA), and benzo[a]pyrene (BaP) as model compounds. Under the optimal conditions, the FPSE-IMS detection limits were 5 ng mL^{-1}, 8 ng mL^{-1}, and 10 ng mL^{-1}, respectively. Satisfactory recoveries were obtained ranging from 80.5% to 100.5%, making the method of FPSE-IMS applicable for the monitoring the water quality on-site, and thus providing early warning in the field.

Kaur et al. in 2019 [51] developed and validated a rapid extraction and clean-up method using selective fabric phase sorptive extraction combined with gas chromatography and mass spectrometry for the determination of broad polarity spectrum emerging pollutants, ethyl paraben, butyl paraben, diethyl phthalate, dibutyl phthalate, lidocaine, prilocaine, triclosan, and bisphenol A in various aqueous samples. Some important parameters of fabric phase sorptive extraction such as extraction time, matrix pH, stirring speed, type, and volume of desorption solvent were investigated and optimized. Under the optimum conditions, the limits of detection were in the range 0.009–0.021 ng/mL. Recoveries ranged from 93 to 99%. The developed method was applied for the determination of the emerging contaminants in tap water, municipal water, ground water, sewage water, and sludge water samples. Three different FPSE sorbent coatings were comparatively studied: sol–gel CW20M (polar), sol–gel PTHF (medium polar), and sol–gel C18 (nonpolar). Sol–gel CW20M-coated FPSE provided the optimum results.

The most recent application comes from Celeiro et al. [52]. This research group in 2020 proposed a novel method based on fabric phase sorptive extraction (FPSE) followed by gas chromatography–tandem mass spectrometry (GC-MS/MS) for the simultaneous determination of 11 UV filters (ethylhexyl salicylate, benzyl salicylate, homosalate, benzophenone-3, isoamylmethoxycinnamate, 4-methylbenzylidenecamphor, methyl anthranilate, etocrylene, 2-ethylhexylmethoxycinnamate, 2-ethylhexyl p-dimethylaminobenzoate, and octocrylene), in natural and recreational waters. Different types and sizes of sol–gel coated FPSE membranes, sample volumes, extraction times, and types and volumes of desorption solvent were optimized. The optimal conditions involved the use of a (2.0 × 2.5) cm^2 FPSE device with PDMS-based coating for the extraction of 20 mL of water for 20 min. Back-extraction was performed by ethyl acetate. Recovery rates under optimum conditions were about 90%. LODs were at the low ng L^{-1} in all cases. The proposed validated FPSE-GC-MS/MS method was applied to different real samples, including environmental water (lake, river, seawater) and recreational water (swimming pool).

8. Trend and Future Perspectives

As it is shown in Figure 3, publications based on FPSE constantly increased, and it can be predicted that it will expand to more analytes as well as more sample matrices in the future. However, there is a slight decrease in the number of published papers in 2020, which may be attributed to the global pandemic that slowed down everything, including scientific research. Among the published applications, the vast majority use liquid chromatographic determination with various detection and identification techniques (Figure 4). However, this tendency has the potential to be altered in the future, since more applications are anticipated covering all chemistries of analytes. New sorptive membranes in new formats can help this direction. Additionally, the ability for implementation in automated systems meets the new analytical performance criteria, and more on-line approaches are expected to be developed.

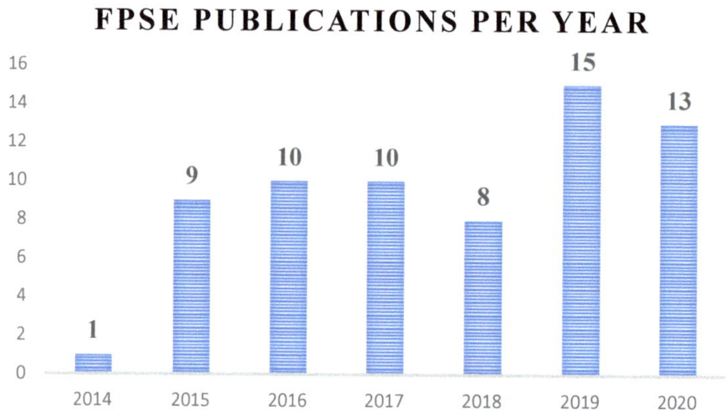

FPSE PUBLICATIONS PER YEAR

Figure 3. Graphical representation of number of papers published on FPSE per year since 2014.

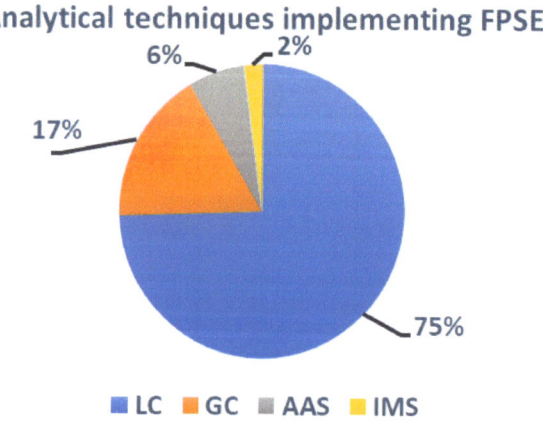

Analytical techniques implementing FPSE

Figure 4. Graphical representation of different analytical instruments used (%) subsequent to FPSE.

9. Conclusions

Fabric phase sorptive extraction has emerged as a new generation sample preparation technique with many new attributes that were not offered before by a single extraction/microextraction technique. Although FPSE is not commercially available yet, it has successfully established itself as an inevitable laboratory consumable within a short period. Many academic research groups across the world have demonstrated the performance superiority, compliance of green analytical principles, substantial minimization of sample preparation workflow, extended pH working range, reusability, and field deployability of FPSE membranes in numerous applications using diversified sample matrices which will undoubtedly provoke new analysts to explore this powerful technology. A broad range of sorbents chemistries offered by FPSE encompassing all the sorbent chemistries available on the SPE and SPME platform will provide an analyst more liberty to select the appropriate sorbent for a given application. Ability to use the same FPSE membrane in SPME mode or SPE mode is indeed a unique concept in the rapidly growing sample preparation technology space. The mathematical model-driven sorbent selection strategy proposed by FPSE also manifests another green component that was not considered before and deserves appreciation.

Author Contributions: Conceptualization, methodology, writing—original draft preparation, writing—review and editing, A.K. and V.S., equal contribution. All authors have read and agreed to the published version of the manuscript.

Funding: This research received no external funding

Conflicts of Interest: The authors declare no conflict of interest.

References

1. Ghorbani, M.; Aghamohammadhassan, M.; Chamsaz, M.; Akhlaghi, H.; Pedramrad, T. Dispersive solid phase microextraction. *Trends Anal. Chem.* **2019**, *118*, 793–809. [CrossRef]
2. He, M.; Wang, Y.; Zhang, Q.; Zang, L.; Chen, B.; Hu, B. Stir bar sorptive extraction and its application. *J. Chromatogr. A* **2021**, *1637*, 461810. [CrossRef]
3. Rezaei, S.M.; Makarem, S.; Alexovič, M.; Tabani, H. Simultaneous separation and quantification of acidic and basic dye specimens via a dual gel electro-membrane extraction from real environmental samples. *J. Iran. Chem. Soc.* **2021**. [CrossRef]
4. Kabir, A.; Furton, K.G. *Fabric Phase Sorptive Extractors*; United States Patents and Trademark Office: Alexandiria, VA, USA, 2016.
5. Kabir, A.; Furton, K.G.; Malik, A. Innovations in Sol–gel microextraction phases for solvent-free sample preparation in analytical chemistry. *Trends Anal. Chem.* **2013**, *45*, 197–218. [CrossRef]
6. Lakade, S.S.; Borrull, F.; Furton, K.G.; Kabir, A.; Marce, R.M.; Fontanals, N. Dynamic fabric phase sorptive extraction for a group of pharmaceuticals and personal care products from environmental waters. *J. Chromatogr. A* **2015**, *1456*, 19–26. [CrossRef]
7. Kabir, A.; Mesa, R.; Jurmain, J.; Furton, K.G. Fabric Phase Sorptive Extraction Explained. *Separations* **2017**, *4*, 21. [CrossRef]
8. Lucena, R. Extraction and stirring integrated techniques: Examples and recent advances. *Anal. Bioanal. Chem.* **2012**, *403*, 2213–2223. [CrossRef] [PubMed]
9. Kumar, R.; Gaurav, H.; Malik, A.K.; Kabir, A.; Furton, K.G. Efficient analysis of selected estrogens using fabric phase sorptive extraction and high performance liquid chromatography-fluorescence detection. *J. Chromatogr. A* **2014**, *1359*, 16–25. [CrossRef] [PubMed]
10. Mesa, R.; Kabir, A.; Samanidou, V.; Furton, K.G. Simultaneous determination of selected estrogenic endocrine disrupting chemicals and bisphenol A residues in whole milk using fabric phase sorptive extraction coupled to HPLC-UV detection and LC-MS/MS. *J. Sep. Sci.* **2019**, *42*, 598–608. [CrossRef]
11. Kaur, R.; Kaur, R.; Rani, S.; Malik, A.K.; Kabir, A.; Furton, K.G.; Samanidou, V.F. Rapid Monitoring of Organochlorine Pesticide Residues in Various Fruit Juices and Water Samples Using Fabric Phase Sorptive Extraction and Gas Chromatography-Mass Spectrometry. *Molecules* **2019**, *24*, 1013. [CrossRef]
12. Kumar, R.; Gaurav; Kabir, A.; Furton, K.G.; Malik, A.K. Development of a fabric phase sorptive extraction with high-performance liquid chromatography and ultraviolet detection method for the analysis of alkyl phenols in environmental samples. *J. Sep. Sci.* **2015**, *38*, 3228–3238. [CrossRef] [PubMed]
13. Samanidou, V.; Kaltzi, I.; Kabir, A.; Furton, K.G. Simplifying sample preparation using fabric phase sorptive extraction technique for the determination of benzodiazepines in blood serum by high-performance liquid chromatography. *Biomed. Chromatogr.* **2016**, *30*, 829–836. [CrossRef]

14. Roldan-Pijuan, M.; Cardenas, R.L.S.; Valcarcel, M.; Kabir, A.; Kenneth, G. Furton Stir fabric phase sorptive extraction for the determination of triazine herbicides in environmental water by using ultra-high performance liquid chromatography-UV detection. *J. Chromatogr. A* **2014**. under review.

15. Montesdeoca-Esponda, S.; Sosa-Ferrera, Z.; Kabir, A.; Furton, K.G.; Santana-Rodriguez, J.J. Fabric phase sorptive extraction followed by UHPLC-MS/MS for the analysis of benzotriazole UV stabilizers in sewage samples. *Anal. Bioanal. Chem.* **2015**, *407*, 8137–8150. [CrossRef]

16. Racamonde, I.; Rodil, R.; Quintana, J.B.; Sieira, B.J.; Kabir, A.; Furton, K.G.; Cela, R. Fabric phase sorptive extraction: A new sorptive microextraction technique for the determination of non-steroidal anti-inflammatory drugs from environmental water samples. *Anal. Chim. Acta* **2015**, *865*, 22–30. [CrossRef]

17. Samanidou, V.; Galanopoulos, L.-D.; Kabir, A.; Furton, K.G. Fast extraction of amphenicols residues from raw milk using novel fabric phase sorptive extraction followed by high-performance liquid chromatography-diode array detection. *Anal. Chim. Acta* **2015**, *855*, 41–50. [CrossRef]

18. Lakade, S.S.; Borrull, F.; Furton, K.G.; Kabir, A.; Fontanals, N.; Maria Marcé, R. Comparative Study of Different Fabric Phase Sorptive Extraction Sorbents to Determine Emerging Contaminants from Environmental Water Using Liquid Chromatography-Tandem Mass Spectrometry. *Talanta* **2015**, *144*, 1342–1351. [CrossRef]

19. Anthemidis, A.; Kazantzi, V.; Samanidou, V.; Kabir, A.; Furton, K.G. An automated flow injection system for metal determination by flame atomic absorption spectrometry involving on-line fabric disk sorptive extraction technique. *Talanta* **2016**, *156–157*, 64–70. [CrossRef] [PubMed]

20. Huang, G.; Dong, S.; Zhang, M.; Zhang, H.; Huang, T. Fabric phase sorptive extraction: Two practical sample pretreatment techniques for brominated flame retardants in water. *Water Res.* **2016**, *101*, 547–554. [CrossRef]

21. Guedes-Alonso, R.; Ciofi, L.; Sosa-Ferrera, Z.; Juan Santana-Rodriguez, J.; del Bubba, M.; Kabir, A.; Furton, K.G. Determination of androgens and progestogens in environmental and biological samples using fabric phase sorptive extraction coupled to ultra-high performance liquid chromatography tandem mass spectrometry. *J. Chromatogr. A* **2016**, *1437*, 116–126. [CrossRef] [PubMed]

22. Karageorgou, E.; Manousi, N.; Samanidou, V.; Kabir, A.; Furton, K.G. Fabric phase sorptive extraction for the fast isolation of sulfonamides residues from raw milk followed by high performance liquid chromatography with ultraviolet detection. *Food Chem.* **2016**, *196*, 428–436. [CrossRef] [PubMed]

23. Aznar, M.; Alfaro, P.; Nerin, C.; Kabir, A.; Furton, K.G. Fabric phase sorptive extraction: An innovative sample preparation approach applied to the analysis of specific migration from food packaging. *Anal. Chim. Acta* **2016**, *936*, 97–107. [CrossRef] [PubMed]

24. Alcudia-León, M.C.; Lucena, R.; Cárdenas, S.; Valcárcel, M.; Kabir, A.; Furton, K.G. Integrated sampling and analysis unit for the determination of sexual pheromones in environmental air using fabric phase sorptive extraction and headspace-gas chromatography–mass spectrometry. *J. Chromatogr. A* **2017**, *1488*, 17–25. [CrossRef] [PubMed]

25. Locatelli, M.; Kabir, A.; Innosa, D.; Lopatriello, T.; Furton, K.G. A fabric phase sorptive extraction-High performance liquid chromatography-Photo diode array detection method for the determination of twelve azole antimicrobial drug residues in human plasma and urine. *J. Chromatogr. B* **2017**, *1040*, 192–198. [CrossRef]

26. Heena; Kaur, R.; Rani, S.; Malik, A.K.; Kabir, A.; Furton, K.G. Determination of cobalt(II), nickel(II) and palladium(II) Ions via fabric phase sorptive extraction in combination with high-performance liquid chromatography-UV detection. *Sep. Sci. Technol.* **2017**, *52*, 81–90. [CrossRef]

27. Anthemidis, V.K. Anthemidis, V.K. A Fiber fabric sorbent extraction for on-line toxic metal determination in energy beverages. In *Metrology Promoting Harmonization and Standardization in Food and Nutrition*, 3rd ed.; IMEKOFOODS: Thessaloniki, Greece, 2017; pp. 377–380.

28. Samanidou, V.; Michaelidou, K.; Kabir, A.; Furton, K.G. Fabric phase sorptive extraction of selected penicillin antibiotic residues from intact milk followed by high performance liquid chromatography with diode array detection. *Food Chem.* **2017**, *224*, 131–138. [CrossRef]

29. Saini, S.; Kabir, A.; Rao, A.; Malik, A.; Furton, K. A Novel Protocol to Monitor Trace Levels of Selected Polycyclic Aromatic Hydrocarbons in Environmental Water Using Fabric Phase Sorptive Extraction Followed by High Performance Liquid Chromatography-Fluorescence Detection. *Separations* **2017**, *4*, 22. [CrossRef]

30. Samanidou, V.; Filippou, O.; Marinou, E.; Kabir, A.; Furton, K.G. Sol–gel-graphene-based fabric-phase sorptive extraction for cow and human breast milk sample cleanup for screening bisphenol A and residual dental restorative material before analysis by HPLC with diode array detection. *J. Sep. Sci.* **2017**, *40*, 2612–2619. [CrossRef] [PubMed]

31. Aznar, M.; Úbeda, S.; Nerin, C.; Kabir, A.; Furton, K.G. Fabric phase sorptive extraction as a reliable tool for rapid screening and detection of freshness markers in oranges. *J. Chromatogr. A* **2017**, *1500*, 32–42. [CrossRef]

32. Santana-Viera, S.; Guedes-Alonso, R.; Sosa-Ferrera, Z.; Santana-Rodríguez, J.J.; Kabir, A.; Furton, K.G. Optimization and application of fabric phase sorptive extraction coupled to ultra-high performance liquid chromatography tandem mass spectrometry for the determination of cytostatic drug residues in environmental waters. *J. Chromatogr. A* **2017**, *1529*, 39–49. [CrossRef]

33. Yang, M.; Gu, Y.; Wu, X.; Xi, X.; Yang, X.; Zhou, W.; Zeng, H.; Zhang, S.; Lu, R.; Gao, H.; et al. Rapid analysis of fungicides in tea infusions using ionic liquid immobilized fabric phase sorptive extraction with the assistance of surfactant fungicides analysis using IL-FPSE assisted with surfactant. *Food Chem.* **2018**, *239*, 797–805. [CrossRef]

34. Rekhi, H.; Kaur, R.; Rani, S.; Malik, A.K.; Kabir, A.; Furton, K.G. Direct Rapid Determination of Trace Aluminum in Various Water Samples with Quercetin by Reverse Phase High-Performance Liquid Chromatography Based on Fabric Phase Sorptive Extraction Technique. *J. Chromatogr. Sci.* **2018**, *56*, 452–460. [CrossRef] [PubMed]

35. Kabir, A.; Furton, K.G.; Tinari, N.; Grossi, L.; Innosa, D.; Macerola, D.; Tartaglia, A.; Di Donato, V.; D'Ovidio, C.; Locatelli, M. Fabric phase sorptive extraction-high performance liquid chromatography-photo diode array detection method for simultaneous monitoring of three inflammatory bowel disease treatment drugs in whole blood, plasma and urine. *J. Chromatogr. B* **2018**, *1084*, 53–63. [CrossRef] [PubMed]

36. Kazantzi, V.; Samanidou, V.; Kabir, A.; Furton, K.; Anthemidis, A. On-Line Fabric Disk Sorptive Extraction via a Flow Preconcentration Platform Coupled with Atomic Absorption Spectrometry for the Determination of Essential and Toxic Elements in Biological Samples. *Separations* **2018**, *5*, 34. [CrossRef]

37. Locatelli, M.; Tinari, N.; Grassadonia, A.; Tartaglia, A.; Macerola, D.; Piccolantonio, S.; Sperandio, E.; D'Ovidio, C.; Carradori, S.; Ulusoy, H.I.; et al. FPSE-HPLC-DAD method for the quantification of anticancer drugs in human whole blood, plasma, and urine. *J. Chromatogr. B* **2018**, *1095*, 204–213. [CrossRef]

38. Tartaglia, A.; Locatelli, M.; Kabir, A.; Furton, K.G.; Macerola, D.; Sperandio, E.; Piccolantonio, S.; Ulusoy, H.I.; Maroni, F.; Bruni, P.; et al. Comparison between Exhaustive and Equilibrium Extraction Using Different SPE Sorbents and Sol–gel Carbowax 20M Coated FPSE Media. *Molecules* **2019**, *24*, 382. [CrossRef]

39. Pérez-Mayán, L.; Rodríguez, I.; Ramil, M.; Kabir, A.; Furton, K.G.; Cela, R. Fabric phase sorptive extraction followed by ultra-performance liquid chromatography-tandem mass spectrometry for the determination of fungicides and insecticides in wine. *J. Chromatogr. A* **2018**, *1584*, 13–23. [CrossRef]

40. Lioupi, A.; Kabir, A.; Furton, K.; Samanidou, V. Fabric phase sorptive extraction for the isolation of five common antidepressants from human urine prior to HPLC-DAD analysis. *J. Chromatogr. B* **2019**, *1118–1119*, 171–179. [CrossRef]

41. Locatelli, M.; Furton, K.G.; Tartaglia, A.; Sperandio, E.; Ulusoy, H.I.; Kabir, A. An FPSE-HPLC-PDA method for rapid determination of solar UV filters in human whole blood, plasma and urine. *J. Chromatogr. B* **2019**, *1118–1119*, 40–50. [CrossRef] [PubMed]

42. Taraboletti, A.; Goudarzi, M.; Kabir, A.; Moon, B.-H.; Laiakis, E.; Lacombe, J.; Ake, P.; Shoishiro, S.; Brenner, D.; Fornace, A., Jr.; et al. Fabric Phase Sorptive Extraction—A metabolomic pre-processing approach for ionizing radiation exposure assessment. *J. Proteome Res.* **2019**, *18*, 3020–3031. [CrossRef] [PubMed]

43. Alampanos, V.; Kabir, A.; Furton, K.G.; Samanidou, V.; Papadoyannis, I. Fabric phase sorptive extraction for simultaneous observation of four penicillin antibiotics from human blood serum prior to high performance liquid chromatography and photo-diode array detection. *Microchem. J.* **2019**, *149*, 103964. [CrossRef]

44. Zilfidou, E.; Kabir, A.; Furton, K.; Samanidou, V. Fabric Phase Sorptive Extraction: Current State of the Art and Future Perspectives. *Separations* **2018**, *5*, 40. [CrossRef]

45. Tartaglia, A.; Kabir, A.; Ulusoy, S.; Sperandio, E.; Piccolantonio, S.; Ulusoy, H.I.; Furton, K.G.; Locatelli, M. FPSE-HPLC-PDA analysis of seven paraben residues in human whole blood, plasma, and urine. *J. Chromatogr. B* **2019**, *1125*, 121707. [CrossRef]

46. Otoukesh, M.; Nerin, C.; Aznar, M.; Kabir, A.; Furton, K.G.; Es'haghi, Z. Determination of adhesive acrylates in recycled polyethylene terephthalate by fabric phase sorptive extraction coupled to ultra performance liquid chromatography—Mass spectrometry. *J. Chromatogr. A* **2019**, *1602*, 56–63. [CrossRef]

47. Lastovka, A.V.; Rogachev, A.D.; Il'ina, I.V.; Kabir, A.; Volcho, K.P.; Fadeeva, V.P.; Pokrovsky, A.G.; Furton, K.G.; Salakhutdinov, N.F. Comparison of dried matrix spots and fabric phase sorptive extraction methods for quantification of highly potent analgesic activity agent (2R,4aR,7R,8aR)-4,7-dimethyl-2-(thiophen-2-yl)octahydro-2H-chromen-4-ol in rat whole blood and plasma using LC–MS/MS. *J. Chromatogr. B* **2019**, *1132*, 121813. [CrossRef]

48. Gulle, S.; Ulusoy, H.I.; Kabir, A.; Tartaglia, A.; Furton, K.G.; Locatelli, M.; Samanidou, V.F. Application of a fabric phase sorptive extraction-high performance liquid chromatography-photodiode array detection method for the trace determination of methyl paraben, propyl paraben and butyl paraben in cosmetic and environmental samples. *Anal. Methods* **2019**, *11*, 6136–6145. [CrossRef]

49. Kaur, R.; Kaur, R.; Rani, S.; Malik, A.K.; Kabir, A.; Furton, K.G. Application of fabric phase sorptive extraction with gas chromatography and mass spectrometry for the determination of organophosphorus pesticides in selected vegetable samples. *J. Sep. Sci.* **2019**, *42*, 862–870. [CrossRef]

50. Sun, T.; Wang, D.; Tang, Y.; Xing, X.; Zhuang, J.; Cheng, J.; Du, Z. Fabric-phase sorptive extraction coupled with ion mobility spectrometry for on-site rapid detection of PAHs in aquatic environment. *Talanta* **2019**, *195*, 109–116. [CrossRef]

51. Kaur, R.; Kaur, R.; Grover, A.; Rani, S.; Malik, A.K.; Kabir, A.; Furton, K.G. Fabric phase sorptive extraction/GC-MS method for rapid determination of broad polarity spectrum multi-class emerging pollutants in various aqueous samples. *J. Sep. Sci.* **2019**, *42*, 2407–2417. [CrossRef] [PubMed]

52. Celeiro, M.; Acerbi, R.; Kabir, A.; Furton, K.G.; Llompart, M. Development of an analytical methodology based on fabric phase sorptive extraction followed by gas chromatography-tandem mass spectrometry to determine UV filters in environmental and recreational waters. *Anal. Chim. Acta X* **2020**, *4*, 100038. [CrossRef] [PubMed]

Article

FT-IR/ATR Solid Film Formation: Qualitative and Quantitative Analysis of a Piperacillin-Tazobactam Formulation

Ioanna Chrisikou [1,2], Malvina Orkoula [1] and Christos Kontoyannis [1,2,*]

1   Department of Pharmacy, University of Patras, University Campus, GR-26504 Rio Achaias, Greece; ioannach94@gmail.com (I.C.); malbie@upatras.gr (M.O.)
2   Institute of Chemical Engineering Sciences, Foundation of Research and Technology-Hellas (ICE-HT/FORTH), GR-26504 Platani Achaias, Greece
*   Correspondence: kontoyan@upatras.gr; Tel.: +30-2610-962328

Academic Editors: Victoria Samanidou and Irene Panderi
Received: 23 November 2020; Accepted: 18 December 2020; Published: 21 December 2020

Abstract: FT-IR/ATR analytical technique is one of the most applicable techniques worldwide. It is closely associated with easy-to-use equipment, rapid analysis, and reliable results. This study reports the simultaneous qualitative and quantitative analysis of two active pharmaceutical ingredients (APIs), of a piperacillin and tazobactam formulation using a film formation method. This method requires film formation on the ATR crystal, resulting from solvent evaporation of a small amount of liquid sample. Good contact between the film and the crystal led to the identification of both APIs, although tazobactam was of low content in the formulation mixture. The quantification of the APIs in the commercial mixture was also achieved, using a single calibration line with a correlation coefficient equal to 0.999, not only after film formation but also in the initial dry formulation before reconstitution. The present spectroscopic technique combined with the proposed relatively simple sample treatment outweighs chromatographic protocols already applied, which require specialized staff and are costly, time-consuming, and not environmentally friendly. Taking all the above into consideration, it turns out that such an approach has the potential to be used for off-line quality control procedures in manufacture or, in terms of portable equipment and automated software, anywhere for on-site analysis, even in a hospital workflow.

Keywords: FT-IR/ATR; sample preparation; film formation; piperacillin; tazobactam

1. Introduction

Fourier Transform Infrared Spectroscopy (FT-IR) is a well-established analytical technique. Mid-IR region (4000–600 cm^{-1}) offers the "fingerprint" of the analyte, as it is rich in information about the structure of the functional groups of the sample tested [1]. Apart from that, it can be applied for quantitative analysis as the energy absorbed at a particular wavelength is in proportion with the number of bonds absorbing the associated quanta of energy [2]. FT-IR/ATR (Attenuated Total Reflectance), based on internal total reflectance, constitutes an alternative mode of the customary technique, which has numerous applications in recent years in a variety of fields including medicine, food industry, environmental, and forensic science [3–7]. In a typical ATR experiment, the incident infrared radiation interacts with the analyte molecules on the crystal surface via an evanescent wave created by total reflectance in the crystal. The attenuation of the evanescent field by the absorption in the sample is then detected. Since this field decays exponentially the depth of penetration in the sample corresponds to few microns (usually 1–2 um) [1,2]. Worth mentioning is the fact that the quality of the spectra obtained, depends strongly on procedures like sample preparation and handling. The good contact between the sample and the crystal surface plays a vital role in excluding the possibility of a noisy spectrum acquisition. The quality of the obtained spectra in the case of solids or powders depends on the applied downforce, whereas when handling liquid samples, the determinant factor is the number of analyte molecules being in contact with the surface of the crystal, so as the solvent contribution to be minimized [7]. In the pharmaceutical field, ATR has been used for qualitative as well as quantitative analysis of either solid or liquid samples. Graham Lawson and his team used this technique to develop a rapid method of quantifying counterfeit tablet formulations. Their method was developed without the need of solvents as the samples under test were in the form of powder or tablets that had been crushed prior to analysis [8]. Kerry J. Hartauer and J. Keith Guillory suggested an alternative procedure for the simultaneous determination of trimethoprim and sulfamethoxazole in a formulation intended for intravenous administration. Their method insisted that the sample solutions be poured into a Circle Cell to cover the instrument crystal completely and 5 min of purge to be allowed before recording the spectra [9]. The percentage of methamphetamine contained in illicit drug mixtures was determined rapidly and effectively without any significant pre-preparation required [10]. Besides that, Alaa A. Makki and his team applied ATR spectroscopy for the individual analysis of three commercially available anticancer drugs, TEVA®, MYLAN® and CERUBIDINE®, respectively containing doxorubicin, epirubicin, and daunorubicin. This study was conducted in order for ATR spectroscopy to be assessed as a tool for the analysis of therapeutic solutions in a hospital workflow. All drug samples measured were in liquid form, while the pre-measurement sample treatment included the film formation of a small (2 μL) sample volume on the instrument crystal through solvent evaporation [11]. Although film formation technique has been reported as an ATR sample preparation method for the analysis of active pharmaceutical ingredients in formulations, there is no published application of such a technique for not only the identification but also the quantification of two different APIs in a single formulation being initially in its solid form. This paper reports on the employment of the film-formation, preparation technique for the qualitative and quantitative analysis of a piperacillin (PIP) and tazobactam (TAZ) intravenous formulation with a mass ratio of APIs equal to 8:1 (89% PIP-11% TAZ). Piperacillin is a broad-spectrum b-lactam antibiotic, which inhibits the synthesis of the bacterial cell wall [12,13], while tazobactam is a beta-lactamase inhibitor, which increases and expands the antimicrobial spectrum of piperacillin. In particular, it protects the antibiotic from degradation caused by beta-lactamase enzymes [12,14]. Up to date, in most cases, PIP and TAZ in formulations have been simultaneously analyzed through exploitation of time-consuming LC (liquid chromatography) protocols [15–17], a procedure that is required by pharmacopoeia for the identification of each API and its impurities [18]. Ultra high performance liquid chromatography tandem mass spectrometry (UHPLC-MS/MS) has also been applied for the simultaneous determination of the two APIs, not in the formulation but in different biological matrices (serum, urine, renal replacement therapy effluent) in regards to therapeutic drug monitoring (TDM) after administration [19,20]. Although this method requires microsample volumes and the run time does not exceed 5 min, the analytical procedure requires more expensive equipment and more specialized staff and cannot be applied directly to the as-received initial dry formulation, before reconstitution. UV/Vis spectroscopy has also been used, but its application proved to be complex since it requires selection of appropriate derivative order and smoothing factor [21,22].

2. Results

2.1. Identification of PIP and TAZ

2.1.1. Powder Spectra

Initially, FT-IR/ATR spectra of the powder of both the pure active substances were obtained (Figure 1) and some of the most significant peaks of them were attributed to the vibration modes of their molecules (Table 1) [1].

Molecules **2020**, *25*, 6051

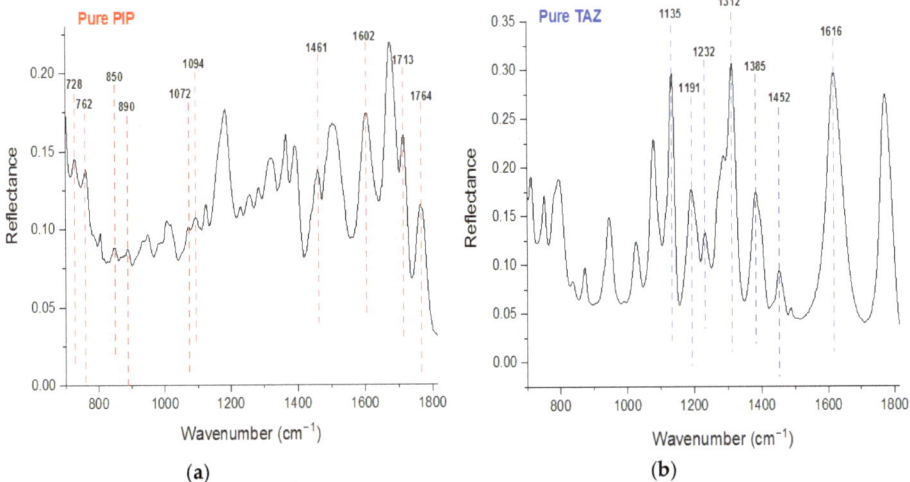

Figure 1. (a) Quantitative Fourier Transform-Infrared/Attenuated Total Reflectance (FT-IR/ATR) powder spectrum of pure piperacillin (PIP) and (b) FT-IR/ATR powder spectrum of pure tazobactam (TAZ).

Table 1. Vibration modes of some prominent peaks of pure active pharmaceutical ingredients (APIs).

API	Wavenumber (cm^{-1})	Vibration Mode
PIP	728, 762, 850, 890	Aromatic CH wagging
	1072, 1094	In-plane movement of aromatic carbons
	1461, 1602	Semi-circle stretching and Quadrum stretching of aromatic ring
	1713, 1764	Carbonyl group (CO) stretching
TAZ	1135, 1191, 1312	In-phase and Out-of-phase stretching of C-SO$_2$-C
	1232, 1452	N=N stretching
	1385, 1616	Carbonyl group (CO) stretching

The FT-IR/ATR spectrum of the formulation was then obtained. A great similarity between the spectrum of the commercial mixture and that of piperacillin was observed, as shown in Figure 2. This was expected due to the high content of piperacillin in the formulation.

On the other hand, TAZ peaks are extensively overlapped by those of PIP, and the presence of TAZ in the formulation becomes evident through small differences between formulation and PIP spectra, such as strengthening of PIP peaks at certain positions or change in intensity ratio between neighboring PIP peaks. All these observations are summarized in Table 2.

2.1.2. Film Spectra

Subsequently, solutions of pure PIP (50 mg/mL), TAZ (30 mg/mL) and the formulation (44.44 mg PIP/mL, 5.56 mg TAZ/mL) were prepared and an aliquot, equal to 5 μL was placed on ATR crystal. Spectra were obtained 20 min later from the film formed thereon after evaporation of the solvent. Worth mentioning is the fact that the concentration of each solution was of such a value that adequate sample quantity could be in good contact with ATR crystal after solvent had been evaporated and a satisfying signal to noise ratio could be obtained.

Film spectra found to be of better quality than those of the powder. The term better quality is referred to the higher intensity of the peaks detected in spectra obtained from the film, as well as the better signal-to-noise ratio. This is achieved due to the better contact between the analyte and the ATR crystal. Apart from that, no peak shift is observed in the case of the film spectra, in comparison with the respective spectra obtained from the powder of the substance under test. This is valid for both PIP and TAZ and means that after solvent evaporation, no other polymorph or hydrated form of the active substances is precipitated. Figure 3 is the proof of the two previous conclusions.

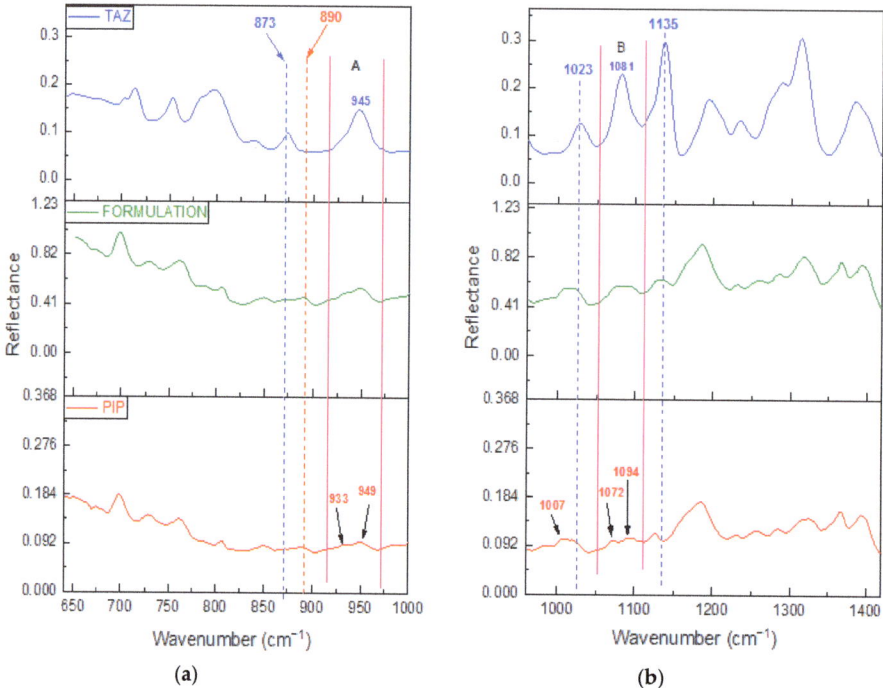

Figure 2. FT-IR/ATR Powder Spectra of PIP, TAZ, and the formulation: (**a**) from 640 to 1000 cm^{-1} and (**b**) from 960 to 1420 cm^{-1}. Differences between PIP and formulation spectra due to TAZ presence in the commercial mixture are with blue dotted lines and in the spectral regions A (915–971 cm^{-1}) and B (1052–1111 cm^{-1}). The characteristic peak of PIP at 890 cm^{-1} is with the red dotted line.

Table 2. Spectral differences between PIP and formulation FT-IR/ATR spectra obtained from powder samples of the substances, due to TAZ presence.

Wavenumber (cm^{-1})	Observation
873	Intensity enhancement and change in intensity ratio against the characteristic PIP peak at 890 cm^{-1}
945	Intensity enhancement and change in intensity ratio between neighboring peaks at 933 and 949 cm^{-1} (area A)
1023	Intensity enhancement and change in intensity ratio against the neighboring peak at 1007 cm^{-1}
1081	Intensity enhancement and change in intensity ratio between neighboring peaks at 1072 and 1094 cm^{-1} (area B)
1135	Detection of TAZ shoulder

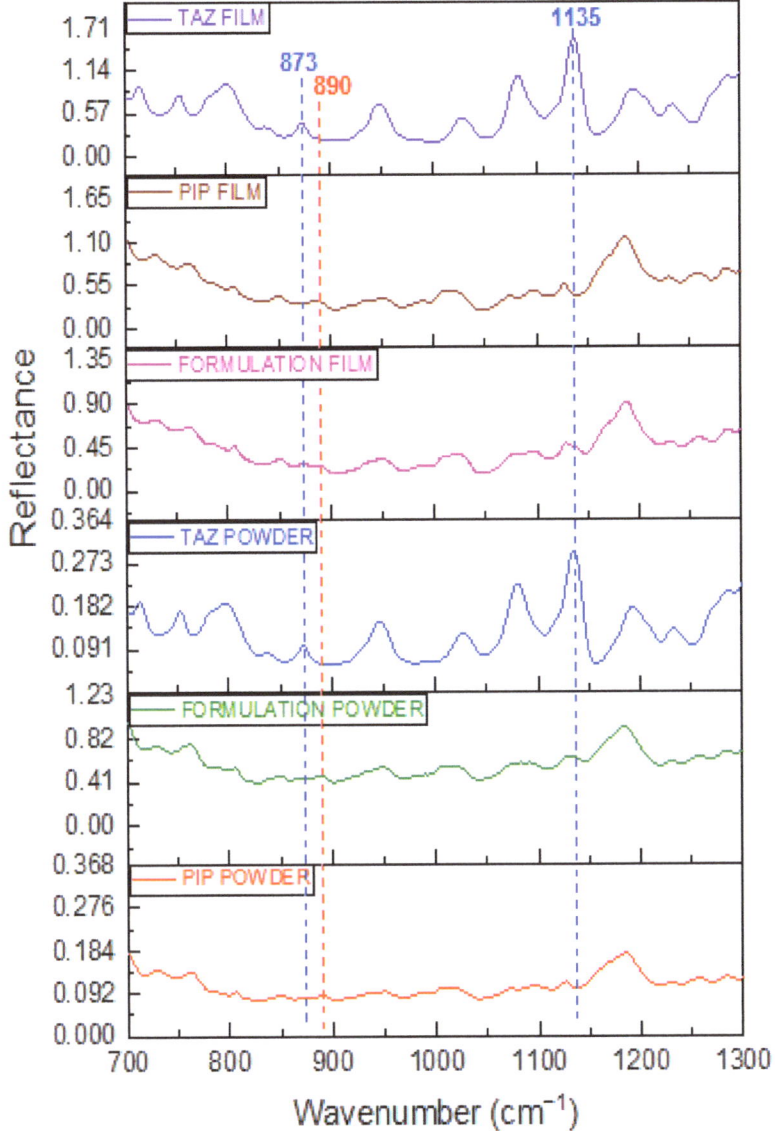

Figure 3. FT-IR/ATR spectra of PIP, TAZ, and Formulation Film (brown, purple, and pink line respectively), as well as that of PIP, TAZ, and Formulation Powder (red, blue, and green line respectively) at spectral region from 700 to 1300 cm^{-1}.

Emphasis should be placed on peaks at 873 and 1135 cm^{-1}, which are the main evidence of tazobactam presence in the formulation, since non-overlapping characteristic peaks of the API do not exist. Particularly, either the strengthening of PIP peak at 873 cm^{-1} or the shoulder detected at 1135 cm^{-1}, due to TAZ, are much more distinct in case of film spectra acquisition. Thus, identification of TAZ API is much easier.

2.2. Quantification of PIP and TAZ

Solutions of a wide range of standard PIP-TAZ mixtures were prepared and their ATR spectra were obtained, using the film formation technique (Figure 4). The purpose of that procedure was the construction of a calibration line for the quantification of % *w/w* content of each API in the solid formulation. Mass ratio of the standard mixtures prepared, ranged from 5% TAZ-95% PIP to 50% TAZ-50% PIP, taking into account that the composition of the commercially available formulations is 11% TAZ-89% PIP.

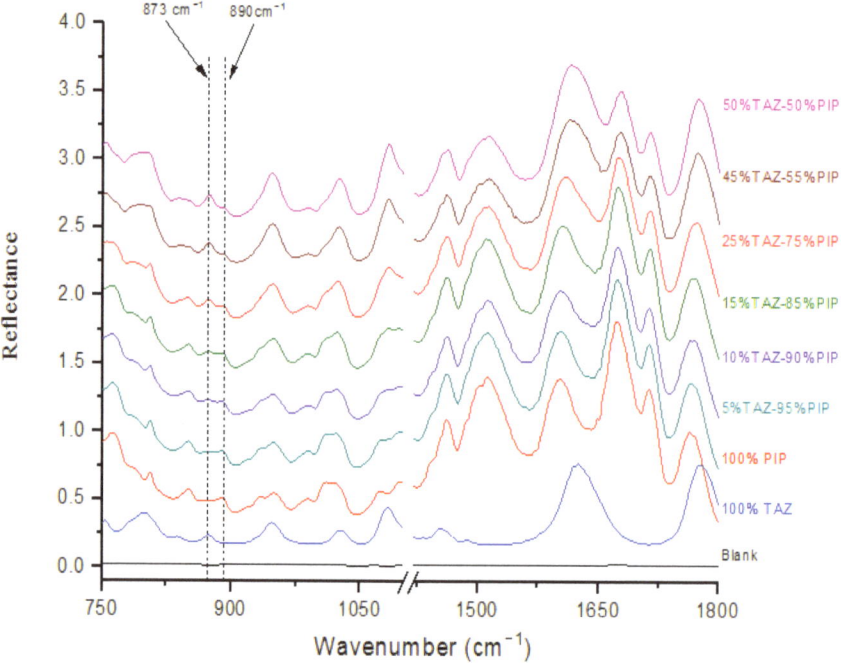

Figure 4. FT-IR/ATR Film Spectra obtained from solutions of standard solid mixtures with a mass ratio of PIP-TAZ equal to 50% TAZ-50% PIP (pink line), 45% TAZ-55% PIP (brown line), 25% TAZ-75% PIP (orange line), 15% TAZ-85% PIP (green line), 10% TAZ-90% PIP (purple line) and 5% TAZ-95% PIP (navy line). Additionally, ATR Film Spectra obtained from pure PIP (red line), pure TAZ (blue line) and blank solution (water film, black line). Note that the spectra have been shifted along y axis for clarity.

As TAZ concentration in the solid mixture increases, the peak at 873 cm^{-1} becomes gradually more evident and it ends up outweighing its neighboring peak at 890 cm^{-1}.

The mean peak height ratio, $I(873)/I(890)$ and relative standard deviation (RSD) values resulting from three replicate measurements were calculated for each standard (Table 3).

At this point, low RSD values in Table 3 are indicative of the reproducibility of replicate sample measurements as well as the high sample homogeneity achieved, given that the instrumental conditions remained constant.

The calibration graph plotted is reproduced in Figure 5 and was constructed through least square regression method, using the ratio $\frac{I(873)}{I(890)}$ (y axis) against ratio $\frac{100}{C_{PIP}}$ (x axis) [See Appendix A].

Table 3. Mean peak height ratio, *I*(873)/*I*(890) and relative standard deviation (RSD) values corresponding to the three replicate film measurements, for each one of the standard mixtures tested.

% Mass Ratio TAZ-PIP	*I*(873)/*I*(890)	RSD%
50:50	2.925	3.14
45:55	2.493	2.19
25:75	1.138	3.48
15:85	0.699	2.22
10:90	0.517	3.89
5:95	0.367	1.90

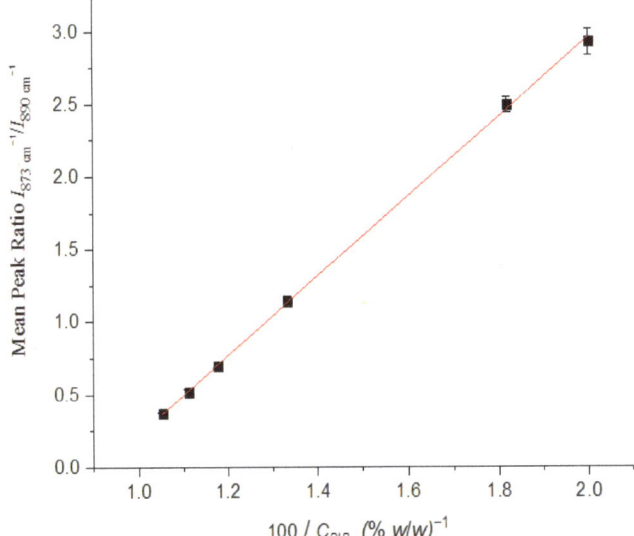

Figure 5. Calibration line for both PIP and TAZ for solid mixtures of them, with APIs' mass ratio ranging from 5% TAZ-95% PIP to 50% TAZ-50% PIP.

The equation describing the calibration line was:

$$\frac{I\left(873 \text{ cm}^{-1}\right)}{I\left(890 \text{ cm}^{-1}\right)} = (2.74 \pm 0.024) \times \left(\frac{100}{C_{PIP}}\right) + (-2.52 \pm 0.026), \left(R^2 = 0.999\right), \tag{1}$$

Since minimum amount of tazobactam corresponds to maximum amount of piperacillin, the limit of detection (LOD) of tazobactam was calculated after many blank (PIP film) measurements. Using Equation (1), LOD was found to be 1.618% *w/w* and LOQ (limit of quantification) equal to 4.854% *w/w*. LOD value was confirmed by visual evaluation method [23].

In order to test the performance of the calibration line a solution of 11% TAZ-89% PIP was treated as unknown sample. The PIP concentration was found to be (89.37 ± 0.40)% *w/w* and that of TAZ was equal to (10.63 ± 0.40)% *w/w*. The relative error values of the results mentioned above were found to be 0.41% and 3.33%, respectively.

In order to test if the calibration line could be applied directly to solid formulation before reconstitution, the as-received formulation powder was treated as unknown sample and its ATR spectrum was recorded one month after the construction of the calibration line. The PIP concentration was found to be (88.68 ± 0.39)% *w/w* and that of TAZ was equal to (11.32 ± 0.39)% *w/w*. The relative error values of the results mentioned above were found to be 0.36% and 2.91%, respectively.

3. Discussion

The aim of the present study was to apply the sample preparation method of film formation, on the ATR crystal after solvent evaporation, in order for piperacillin and tazobactam, two different APIs coexisting in a commercial drug mixture, to be identified and quantified. The above-mentioned methodology proved to be a valuable tool that exceeded our expectations. This sample preparation technique ensured a good contact between the sample and the ATR crystal, thus making it possible to enhance the spectral detail. The amount of liquid formulation placed on the crystal had such a value that the film formed thereon was thin enough to permit the acquisition of a complete spectral image of the sample. Given that the penetration depth of the IR beam is restricted to a few microns, it was achieved to obtain whole analyte information. Such a sample preparation method can also overcome the usual problems with solvent IR absorption, which are especially significant for aqueous solutions, irrespective of the concentration of analytes dissolved in it. An additional advantage of the technique used is that by handling a liquid sample, no pressure needs to be employed on the sample surface, thus protecting crystal materials prone to crack.

Although the formulation had a quite low content of TAZ API and the characteristic peaks of the latter were extensively overlapped by those of PIP, it was effectively identified. The quantification of APIs' concentrations in solid drug took place after selection of a suitable pair of bands, which could yield reliable results. Variations in signal intensity of some characteristic peaks of each API between replicate samples was not only due to non-uniform distribution of analyte particles on the surface of the instrument's crystal, but also to different orientation of them, thus affecting their interaction with radiation. The calibration curve obtained presents a linear dynamic range from 5% to 50% *w/w* TAZ and a correlation coefficient about 0.999.

Up to date, piperacillin and tazobactam have been in most cases analyzed through chromatographic techniques, which are quite sensitive and precise but gather some important disadvantages. Apart from being time-consuming, HPLC requires large volume of solvents, thus excluding itself from the category of "green" techniques. UHPLC, on the other hand, takes only 5 min to be applied and few µL of sample, but the whole procedure still requires a mobile phase and a more specialized user. Also, the proposed methodology can be applied directly to dry formulation before reconstitution i.e., no sample preparation is involved, while chromatographic techniques can be applied only for liquid samples. That was proved when the performance of the calibration line was tested not only for the liquid reconstituted sample after drying on the ATR crystal but also for the solid formulation.

Considering the above, such an analytical technique combined with the proposed relatively simple sample treatment should be considered as an alternative when developing quality-control procedures for solid or liquid pharmaceuticals. The advantages it gathers make it potentially ideal for in-line control in manufacture in order for simultaneous quantification of multiple products to be performed. Besides, the methodology presented here is easy to be applied, not only for this formulation but for numerous intravenous administered formulations containing various APIs such as Vancomycin, Voriconazole, Propofol, since it requires limited training and no particular analytical knowledge. Thus, it can be incorporated in a hospital workflow. The identification and quantification of the drug formulation in such cases could be performed either after reconstitution of the formulation (using a negligible amount of the liquid drug from its glass bottle by syringe) or before its administration (using the liquid drug from the intravenous infusion bag where it would have been further diluted) or even by directly checking the dry formulation before reconstitution. A nurse or the hospital pharmacist could then place either the dry formulation directly on the ATR crystal and obtain with no time delay the spectrum of formulation (no sample treatment is involved) or use the liquid sample after reconstitution and film formation so as to record the spectrum. Such an approach is quite promising as it could confirm the identity as well as the concentration of a formulation before its administration to a patient, thus eliminating errors that are irreversible and can cost human lives.

4. Materials and Methods

In the present study, pure piperacillin and tazobactam APIs, were purchased in the form of sodium salts, from Glentham Life Sciences (Corsham, UK), which is a UK based supplying company. The generic formulation used was Zobactam® (Vocate Pharmaceuticals S.A., Athens, Greece), a combination of piperacillin and tazobactam in a mass ratio of 8:1, respectively. The formulation was in the form of powder for injectable solution with no excipients and had been supplied by Aenorasis S.A., a Greece based medical equipment company (Athens, Greece). In order to complete reconstitution of 4.5 g of the drug to be achieved, 20 mL of water for injection were used.

4.1. Preparation of Standard Mixtures

Pure APIs were used to prepare 6 dry mixtures. The mass ratio of TAZ:PIP ranged from 50:50 to 5:95. Using an analytical balance (Kern Inc., Grove City, OH, USA, ABJ 220-4NM) and depending on the desired ratio, an appropriate amount of each API was obtained in order for a total mass of 100 mg of the standard mixture to be prepared (e.g., for the dry mixture 25:75, 25 mg TAZ and 75 mg PIP were obtained). Each mixture was placed in a special plastic container (NALGENE®, Rochester, NY, USA) having a magnetic rod and was homogenized by placing the latter on a magnetic stirrer (HANNA instruments, HI 190M (Woonsocket, RI, USA)) for 5 min.

4.2. Preparation of Solutions of Standard Mixtures

The total mass of the prepared dry mixtures was transferred to 20 mL glass vials and dissolved by adding 2 mL of water for injection. The solvent was added using a 1 mL automatic pipette (Labnet International Inc., Edison, NJ, USA, Biopette™, Autoclavable Pipettes). In order for complete dissolution to be achieved, Vortex (IKA®, Staufen, Germany, MS2, Minishaker) as well as ultrasonic bath (Branson Ultrasonics, Danbury, CT, USA, 2510E-MT) at a frequency of 60 HZ were used for about 30 s and 15 min, respectively.

4.3. Development of Film Forming Methodology

The instrumentation used was comprised of a cavity, the circular bottom of which had an internal diameter of 4.00 mm, an external diameter of 10.76 mm and a depth of 1.60 mm and was in direct contact with the 9 bounce ATR crystal, as shown in Figure 6. The sample tested was placed in the center of this cavity in order for the radiation to interact with the analyte molecules precipitated thereon after evaporation. Of great significance was the amount of liquid sample placed in the cavity, so as not only intense peaks to be detected and a good signal-to-noise ratio to be achieved, but also no sample particles to be precipitated outside the bottom of it (scanning region). After trials it was concluded that the capacity of the cavity was of about 70 µL and the evaporation time was more than 2 h. An amount of 10 µL of the sample was tested, but it was noticed that not all analyte particles were precipitated within the scanning region and the evaporation time exceeded 30 min. Finally, 5 µL was the sample quantity considered as the most suitable for this methodology. Firstly, due to the fact that it was adequate to cover the whole surface of the scanning region so as more analyte particles to be precipitated thereon and not outside it, thus more intense peaks to be detected. Secondly, the evaporation time, in this case, was equal to 20 min.

4.4. FT-IR/ATR Spectra Acquisition

ATR spectra were obtained using a FTIR-ATR spectrometer (PerkinElmer Spectrum 100) with a 9 bounce Diamond/ZnSe crystal (Waltham, MA, USA). The background spectrum within the instrument was recorded prior to the start of each measurement. The powder samples of APIs and the formulation were measured by placing about 30 mg of each on the surface of instrument's crystal using a spatula. Then, by the aid of the ATR pressure arm, 60% of the total pressure applicability was used so as good contact between analyte molecules and the crystal to be achieved. The pressure applied was constant for all solid samples. As for liquid samples, 5µL of each solution of standard dry mixtures was placed on ATR crystal using a 10 µL automatic pipette (Scilogex, MicroPette Plus Autoclavable pipettor, YM3K012623) and the spectrum of the sample film was acquired 20 min later, as soon as the solvent had been evaporated. This was repeated for 3 different solution portions from each sample. Each measurement was the result of 25 scans in the mid-infrared region (4000–600 cm^{-1}) and instrument's resolution was equal to 4 cm^{-1}.

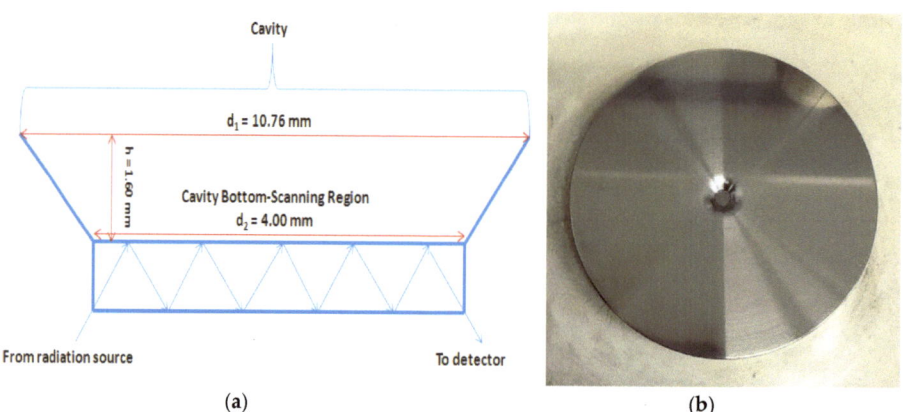

(a) (b)

Figure 6. (a) Scheme of ATR Instrumentation-Sample Position Structure; (b) Photo of ATR Instrumentation-Sample Position Structure.

4.5. Spectra Processing

Spectral data were processed using Origin (OriginPro8, OriginLabPro®, Northampton, MA, USA) and recorded as absorbance data. OriginPro8 integration method of measuring the absolute height (or intensity) of selected peaks was used in conjunction with different combinations of fingerprint APIs' peaks so as to assess the potential to quantify the level of APIs present. The calibration graph was prepared not for individual peaks, but for a combination of them. The one and only baseline corresponding to both of the neighboring peaks of interest (873 cm^{-1} for TAZ and 890 cm^{-1} for PIP) ranged from 862 to 905 cm^{-1}.

5. Conclusions

Identification and quantitative determination of intravenously administered TAZ-PIP formulation, before and after reconstitution, was accomplished successfully by FT-IR/ATR. The initial powder formulation was analyzed by placing the powder directly on the ATR crystal, while the resulted solution, after reconstitution, was analyzed after the evaporation of the water and the subsequent film formation on the ATR crystal, using less than 2 µL. The latter approach, despite being more time-consuming, due to required time for the film formation, has the advantage of being aseptic and practically non-destructive for the sample since the minute amount needed for the measurement can be drawn with a syringe directly from the bottle or from the infusion bag. Usage of FT-IR/ATR, through film formation, is less complicated than most of the other analytical techniques and thus, this methodology can be introduced into the workflow of a hospital for the reliable, easy to apply, quantitative, and qualitative analysis of all intravenously administered formulations.

Author Contributions: Conceptualization, I.C., M.O. and C.K.; methodology, I.C., M.O. and C.K.; investigation, I.C.; validation, I.C., M.O. and C.K.; data curation, I.C.; writing—original draft preparation, C.K.; review and editing, M.O. and C.K.; supervision, M.O. and C.K.; project administration, C.K. All authors have read and agreed to the published version of the manuscript.

Funding: This research received no external funding.

Acknowledgments: The generic formulation used (Zobactam®) was donated by Aenorasis S.A., a Greece based medical equipment company.

Conflicts of Interest: The authors declare no conflict of interest.

Appendix A

Appendix A is a section containing the mathematical proof of how the final equation of the calibration curve was developed. Spectral observations arising from film spectra of both pure active substances and the commercial formulation (Figures 1 and 2) confirmed that peak at 873 cm^{-1} is a result of not only PIP but also TAZ presence. Thus, signal intensity at that position is expected to be given from the Equation:

$$I\left(873 \text{ cm}^{-1}\right) = k_1 \times C_{PIP} + k_2 \times C_{TAZ}, \tag{A1}$$

where k_1 and k_2 are standards correlating the value $I\left(873 \text{ cm}^{-1}\right)$ with C_{PIP} and $I\left(873 \text{ cm}^{-1}\right)$ with C_{TAZ} respectively.

Due to the fact that peak at 890 cm^{-1} is characteristic of piperacillin, its intensity is expected to be given from the equation:

$$I\left(890 \text{ cm}^{-1}\right) = k_3 \times C_{PIP}, \tag{A2}$$

where k_3 is a standard correlating the value $I\left(890 \text{ cm}^{-1}\right)$ with C_{PIP}.

If (A1) is divided by (A2), then it turns out that:

$$\frac{I\left(873 \text{ cm}^{-1}\right)}{I\left(890 \text{ cm}^{-1}\right)} = \frac{k_1}{k_3} + \frac{k_2}{k_3} \times \frac{C_{TAZ}}{C_{PIP}} \tag{A3}$$

Given that the formulation consists exclusively of the APIs it is valid that:

$$C_{TAZ} = 100 - C_{PIP}, \tag{A4}$$

Finally, combining (A3) with (A4) a new equation is formed, where:

$$\frac{I\left(873 \text{ cm}^{-1}\right)}{I\left(890 \text{ cm}^{-1}\right)} = \frac{k_1 - k_2}{k_3} + \frac{k_2}{k_3} \times \frac{100}{C_{PIP}} \tag{A5}$$

References

1. Larkin, P. *Infrared and Raman Spectroscopy: Principles and Spectral Interpretation*; Elsevier Science: Philadelphia, PA, USA, 2017.
2. Skoog, D.A.; Holler, F.J.; Nieman, T.A. *Instrumental Analysis Principles*, 7th ed.; Cengage Learning: Boston, MA, USA, 2016; pp. 389–436.
3. Hans, K.M.C.; Müller, S.; Sigrist, M.W. Infrared attenuated total reflection (IR-ATR) spectroscopy for detecting drugs in human saliva. *Drug Test. Anal.* **2012**, *4*, 420–429. [CrossRef] [PubMed]
4. Ghauch, A.; Deveau, P.A.; Jacob, V.; Baussand, P. Use of FTIR spectroscopy coupled with ATR for the determination of atmospheric compounds. *Talanta* **2006**, *68*, 1294–1302. [CrossRef] [PubMed]
5. Benramdane, L.; Bouatia, M.; Idrissi, M.O.B.; Draoui, M. Infrared Analysis of Urinary Stones, Using a Single Reflection Accessory and a KBr Pellet Transmission. *Spectrosc. Lett.* **2008**, *41*, 72–80. [CrossRef]
6. Wilson, R.H.; Tapp, H.S. Mid-infrared spectroscopy for food analysis: Recent new applications and relevant developments in sample presentation methods. *TrAC Trends Anal. Chem.* **1999**, *18*, 85–93. [CrossRef]

7. Durak, T.; Depciuch, J. Effect of plant sample preparation and measuring methods on ATR-FTIR spectra results. *Environ. Exp. Bot.* **2020**, *169*, 103915. [CrossRef]

8. Lawson, G. Counterfeit Tablet Investigations: Can ATR FT/IR Provide Rapid Targeted Quantitative Analyses? *J. Anal. Bioanal. Tech.* **2014**, *5*, 5. [CrossRef]

9. Hartauer, K.J.; Guillory, J.K. Quantitative Fourier Transform-Infrared/Attenuated Total Reflectance (FT-IR/ATR) Analysis of Trimethoprim and Sulfamethoxazole in a Pharmaceutical Formulation Using Partial Least Squares. *Pharm. Res.* **1989**, *6*, 608–611. [CrossRef] [PubMed]

10. Hughes, J.; Ayoko, G.; Collett, S.; Golding, G. Rapid quantification of methamphetamine: Using attenuated total reflectance fourier transform infrared spectroscopy (ATR-FTIR) and chemometrics. *PLoS ONE* **2013**, *8*, e69609. [CrossRef] [PubMed]

11. Makki, A.A.; Bonnier, F.; Respaud, R.; Chtara, F.; Tfayli, A.; Tauber, C.; Bertrand, D.; Byrne, H.J.; Mohammed, E.; Chourpa, I. Qualitative and quantitative analysis of therapeutic solutions using Raman and infrared spectroscopy. *Spectrochim. Acta Part A Mol. Biomol. Spectrosc.* **2019**, *218*, 97–108. [CrossRef] [PubMed]

12. Shorr, R.I.; Hoth, A.B.; Rawls, N. (Eds.) *Drugs for the Geriatric Patient*; W.B. Saunders: Philadelphia, PA, USA, 2007; pp. 930–1062.

13. Drawz, S.M.; Bonomo, R.A. Three Decades of β-Lactamase Inhibitors. *Clin. Microbiol. Rev.* **2010**, *23*, 160. [CrossRef] [PubMed]

14. Kuriyama, T.; Karasawa, T.; Williams, D.W. Chapter Thirteen—Antimicrobial Chemotherapy: Significance to Healthcare. In *Biofilms in Infection Prevention and Control*; Percival, S.L., Williams, D.W., Randle, J., Cooper, T., Eds.; Academic Press: Boston, MA, USA, 2014; pp. 209–244.

15. Rama Krishna Veni, P.; Sharmila, N.; Narayana, K.J.P.; Hari Babu, B.; Satyanarayana, P.V.V. Simultaneous determination of piperacillin and tazobactam in pharmaceutical formulations by RP-HPLC method. *J. Pharm. Res.* **2013**, *7*, 127–131. [CrossRef]

16. Atmakuri, L.R.; Krishna, K.; Kumar, C.H.; Raja, T.A. Simultaneous determination of piperacillin and tazobactum in bulk and pharmaceutical dosage forms by rphplc. *Int. J. Pharm. Pharm. Sci.* **2011**, *3*, 134–136.

17. Pai, S.; Rao, G.K.; Murthy, M.S.; Prathibha, H. Simultaneous estimation of piperacillin and tazobactam in injection formulations. *Indian J. Pharm. Sci.* **2006**, *68*, 799–801. [CrossRef]

18. Council of Europe, European Directorate for the Quality of Medicines & Healthcare. *European Pharmacopoeia 10.4*; Council of Europe: Strasburg, France, 2020; p. 1168.

19. Naicker, S.; Guerra Valero, Y.C.; Ordenez Meija, J.L.; Lipman, J.; Roberts, J.A.; Wallis, S.C.; Parker, S.L. A UHPLC-MS/MS method for the simultaneous determination of piperacillin and tazobactam in plasma (total and unbound), urine and renal replacement therapy effluent. *J. Pharm. Biomed. Anal.* **2018**, *148*, 324–333. [CrossRef] [PubMed]

20. Zander, J.; Maier, B.; Suhr, A.; Zoller, M.; Frey, L.; Teupser, D.; Vogeser, M. Quantification of piperacillin, tazobactam, cefepime, meropenem, ciprofloxacin and linezolid in serum using an isotope dilution UHPLC-MS/MS method with semi-automated sample preparation. *Clin. Chem. Lab. Med.* **2015**, *53*, 781–791. [CrossRef] [PubMed]

21. Karpova, S.P.; Blazheyevskiy, M.; Mozgova, O. Development and validation of UV spectrophotometric area under curve method quantitative estimation of piperacillin. *Int. J. Pharm. Sci. Res.* **2018**, *9*, 3556–3560.

22. Toral, M.I.; Nova-Ramírez, F.; Nacaratte, F. Simultaneous determination of piperacillin and tazobactam in the pharmaceutical formulation Tazonam® by derivative spectrophotometry. *J. Chil. Chem. Soc.* **2012**, *57*, 1189–1193. [CrossRef]

23. ICH Harmonized Tripartite Guideline Q2 (R1) Validation of Analytical Procedures. Available online: https://www.ema.europa.eu/en/ich-q2-r1-validation-analytical-procedures-text-methodology (accessed on 20 March 2020).

Sample Availability: Samples of the compounds PIP and TAZ are available from the authors.

 *molecules*

Article

Sample Preparation of Posaconazole Oral Suspensions for Identification of the Crystal Form of the Active Pharmaceutical Ingredient

Michail Lykouras [1] , **Stefani Fertaki** [1,2] , **Malvina Orkoula** [1,2] **and Christos Kontoyannis** [1,2,*]

[1] Department of Pharmacy, University of Patras, GR-26504 Rio Achaias, Greece; michalislyk@gmail.com (M.L.); sfertaki@gmail.com (S.F.); malbie@upatras.gr (M.O.)
[2] Institute of Chemical Engineering Sciences, Foundation of Research and Technology-Hellas (ICE-HT/FORTH), GR-26504 Platani Achaias, Greece
* Correspondence: kontoyan@upatras.gr; Tel.: +30-2610-962328

Academic Editors: Victoria Samanidou and Irene Panderi
Received: 21 November 2020; Accepted: 17 December 2020; Published: 19 December 2020

Abstract: Determination of the polymorphic form of an active pharmaceutical ingredient (API) in a suspension could be really challenging because of the water phase and the low concentration of the API in this formulation. Posaconazole is an antifungal drug available also as an oral suspension. The aim of this study was to develop a sample-preparation method for polymorphic identification of the dispersed API by increasing the concentration of the API but with no compromise of polymorph stability. For this purpose, filtration, drying and centrifugation were tested for separating the API from the suspending medium. Centrifugation was selected because it succeeded in separating Posaconazole API with no polymorph transformation during the process. During this study, it was found that Posaconazole in oral suspensions is Form-S. However, when slower scanning rates were used for acquiring an XRPD pattern with better signal/noise ratio, Posaconazole was converted to Form I due to water loss. In order to protect the sample from conversion, different approaches were tested to secure an airtight sample including a commercially available XRPD sample holder with a dome-like transparent cap, standard polymethylmethacrylate (PMMA) sample holders covered with Mylar film, transparent pressure-sensitive tape and a transparent food membrane. Only usage of the transparent food membrane was found to protect the API from conversion for a period of at least two weeks and resulted in a Posaconazole Form-S XRPD pattern with no artificial peaks.

Keywords: posaconazole; oral suspension; polymorphism; form-S; sample preparation; centrifugation; XRPD

1. Introduction

Solid Active Pharmaceutical Ingredients (APIs) might exist in various crystal forms with different physical properties, although their chemical properties are identical. The multiple crystal forms are known as polymorphs, and the phenomenon is called polymorphism [1]. The crystal forms created by the combination of API molecules and solvent molecules in different stoichiometric proportions are known as solvates. When the solvent is water, the crystal form is a hydrate. Pseudopolymorphism is the term that describes the behaviour of solvates and hydrates [2]. Another form of solid APIs is the amorphous solid lacking long-range order or well-defined conformation of API molecules [3].

The phenomenon of polymorphism is of high importance for the pharmaceutical industry because the phase transition of an API could lead to undesirable properties affecting the stability of the final product, influencing the dissolution properties and the bioavailability of the API, and hence generating different pharmacological action [4]. Therefore, identification of the polymorphic or pseudopolymorphic form of the API in the final pharmaceutical product is considered critical for the development of the product.

Chromatographic techniques could not be applied for the qualitative determination of the crystal form of an API because different polymorphs cannot be identified by these methods. Among the analytical techniques used for polymorphic identification of a solid API, methods at the molecular level, particulate level and bulk level are included. The first class of methods involves spectroscopic methods, like Infrared Spectroscopy (IR), Raman Spectroscopy and solid-state Nuclear Magnetic Resonance (ss-NMR). The second category involves X-ray Powder Diffraction (XRPD); Light Microscopy; Scanning Electron Microscopy (SEM); and thermoanalytical methods, such as Differential Scanning Calorimetry (DSC), Thermogravimetric Analysis (TGA) and Dynamic Vapour Sorption (DVS). The last class involves the method of Karl Fisher Titration and the Brunauer, Emmett and Teller (BET) method [1].

Oral suspensions are pharmaceutical formulations widely preferred for paediatric or geriatric administration and are also characterized as dispersions because the solid API and possibly some excipients are dispersed as solid particles in a liquid vehicle or suspending medium [5]. Due to the liquid suspending medium, which is frequently water, the APIs in oral suspensions are prone to polymorphic transition causing physical instability of the suspension. This lack of physical stability can be expressed as crystal growth or caking, i.e., the loss of a suspension's ability to be resuspended uniformly [6]. Therefore, identification of the API's polymorph in an oral suspension is very important so that the stability of that pharmaceutical formulation is ensured.

However, determination of the polymorphic form of the API in an oral suspension is often challenging. The liquid phase of oral suspensions affecting the signal/noise ratio, the low concentration of the API in those pharmaceutical formulations and the lack of homogeneity of the samples are frequently the basic obstacles in the analysis methods used for their polymorphic characterization [7]. These challenges in the polymorphic characterization of the API in pharmaceutical oral suspensions in combination with the lack of suitable methods in the literature for overcoming these difficulties render the development of a novel sample-preparation method to be of high importance. The novel method to be developed should be capable of eliminating the interference of water contained in pharmaceutical oral suspensions without causing phase conversion of the API and at the same time differentiating the possible different polymorphs of the API in those formulations successfully.

For this purpose, Posaconazole oral suspensions were used. Posaconazole is a broad-spectrum member of the second generation antifungal triazole drugs and is used against invasive fungal infections [8,9]. It is commercially available as an oral suspension (40 mg/mL), as a concentrate for solution for intravenous infusion (300 mg) and as gastro-resistant tablets (100 mg) [10]. Although 12 different forms of Posaconazole have been already described [11–18], Posaconazole Form I is used for the production of Posaconazole oral suspension [19].

The quite low concentration of Posaconazole in oral suspensions (40 mg/mL) and the nature of the suspensions are obstacles in identification of the crystal form of the API in these formulations. Therefore, the aim of this study was to develop a sample-preparation method to artificially increase the concentration of the API in the suspension without causing a polymorphic conversion and to subsequently determine the crystal form of Posaconazole in the concentrated sample.

2. Results

2.1. Posaconazole Oral Suspension Analysis via XRPD

XRPD, Raman spectroscopy and Attenuated Total Reflectance Fourier-Transform Infrared (ATR/FT-IR) spectroscopy were tested for identification of the Posaconazole crystal form in Posaconazole oral suspension. However, the API could be barely detected via ATR/FT-IR spectroscopy in the as-received suspension and both techniques could not be used for differentiating among the different polymorphs of Posaconazole due to minor spectral differences. Hence, XRPD was selected for analysis of the Posaconazole oral suspension. The XRPD patterns of Posaconazole Form I, Posaconazole oral suspension's placebo and Posaconazole oral suspension were recorded and compared to each other (Figure 1). Although Posaconazole Form I has been used to produce Posaconazole oral suspensions [19], the XRPD pattern of the oral suspensions revealed that these formulations were a mixture of Posaconazole oral suspension's placebo and another polymorph or pseudopolymorph, or a combination of different forms of Posaconazole. The liquid phase of the oral suspension and the low concentration of the API in the suspension (40 mg/mL) resulted in a low signal/noise ratio in the XRPD pattern of Posaconazole oral suspension. These issues hindered identification of the crystal form of the API in the oral suspension, and it was not possible to determine whether the API in the oral suspension was found in one or more crystal forms at the same time.

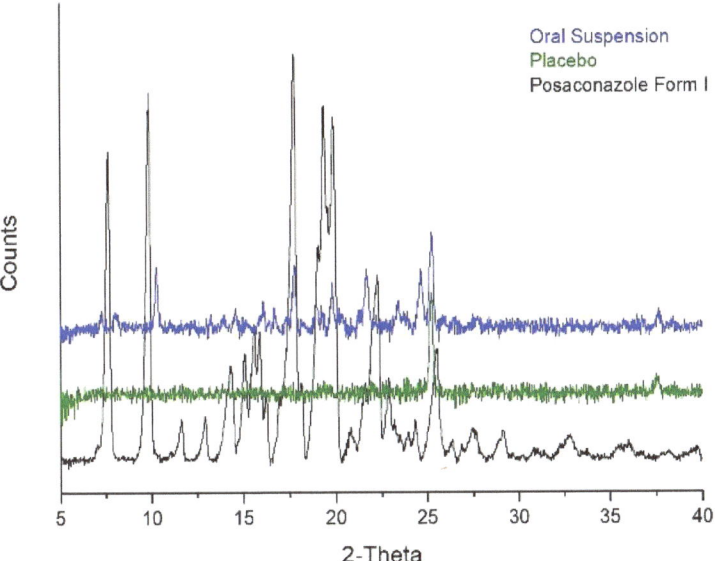

Figure 1. The XRPD patterns of Posaconazole oral suspension (**blue line**), Posaconazole oral suspension's placebo (**green line**) and Posaconazole Form I (**black line**).

2.2. Development of the Sample-Preparation Method for Increasing Posaconazole's Concentration in Oral Suspensions

In order to overcome the problem of low signal/noise ratio, three different methods for increasing Posaconazole's concentration in the sample were tested. Filtering, drying or centrifuging Posaconazole oral suspensions were suggested so that the suspending medium would be removed and the separated precipitate would contain Posaconazole API and the solid excipients of the oral suspensions. Filtration under vacuum failed to remove the suspending medium probably due to the presence of xanthan gum, which is used as a suspending agent of the API, which clogged the pores of the filters.

The method of drying succeeded in removing the liquid phase of the suspension and in increasing Posaconazole's concentration in the solid residue. More specifically, the final concentration of Posaconazole was approximately 12.5% *w/w*, which was more than triple in comparison to the initial concentration. However, the recorded XRPD pattern of the concentrated sample did not resemble the initial XRPD pattern of the oral suspension and highly resembled the pattern of Posaconazole Form I (Figure 2). Consequently, this sample-preparation method failed in retaining the polymorphic form of Posaconazole as it was in the oral suspension.

Molecules **2020**, *25*, 6032

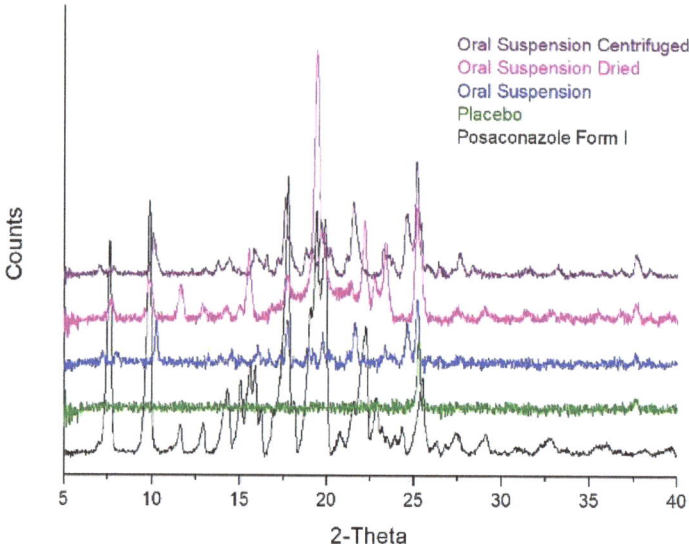

Figure 2. Development of the sample-preparation method for increasing Posaconazole's concentration in oral suspensions: XRPD patterns of Posaconazole oral suspension's precipitate after drying (**magenta line**) and after centrifugation (**purple line**), Posaconazole oral suspension (**blue line**), Posaconazole oral suspension's placebo (**green line**) and Posaconazole Form I (black line).

The final method tested for increasing the concentration of the API in the oral suspension was centrifugation. The precipitate after centrifugation was separated, containing Posaconazole API and the undissolved excipients. The concentration of the API in the precipitate was determined at approximately 20% *w/w*, i.e., it was increased more than 5 times in comparison to Posaconazole's concentration in the oral suspension. The XRPD pattern of the precipitate highly resembled the pattern of the oral suspension, and its signal/noise ratio was increased (Figure 2). The acquired XRPD pattern was compared to the XRPD patterns of all known Posaconazole Forms [11–18], and it was found that Posaconazole in the collected precipitate was Form-S.

2.3. Setting the Scanning Rate of the XRPD Analysis

The effect of scanning rate on the signal/noise ratio of the XRPD pattern of the concentrated sample was investigated by applying scanning rates of 0.5 s/step, 1.0 s/step, 2.0 s/step and 4.0 s/step on fresh separated precipitates for each scanning rate. The XRD pattern which was acquired with a faster 0.5 s/step scanning rate (15 min scan) was used as the reference pattern against which all other XRD patterns were compared. When the 2-times slower scanning rate was applied (1.0 s/step, 30 min scan), it was observed that the quality of the pattern (signal/noise ratio) was improved; however, the pattern in the region 22–40 2-theta did not match the pattern recorded with the 0.5 s/step scanning speed. Although the pattern in the region 4–22 2-theta was identical to the sum of the centrifuged placebo and Posaconazole Form-S, in the region 22–40 2-theta, additional peaks attributed to Posaconazole Form I were also detected (Figure 3). Similar findings were also observed when a 2.0 s/step scanning rate (60 min scan) was applied. In the region 20–40 2-theta, additional peaks have been detected that could be attributed to Posaconazole Form I. As a result, the first half of the pattern highly resembled the sum of Posaconazole Form-S and the centrifuged placebo, while the second half matched the sum of Posaconazole Form I, Posaconazole Form-S and the centrifuged Placebo (Figure 3). When using the 8-times slower 4.0 s/step scanning rate (120 min scan), peaks of Posaconazole Form I could be detected across the XRD pattern (Figure 3). Therefore, it was concluded that, although the signal/noise ratio was improved, Posaconazole Form-S was susceptible to polymorphic conversion to Posaconazole Form I when slow scanning rates are used.

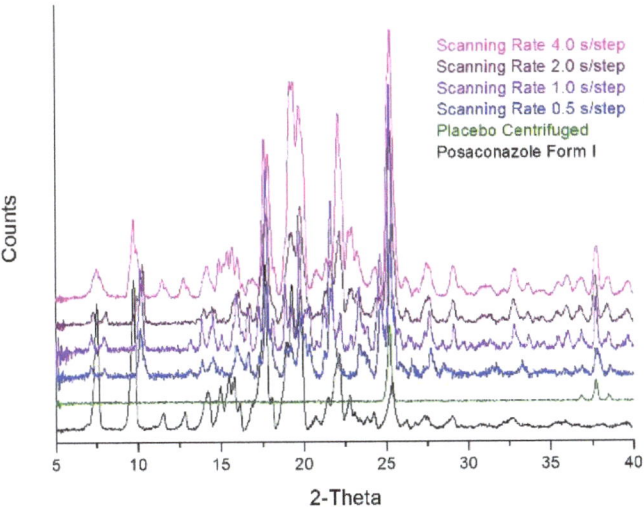

Figure 3. Setting the scanning rate of the XRPD analysis of the oral suspension's precipitate: XRPD patterns of Posaconazole Form I (**black line**), precipitate of Posaconazole oral suspension's placebo after centrifugation (**green line**) and precipitate of Posaconazole oral suspension after centrifugation recorded with scanning rate of 0.5 s/step (**blue line**), 1.0 s/step (**violet line**), 2.0 s/step (**purple line**) and 4.0 s/step (**magenta line**).

2.4. Stability of Posaconazole API in the Oral Suspension's Precipitate

Because of the polymorphic conversion observed during XRPD analysis using a slower scanning rate, the stability of Posaconazole Form-S in the precipitate was studied at ambient temperature. This investigation led to the detection of a quick conversion to Posaconazole Form I. More specifically, immediately after centrifugation and for the next 30 min, the XRPD pattern of the precipitate was practically identical to the sum of Posaconazole Form-S and Posaconazole placebo. After 60 min since isolation of the precipitate by centrifugation, the conversion to Form I had already started and it was completed after 90 min. No other conversion was observed after this time point (Figure 4). These results were quite in accordance with the XRPD patterns recorded at different scanning rates. However, the XRPD pattern recorded with a 1.0 s/step scanning rate after the 22 2-theta region was a mixture of Posaconazole Form-S and Form I, although the duration of the analysis was only 30 min, while in stability tests, at ambient temperature, Posaconazole remained in Form-S for that period of time (Figure 4). An explanation to this could be the higher temperature in the X-ray diffractometer chamber during analysis and consequently the higher water loss from the precipitate leading to polymorphic conversion to Posaconazole Form I. Based on the above observation, it is highly possible that Posaconazole Form-S is likely to be a hydrate form.

Molecules **2020**, *25*, 6032

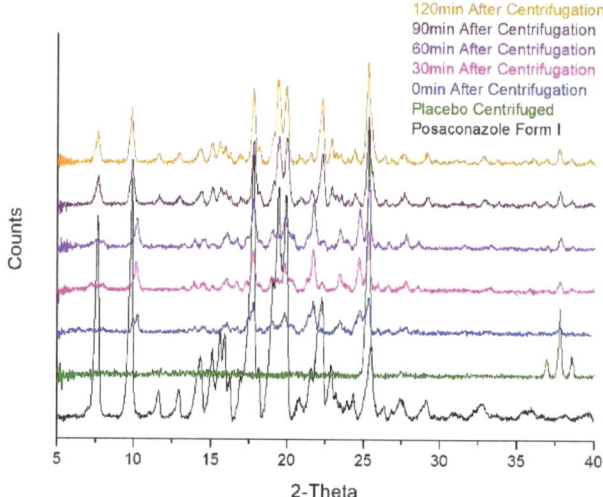

Figure 4. Stability of Posaconazole Form-S in the oral suspension's precipitate: XRPD patterns of Posaconazole Form I (**black line**), precipitate of Posaconazole oral suspension's placebo (**green line**) and precipitate of Posaconazole oral suspension immediately after centrifugation (**blue line**), 30 min (**magenta line**), 60 min (**violet line**), 90 min (**purple line**) and 120 min (**orange line**).

2.5. Development of the Sample-Preparation Method for Delaying the Polymorphic Conversion

In order to prevent water loss from Posaconazole oral suspension's precipitate and subsequent transformation of Form-S to Form I, the following protective materials were applied: a commercially available airtight XRPD sample holder with an X-ray transparent dome-like cap, Mylar film, transparent pressure-sensitive tape and a transparent food membrane. After the application of each protective material, the respective XRPD pattern was recorded and compared against each other and against the pattern of the uncovered centrifuged Posaconazole oral suspension.

The airtight specimen XRPD sample holder with the dome-like transparent cap, although ideal for environmentally sensitive materials, resulted in an XRPD pattern with an extra broad peak in the region 8.3–12.1 2-theta and an intense additional peak at 13.4 2-theta, which interfered with characteristic peaks of Posaconazole Form I and Form-S (Figure 5).

The Mylar film was used to cover the precipitate on a standard PMMA sample holder. The resulting XRPD pattern had a low signal/noise ratio, the peaks were rather broad and neighbouring peaks were merged while some other peaks were missing. Thus, the usage of Mylar film hindered identification of the Posaconazole polymorphic form (Figure 5).

In the XRPD pattern of the precipitate covered with the transparent pressure-sensitive tape, peaks of high intensity at 14.1, 16.9, 18.6 and 25.6 2-theta appeared because of the tape while the peaks of the precipitate were barely visible (Figure 5). As a result, the pressure-sensitive tape cannot be applied as a protective material for detection of the Posaconazole polymorphic form.

The transparent food membrane was the only material which had no interference with the peaks of the precipitate. The XRPD pattern of the sample that was protected with the transparent food membrane was practically identical to the respective pattern recorded without the use of the membrane, i.e., Posaconazole remained as Form-S (Figure 5).

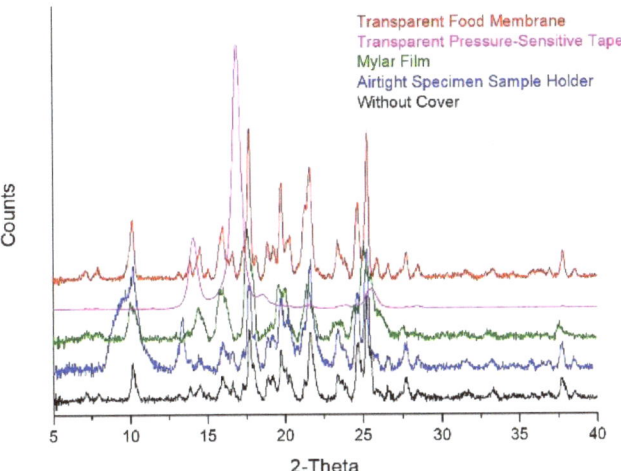

Figure 5. Sample-preparation methods for delaying polymorphic conversion of Posaconazole in the received precipitate after centrifuging Posaconazole oral suspension: XRPD patterns of Posaconazole precipitate without any cover (**black line**) and Posaconazole precipitates placed on an airtight specimen sample holder with dome-like X-ray transparent cap (**blue line**), on a PMMA sample holder covered with a Mylar film (**green line**), on a PMMA sample holder covered with transparent pressure-sensitive tape (**magenta line**) and on a PMMA sample holder enfolded with a transparent food membrane (**red line**).

2.6. Stability of Posaconazole API in the Oral Suspension's Precipitate Covered with Transparent Food Membrane

The ability of the food membrane to protect the sample from polymorphic conversion was tested not only under ambient conditions but also during XRPD recording. The precipitate was covered with the membrane, and the pattern was recorded with gradually increasing scanning rates from 0.5 s/step to 1.0 s/step, 2.0 s/step and finally 4.0 s/step. No difference was observed among the patterns recorded with the four different scanning rates. Posaconazole in all four patterns remained as Form-S. When a slower scanning speed (2.0 s/step or 4.0 s/step) was used, a pattern with higher signal/noise ratio was obtained (Figure 6a).

The stability of Posaconazole Form-S in the precipitate when covered with the transparent food membrane was studied also against time. It was found that the membrane delayed the polymorphic conversion from Posaconazole Form-S to Form I for 2 weeks. After 1 month, both Posaconazole Form-S and Form I were identified in the precipitate by XRPD. However, after 2 months, Posaconazole was converted completely to Form I (Figure 6b). With the use of the membrane, a very significant delay in the polymorphic conversion of Posaconazole Form-S to Form I was achieved compared to the 1 h required for conversion without the membrane. Therefore, the transparent food membrane was efficient in trapping water molecules in the precipitate, delaying Posaconazole polymorphic conversion and giving the opportunity for more time-consuming analysis in order to obtain an XRPD pattern of higher quality.

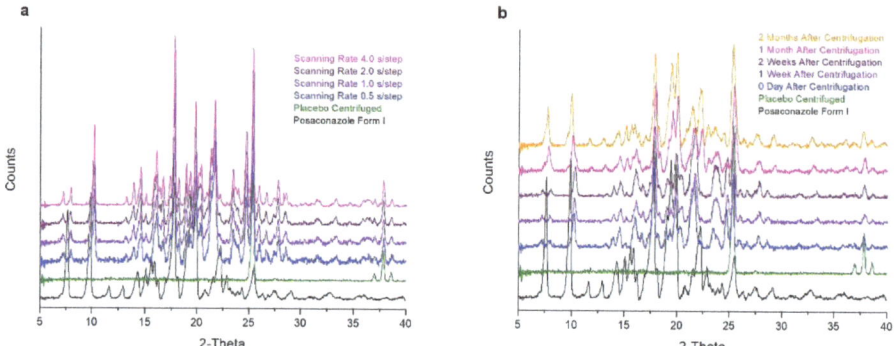

Figure 6. Stabilization of Posaconazole Form-S in the oral suspension's precipitate after covering the sample with transparent food membrane: (**a**) the XRPD patterns of Posaconazole oral suspension's precipitate recorded with different scanning rates at 0.5 s/step (blue line), 1.0 s/step (violet line), 2.0 s/step (purple line) and 4.0 s/step (magenta line); of the precipitate of Posaconazole oral suspension's placebo (green line); and of Posaconazole Form I (black line) and (**b**) the XRPD patterns of Posaconazole oral suspension's precipitate immediately after centrifugation and enfolded with the transparent food membrane (blue line) after 1 week (violet line), 2 weeks (purple line), 1 month (magenta line) and 2 months (orange line); of the precipitate of Posaconazole oral suspension's placebo (green line); and of Posaconazole Form I (black line).

3. Discussion

Posaconazole oral suspension treated with the developed sample-preparation method resulted in an XRPD pattern that was practically identical to the untreated Posaconazole oral suspension. Comparison of the patterns of the precipitate recorded with 2.0 s/step or 4.0 s/step against the initial untreated oral suspension revealed that the signal/noise ratio increased significantly (Figure 7). This improvement in the quality of Posaconazole oral suspension's XRPD pattern gave the opportunity for complete characterization of Posaconazole Form-S and excluded the possibility of impurity from any other polymorphic form of Posaconazole. Hence, using the developed method, the S/N ratio of the XRPD pattern of the suspension increased and the S polymorph was stabilized. This method is fast, reliable, environmentally friendly and capable of identifying the crystal form of the API in oral suspensions.

In this study, the characteristic 2-theta and the respective d-spacing of the XRPD pattern's peaks of Posaconazole Form-S were determined and compared to those corresponding to Posaconazole Form I (Table 1). The characteristic peaks of Posaconazole Form-S are at 7.2, 7.9, 10.2, 13.2, 13.9, 14.5, 15.0, 16.0, 16.6, 17.3, 17.7, 18.1, 18.9, 19.3, 19.8, 20.2, 20.9, 21.3, 21.6, 22.1, 23.4, 23.7, 24.7, 25.9, 26.6, 27.4, 27.7, 28.5, 30.3, 31.0, 31.6, 33.2, 34.5, 35.2, 35.8, 36.5, 36.8, 37.4 and 38.5 2-theta. The existence of those peaks and the absence of the peaks at 7.6, 9.8, 11.1, 11.6, 12.9, 14.3, 15.6, 15.8, 16.2, 17.1, 19.1, 22.9, 23.1, 23.9, 24.3, 25.5, 26.3, 26.9, 29.2, 32.3, 32.8, 33.5, 35.5, 36.0, 37.9, 38.2 and 39.5 2-theta peaks can be used in order to differentiate Posaconazole Form-S from Posaconazole Form I in the oral suspension.

Previous attempts to identify the polymorphism of the API in the reference commercial Posaconazole oral suspension, Noxafil®, were reported [20]. However, the non-controlled treatment of suspension (drying and centrifugation) resulted in XRPD patterns that matched the pattern of Form I [20]. Also, in the scientific discussion of Noxafil® by the European Medicines Agency [19], it is reported that Posaconazole is produced as Form I and in the whole manufacturing process, even if the final oral suspension remains in the same crystal form [19].

Table 1. The characteristic 2-theta diffraction angles and d-spacing of the XRPD patterns of Posaconazole Form I and Posaconazole Form-S.

Posaconazole Form I		Posaconazole Form-S	
2-Theta Angles (Degrees) [1]	d-Spacing (Å)	2-Theta Angles (Degrees) [1]	d-Spacing (Å)
7.6	11.63	7.2	12.35
9.8	8.98	7.9	11.17
11.1	7.94	10.2	8.69
11.6	7.61	13.2	6.70
12.9	6.85	13.9	6.39
14.3	6.20	14.5	6.10
15.1	5.88	15.0	5.89
15.6	5.68	16.0	5.54
15.8	5.59	16.6	5.33
16.2	5.47	17.3	5.11
17.1	5.19	17.7	5.01
17.7	5.00	18.1	4.89
18.1	4.90	18.9	4.69
19.1	4.64	19.3	4.60
19.4	4.57	19.8	4.49
19.9	4.47	20.2	4.39
20.8	4.26	20.9	4.25
21.6	4.11	21.3	4.16
22.2	4.00	21.6	4.10
22.9	3.89	22.1	4.01
23.1	3.84	23.4	3.80
23.5	3.79	23.7	3.75
23.9	3.71	24.7	3.61
24.3	3.66	25.9	3.44
25.5	3.50	26.6	3.35
26.3	3.39	27.4	3.26
26.9	3.31	27.7	3.21
27.5	3.24	28.5	3.13
29.2	3.06	30.3	2.95
31.0	2.88	31.0	2.88
32.3	2.77	31.6	2.83
32.8	2.73	33.2	2.69
33.5	2.67	34.5	2.59
35.5	2.53	35.2	2.55
36.0	2.50	35.8	2.51
36.7	2.45	36.5	2.46
37.9	2.37	36.8	2.44
38.2	2.36	37.4	2.40
39.5	2.28	38.5	2.34

[1] The 2-theta diffraction angles may vary by ±0.2° [21].

This study is the first to report that the Posaconazole API in oral suspensions is Form-S. The reference Noxafil® oral suspension was also analysed using XRPD, and the developed sample-preparation method was applied. It was found that, after centrifugation of Noxafil®, the XRPD pattern of its precipitate was practically identical to the pattern of the precipitate of the generic Posaconazole oral suspension (Figure 8). Moreover, the XRPD pattern of the Noxafil® precipitate highly resembled the pattern of the untreated Noxafil® oral suspension (Figure 8) while Posaconazole Form I was not detected in the suspension.

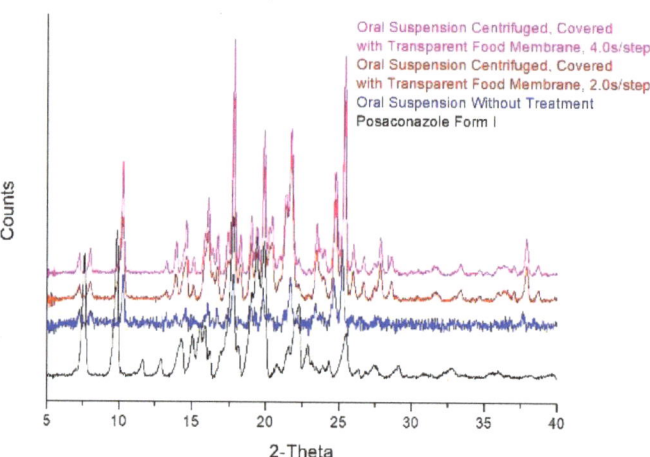

Figure 7. Comparison of Posaconazole oral suspension without treatment and Posaconazole oral suspension treated with the developed method: XRPD patterns of Posaconazole Form I (**black line**), Posaconazole oral suspension without treatment (**blue line**) and Posaconazole oral suspension centrifuged and covered with the transparent food membrane recorded with 2.0 s/step (**red line**) and 4.0 s/step scanning rate (**magenta line**).

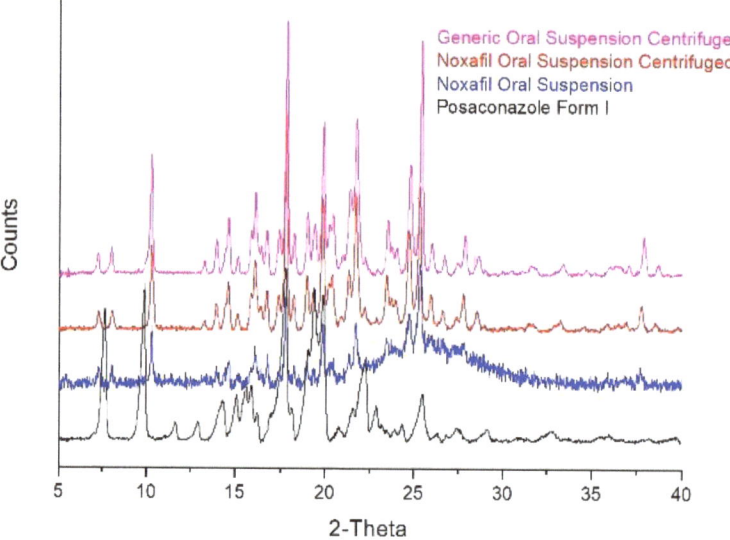

Figure 8. Comparison of the generic Posaconazole oral suspension and the reference Posaconazole oral suspension, Noxafil®, treated with the developed sample-preparation method: The XRPD patterns of Posaconazole Form I (**black line**), Noxafil® Posaconazole oral suspension (**blue line**), Noxafil® Posaconazole oral suspension centrifuged and covered with the transparent food membrane (**red line**) and generic Posaconazole oral suspension centrifuged and covered with the transparent food membrane (**magenta line**).

The developed sample-preparation method is likely to be also efficient in the identification of the crystal form of the API in other suspensions and formulations containing water molecules in which the concentration of the API is rather low, like oral suspensions (other than Posaconazole oral suspension and otic suspensions) and ophthalmic suspensions. Moreover, the use of the proposed method could be extended to identification of the crystal form of the API in semi-solid dispersions, such as o/w creams. Furthermore, application of the transparent food membrane for covering the sample would be useful for determining the polymorphic form of the API in environmentally sensitive samples, which are susceptible to polymorphic conversion.

4. Materials and Methods

4.1. Materials and Samples

Posaconazole API Form I; Posaconazole oral suspensions (40 mg/mL); and Posaconazole oral suspension's placebo, composed of Polysorbate 80, sodium citrate monohydrate, citric acid monohydrate, simethicone, xanthan gum, sodium benzoate, liquid glucose, glycerin, artificial cherry flavour, titanium dioxide and purified water, were kindly provided by the Greek pharmaceutical company GENEPHARM S.A (Pallini, Attica, Greece). The commercial reference Posaconazole oral suspensions, Noxafil®, was acquired from a local drug store.

4.2. X-Ray Powder Diffraction (XRPD)

For identification of the polymorphism of the Posaconazole API in oral suspensions, an X-ray Powder Diffractometer (Bruker AXS D2 Phaser 2nd Gen, Karlsruhe, Germany) was used, equipped with a standard Bragg Brentano geometry with fixed primary and linear LYNXEYE (1D mode) detector. Ceramic X-ray tube KFL Cu-2K, 0.4 mm × 12 mm, with a Ka spectral line (λ = 1.54184 Å) was used as the incident radiation. The tube worked with 300 W (30 kV voltage and 10 mA current). The scan mode was continuous, and a locked coupled scan type was used. The step size was 0.02° (2θ), the scan speed varied while the region of 2–40° (2θ) was scanned. No rotation was applied for recording the XRPD patterns. A 0.6-mm primary divergence slit, a 3-mm air scatter screen, an 8-mm anti-scatter slit, a 2.5° soler slit and a 5° (2θ) opening for the Position Sensitive Detector (PSD), were used. Polymethylmethacrylate (PMMA) XRPD sample holders with a 25-mm diameter and a 0.5-mm-deep circular cavity for spreading the sample were used. The background of the XRPD patterns was subtracted using the software DIFFRAC.SUITE EVA V4.1.1.

4.2.1. Posaconazole Oral Suspension Analysis via XRPD

The Posaconazole API was loaded on the PMMA sample holder by simply spreading the powder using a 25 mm × 75 mm × 1 mm glass slide. The XRPD pattern of Posaconazole Form I was recorded using a scan rate of 0.2 s/step. All other settings were the same as those previously mentioned.

Regarding Posaconazole oral suspension, reference Posaconazole oral suspension, Noxafil® and the suspension's placebo, the sample was shaken evenly and placed on the PMMA sample holder with a Pasteur pipette, avoiding bubbles in the samples. Subsequently, their XRPD patterns were recorded with a scan speed of 1.0 s/step.

4.2.2. Sample-Preparation Methods for Increasing Posaconazole's Concentration in Oral Suspensions

In order to overcome the problem of a low signal/noise ratio in the XRPD pattern of the oral suspension, three different methods were tested to increase the concentration of the API in the oral suspensions by separating the solid components of the suspension from the liquid excipients.

The first method tested was filtration under vacuum: 5 mL of Posaconazole oral suspension (40 mg/mL) were filtered using 0.22-μm GSWP nitrocellulose membrane filters (Merck Millipore Ltd., Cork, Ireland) or borosilicate glass microfibre filter circles MN GF-1 with a retention capacity of 0.7 μm (Macherey-Nagel, Düren, Germany) or cellulose filter circles MN 619 de with a retention capacity of 1–2 μm (Macherey-Nagel, Düren, Germany) on a Büchner funnel or on a crucible funnel and a vacuum pump (KNF Neuberger Inc. Laboport, Trenton, NJ, USA). As the d(4.3) of Posaconazole API was found at 7.42 μm and the d(4.3) of the oral suspension was 6.72 μm (Hydro SV detection unit, Mastersizer 3000, Malvern Panalytical, Malvern, UK), filters with higher retention capacity were not used. The oral suspension could not be filtered in any case, and no XRPD pattern was obtained.

The second method was drying of the oral suspension at room temperature: 5 mL (5.4 g) of Posaconazole oral suspension (40 mg/mL, approximately 3.7% *w/w*) was placed on a borosilicate glass Petri dish (70 mm diameter × 15 mm height) after shaking the suspension's bottle evenly, and the suspension was left to dry overnight. The received solid was sticky and gummy, and its weight was found at approximately 1.6 g. Subsequently, the XRPD pattern of the dried oral suspension was recorded using a scan rate of 0.5 s/step.

The third method was centrifugation of the oral suspension. Different rpms (2000 rpm, 4000 rpm, 6000 rpm and 8000 rpm), temperature (10 °C, 15 °C, 20 °C and 25 °C), times (13 min, 23 min and 33 min) of centrifugation and amount of sample (2.5 mL, 5 mL, 10 mL and 15 mL) were tested and validated before selecting the most appropriate conditions. Among the different rpms, 8000 rpm was selected since no precipitate formed when 2000 rpm and 4000 rpm were used while, at 6000 rpm, the precipitate was not completely recovered. The temperature of 25 °C was found to be the most appropriate, as solid components were found in the supernatant when lower temperatures were used. When the suspension was centrifuged for 13 min, the precipitate was not fully recovered as solid components could be detected in the supernatant. The centrifugations of 23 min and 33 min resulted in practically identical XRPD patterns with no expenses in the recovery ratio. Thus, 23-min centrifugation was selected as the aim was to develop a fast sample-preparation method. Concerning the amount of sample, 5 mL was selected, as all amounts resulted in the same recovery ratio and our aim was to develop a method with the smallest amount possible; however, the PMMA sample holder of the XRPD requires approximately 1 mL to be loaded and this amount of precipitate is recovered when at least 5 mL is centrifuged.

Hence, 5 mL (5.4 g) of Posaconazole oral suspension (40 mg/mL, approximately 3.7% *w/w*) was placed in a 50-mL falcon tube and the sample was centrifuged at 8000 rpm and 25 °C for 23 min using a refrigerated centrifuge (Heraeus Biofuge Stratos, Kendro, Osterode, Germany). The supernatant, which was approximately 4 mL (4.4 g), was removed, and the precipitate was collected. The precipitate was loaded on a PMMA sample holder using a spatula, and the XRPD pattern was recorded using a scan rate of 0.5 s/step. For evaluation of the API's recovery from the Posaconazole oral suspension, artificial Posaconazole Form I water dispersions were prepared in triplicates, they were centrifuged and then left to dry at ambient temperature overnight. Each dispersion was prepared by adding 200 mg of Posaconazole Form I in purified water (15 MΩ·cm at 25 °C, Elix, Merck Millipore, Darmstadt, Germany), and the ratio of the API recovered after centrifugation and drying of the dispersion were determined (Table 2). Full recovery was achieved. Moreover, the ratio of precipitate recovered after centrifugation of the oral suspension was determined and it was found equal to 19.22% ± 0.77% (Table 3). For validating the precision and reproducibility of centrifugation, the oral suspension was centrifuged and analysed in triplicates every 4 h in the same day, and for validating the repeatability and ruggedness of the process, it was done in triplicates for three consecutive days. In addition, this method was applied to three different batches of the oral suspension.

Table 2. Mass of water dispersion, separated supernatant, recovered precipitate and recovered dried precipitate as well as the recovery ratio of the Posaconazole API after centrifugation of Posaconazole water dispersion in triplicates.

	Sample Mass (g)	Supernatant Mass (g)	Precipitate Mass (g)	Dried Precipitate Mass (g)	Recovery Ratio From the Initial API Mass [1] (%)
Sample 1	4.964	4.210	0.751	0.196	97.90
Sample 2	5.023	4.226	0.798	0.200	100.05
Sample 3	4.988	4.258	0.728	0.202	100.80
Average	4.992	4.231	0.759	0.199	99.58
Standard Deviation	0.030	0.024	0.035	0.003	1.51

[1] The initial Posaconazole active pharmaceutical ingredient (API) mass in all three samples was 0.200 g.

Table 3. Mass of oral suspension, separated supernatant and recovered precipitate as well as the recovery ratio of the precipitate after centrifugation of the Posaconazole oral suspension in triplicates.

	Sample Mass (g)	Supernatant Mass (g)	Precipitate Mass (g)	Recovery Ratio [1] (%)
Sample 1	5.454	4.425	1.029	18.87
Sample 2	5.250	4.190	1.056	20.11
Sample 3	5.413	4.399	1.012	18.70
Average	5.372	4.338	1.032	19.22
Standard Deviation	0.108	0.129	0.022	0.77

[1] The recovery ratio was calculated as a 100% ratio of the precipitate mass divided by the mass of the respective sample.

4.2.3. Setting the Scanning Rate of the XRPD Analysis

The separated precipitate after centrifugation was loaded on a PMMA sample holder, and its XRPD pattern was recorded using four different scanning rates; 0.5 s/step, 1.0 s/step, 2.0 s/step and 4.0 s/step scan speeds were applied. For each XRPD pattern, a fresh precipitate was prepared.

4.2.4. Studying the Stability of Posaconazole in Oral Suspension's Precipitate

For determination of the time effect on Posaconazole's polymorphism in the oral suspension's precipitate isolated by centrifugation, its XRPD pattern was recorded immediately after centrifugation (0 min) and after 30 min, 60 min, 90 min and 120 min using a scanning rate of 0.5 s/step.

4.2.5. Sample-Preparation Method for Delaying the Polymorphic Conversion

Different types of covers were used to avoid water loss from the precipitate. For each type of cover, a fresh suspension was centrifuged and the respective precipitate was collected.

A commercially available airtight specimen sample holder with dome-like x-ray transparent cap for environmentally sensitive materials (Bruker, Karlsruhe, Germany) was tested. The precipitate was placed on the airtight specimen sample holder, and the dome-like cap was screwed on the sample holder. The XRPD pattern of the sample was recorded using a 1.0 s/step scanning rate, and no air scatter screen was used. All other settings were the same.

The precipitate was transferred on a PMMA sample holder and covered airtight with a 75 mm × 50 mm biaxially oriented polyethylene terephthalate (PET) film: Mylar film (DuPont Teijin Films, Luxembourg S.A.). The XRPD pattern of the precipitate covered with Mylar film was acquired with a 1.0 s/step scanning speed.

The precipitate was spread on a PMMA sample holder with a spatula, and it was covered with three pieces (40 mm length each) of transparent pressure-sensitive tape (Scotch tape, 12 mm width, 3M, Greece). The XRPD pattern of the precipitate covered with transparent pressure-sensitive tape was recorded using 1.0 s/step scanning rate.

A piece of 120 mm × 130 mm and 0.006-mm-thick polyethylene low-density transparent food membrane (Vileda Freshmate 50 m, Aspropyrgos, Greece) was used to enfold the precipitate loaded on a PMMA sample holder. The XRPD pattern of the sample covered with the transparent food membrane was obtained with a 1.0 s/step scanning rate.

4.2.6. Study of Stability of the Posaconazole API in the Oral Suspension's Precipitate Covered with a Transparent Food Membrane

The Posaconazole oral suspension was centrifuged, and the precipitate was separated, loaded on a PMMA sample holder and covered diligently with a transparent food membrane in order to prevent water loss. The XRPD pattern of the sample was acquired using four different scanning rates: 0.5 s/step, 1.0 s/step, 2.0 s/step and 4.0 s/step.

The same procedure was followed for preparing the sample in order to study the stability of Posaconazole in the precipitate when using the transparent food membrane. The XRPD pattern of the sample was recorded immediately after centrifugation (0 days) and after 1 week, 2 weeks, 1 month and 2 months. Each time, a scanning rate of 0.5 s/step was used.

4.3. Spectral Analysis

The software OriginPro 8 (OriginLab Corporation, Northampton, MA, USA) was used for creation of the graphs and for analysis of the XRPD patterns.

5. Conclusions

To conclude, a sample-preparation method was developed for characterisation of the crystal form of the Posaconazole API in oral suspensions using XRPD. More specifically, the API can be isolated as a precipitate together with the other undissolved excipients from the oral suspensions using centrifugation. However, the Posaconazole API is prone to water loss and is instable, and a polymorphic conversion to the thermodynamically stable Form I is rapid. In order to overcome the stability issue, a transparent food membrane was used to cover the precipitate loaded on the PMMA sample holder and a slower scanning rate of 2.0 s/step or 4.0 s/step was applied to increase the signal/noise ratio. Application of this novel sample-preparation method in Posaconazole oral suspensions revealed that the crystal form of Posaconazole in oral suspensions is Form-S.

Author Contributions: Conceptualization, M.L, M.O. and C.K.; methodology, M.L., M.O. and C.K.; investigation, M.L. and S.F.; validation, M.L., M.O. and C.K.; data curation, M.L. and S.F.; supervision, M.O. and C.K.; project administration, C.K.; writing—original draft preparation, M.L.; writing—review and editing, M.O. and C.K. All authors have read and agreed to the published version of the manuscript.

Funding: The research work was supported by the Hellenic Foundation for Research and Innovation (HFRI) under the HFRI PhD Fellowship grant (Fellowship Number: 1298).

Acknowledgments: The authors gratefully acknowledge the Greek pharmaceutical company GENEPHARM S.A. for the donation of Posaconazole API Form I, Posaconazole oral suspension's placebo and the Posaconazole oral suspension.

Conflicts of Interest: The authors declare no conflict of interest.

References

1. Chieng, N.; Rades, T.; Aaltonen, J. An overview of recent studies on the analysis of pharmaceutical polymorphs. *J. Pharm. Biomed. Anal.* **2011**, *55*, 618–644. [CrossRef] [PubMed]
2. Bechtloff, B.; Nordhoff, S.; Ulrich, J. Pseudopolymorphs in Industrial Use. *Cryst. Res. Technol.* **2001**, *36*, 1315–1328. [CrossRef]
3. Yu, L. Amorphous pharmaceutical solids: preparation, characterization and stabilization. *Adv. Drug Deliv. Rev.* **2001**, *48*, 27–42. [CrossRef]
4. Bauer, J.F. Polymorphism—A Critical Consideration in Pharmaceutical Development, Manufacturing, and Stability. *J. Valid. Technol.* **2008**, *14*, 15–23.
5. Kathpalia, H.; Phadke, C. Novel oral suspensions: A review. *Curr. Drug Deliv.* **2014**, *11*, 338–358. [CrossRef] [PubMed]

6.   Haleblian, J.; McCrone, W. Pharmaceutical Applications of Polymorphism. *J. Pharm. Sci.* **1969**, *58*, 911–929. [CrossRef] [PubMed]

7.   Mazurek, S.; Szostak, R. Quantification of active ingredients in pharmaceutical suspensions by FT Raman spectroscopy. *Vib. Spectrosc.* **2017**, *93*, 57–64. [CrossRef]

8.   Hof, H. A new, broad-spectrum azole antifungal: posaconazole? mechanisms of action and resistance, spectrum of activity. *Mycoses* **2006**, *49*, 2–6. [CrossRef] [PubMed]

9.   Leung, S.; Poulakos, M.; Machin, J. Posaconazole: An Update of Its Clinical Use. *Pharmacy* **2015**, *3*, 210–268. [CrossRef] [PubMed]

10.   Imran, M.; Nayeem, N.; Bawadekji, A. Posaconazole: A Pharmaceutical Review. *J. North Basic Appl. Sci.* **2019**, *4*, 109–123.

11.   Andrews, D.; Leong, W.; Sudhakar, A. Crystalline Antifungal Polymorph. US6958337B2, 25 October 2005.

12.   Wieser, J.; Pichler, A.; Hotter, A.; Griesser, U.; Langes, C. A Crystalline Form of Posaconazole. US20110065722A1, 17 March 2011.

13.   Chaudhari, G.; Krishna, V.; Sanikommu, S.R.; Chaudhari, G.; Verdia, J.; Khan, M.A. Process for Preparation of Posaconazole and Crystalline Polymorphic form V of Posaconazole. WO2011158248A2, 22 December 2011.

14.   Wieser, J.; Pichler, A.; Hotter, A.; Griesser, U.; Langes, C. Crystalline Form of Posaconazole. US20120101277A1, 26 April 2012.

15.   Wieser, J.; Pichler, A.; Hotter, A.; Griesser, U.; Langes, C.; Laschober, C. Pharmaceutical Compositions Containing a Crystalline Form of Posaconazole. US20110105525A1, 5 May 2011.

16.   Badone, D.; Negri, C.; REPETTI, A. A Crystalline Form of Posaconazole. WO2015092595A1, 25 June 2015.

17.   Charyulu, P.V.R.; Gowda, D.J.C.; Rajmahendra, S.; Raman, M. Crystalline Forms of Posaconazole Intermediate And Process for the Preparation of Amorphous Posaconazole. US10457668B2, 29 October 2019.

18.   Reddy, M.S.; Rajan, S.T.; Eswaraiah, S.; Vishnuvardhan, S. Process for the Preparation of Triazole Antifungal Drug, its Intermediates and Polymorphs Thereof. WO2013042138A2, 28 March 2013.

19.   European Medicines Agency. Noxafil, INN-Posaconazole. Available online: https://www.ema.europa.eu/en/documents/scientific-discussion/noxafil-epar-scientific-discussion_en.pdf (accessed on 24 May 2020).

20.   Zografos, S.; Fertaki, S.; Kontoyannis, C. Identification and Quantitative Analysis of Posaconazole API in Noxafil®Oral Suspension. In Proceedings of the 10th Panhellenic Scientific Conference in Chemical Engineering, Patras, Greece, 4–6 June 2015.

21.   European Pharmacopoeia 10.4. 2.9.33. Characterisation of Crystalline and Partially Crystalline Solids by X-Ray Powder Diffraction (XRPD). Available online: https://pheur.edqm.eu/app/10-4/content/default/20933E.htm (accessed on 29 October 2020).

Sample Availability: Samples are not available from the authors.

 *molecules*

Article

Miniaturized Salting-Out Assisted Liquid-Liquid Extraction Combined with Disposable Pipette Extraction for Fast Sample Preparation of Neonicotinoid Pesticides in Bee Pollen

Xijuan Tu [1,2] and Wenbin Chen [1,2,*]

[1] College of Bee Science, Fujian Agriculture and Forestry University, Fuzhou 350002, China; xjtu@fafu.edu.cn
[2] College of Animal Science, Fujian Agriculture and Forestry University, Fuzhou 350002, China
* Correspondence: wbchen@fafu.edu.cn

Academic Editors: Victoria Samanidou and Irene Panderi
Received: 12 November 2020; Accepted: 29 November 2020; Published: 3 December 2020

Abstract: As the main source of nutrients for the important pollinator honeybee, bee pollen is crucial for the health of the honeybee and the agro-ecosystem. In the present study, a new sample preparation procedure has been developed for the determination of neonicotinoid pesticides in bee pollen. The neonicotinoid pesticides were extracted using miniaturized salting-out assisted liquid-liquid extraction (mini-SALLE), followed by disposable pipette extraction (DPX) for the clean-up of analytes. Effects of DPX parameters on the clean-up performance were systematically investigated, including sorbent types (PSA, C18, and silica gel), mass of sorbent, loading modes, and elution conditions. In addition, the clean-up effect of classical dispersive solid-phase extraction (d-SPE) was compared with that of the DPX method. Results indicated that PSA-based DPX showed excellent clean-up ability for the high performance liquid chromatography (HPLC) analysis of neonicotinoid pesticides in bee pollen. The proposed DPX method was fully validated and demonstrated to provide the advantage of simple and rapid clean-up with low consumption of solvent. This is the first report of DPX method applied in bee pollen matrix, and would be valuable for the development of a fast sample preparation method for this challenging and important matrix.

Keywords: sample preparation; disposable pipette extraction; neonicotinoid pesticides; bee pollen; HPLC; salting-out assisted liquid-liquid extraction

1. Introduction

As the most important managed pollinator, the honeybee is crucial to the ecosystem, agriculture, and food production. However, precipitous loss of the honeybee population has been reported in Europe and North America, which raises the concern of a pollination crisis [1]. Multiple stressors have been considered as potential causes of the honeybee decline, including nutrition, pesticide, parasites, and disease [2].

Bee pollen is the major source of protein in the honeybee diet, and it also provides the honeybee with essential nutrients, e.g., lipids, vitamins, and minerals [3]. In addition to nutritional values, phytochemicals in bee pollen were reported to be critical in the up-regulating detoxification and immunity genes of the western honeybee [4]. Thus, the security of bee pollen is important for keeping the honeybee healthy. However, bee pollen has the potential to be contaminated by pesticides due to the widespread use of chemicals in plant protection.

Determination of pesticide residues in bee pollen is important for evaluating the risk of exposure. Because of the complexity of the constituents in bee pollen, sample preparation to extract and clean up target compounds is generally required when chromatography-based technology is employed for the analysis of pesticide residues [5]. Classical solid-liquid extraction [6], solid-phase extraction (SPE) [7,8], matrix solid-phase dispersion (MSPD) [9], and dispersive solid-phase extraction (d-SPE) [10–12] methods have been developed for the determination of pesticides in bee pollen. Despite this progress, it is still a great challenge to develop a simple and rapid sample preparation method for residue analysis in bee pollen matrix. In the present work, we have proposed a new sample preparation procedure based on salting-out assisted liquid-liquid extraction [13] and disposable pipette extraction (DPX) clean-up for the determination of neonicotinoid residues in bee pollen.

DPX is an alternative solid-phase extraction (SPE) method which demonstrates the advantages of reducing labor, time, and solvent consumption in the clean-up of analytes [14,15]. In a typical DPX device, sorbents are assembled in a pipette tip with a screen at the bottom and a barrier on the top [16]. Sample solution may be introduced by drawing in from the bottom of the tip or loading on the top of it [16,17]. The analytes are then washed and eluted from the sorbent with a suitable solvent. The operation of DPX is simple and labor-saving. In addition, it allows for automated and high-throughput preparation when combined with the liquid handling system [18,19]. This sample preparation method has been widely used in the analysis of drugs [14,18–24], pesticide residues [16,17,25–30], heavy metal ions [31], and environmental contaminants [32–35]. In the present work, the parameters of DPX were systematically investigated on the clean-up of neonicotinoid pesticides in bee pollen. To the best of our knowledge, this is the first report on the use of the DPX method for the matrix of bee pollen.

2. Results and Discussion

2.1. Salting-Out Assisted Liquid-Liquid Extraction

A miniaturized salting-out assisted liquid-liquid extraction (mini-SALLE) was used for the rapid extraction of three neonicotinoid pesticides—thiamethoxam, acetamiprid, and thiacloprid—from bee pollen (Figure S1). In the mini-SALLE protocol, 2 mL ACN mixed with 2 mL H_2O was applied as the extraction solution. The bee pollen sample was homogenized with the ACN-H_2O solution before phase separation agent was introduced to trigger the partition of ACN from the mixture. Salts, 0.4 g of $MgSO_4$ and 0.2 g of NaCl, were used to induce phase separation and extract target compounds into the upper ACN phase. The former, $MgSO_4$, has been demonstrated to be an efficient phase-separation agent as the high extraction yield for relatively high-polarity compounds [36]. The latter, NaCl, was used to reduce the co-extracted protein and sugar contents from the matrix [37,38].

The obtained SALLE extract from the bee pollen showed a high background of matrix interferences. Bee pollen is a complex matrix which contains pigments, nutrients such as proteins, carbohydrates, vitamins, lipids, and various phytochemical compounds [3,39,40]. We observed that the final extract after SALLE (upper ACN phase) was still a light yellow color, though the phase partition had retained a substantial number of matrix compounds in the lower H_2O phase. Moreover, as shown in Figure 1a, the diluted SALLE extract showed complicated and high-intensity chromatography peaks. It is important to note that the SALLE extract was diluted 50 times before being injected into the HPLC system to avoid damaging the HPLC column; therefore, the real response intensity of the extract would be tens of that shown in Figure 1a. It is also important to note that the gradient elution used in Figure 1a was previously reported for the separation of multiple phenolic compounds [36]. As shown in Figure 1b, this HPLC condition was also suitable for the separation of target neonicotinoid pesticides and the inner standard. Since bee pollen is a rich source of phenolic compounds [39,40], there are complicated peaks in the chromatogram of the SALLE extract; as a consequence, peaks of analytes could not be distinguished from the high background of interferences. This means that the SALLE extract required further clean-up to reduce interfering compounds.

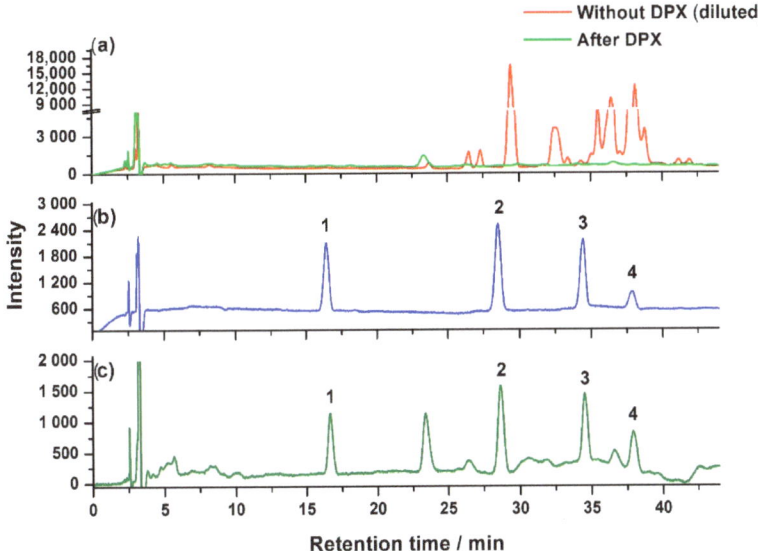

Figure 1. Representative chromatogram of (**a**) diluted SALLE extract without DPX and SALLE extract after PSA-based DPX, (**b**) standards of neonicotinoid pesticides and internal standard, (**c**) spiked bee pollen sample after SALLE and following PSA-based DPX clean-up. Peak 1: thiamethoxam, 2: acetamiprid, 3: thiacloprid, and 4: ethyl 6-chloropyridine-2-carboxylate (internal standard).

2.2. DPX Clean-Up

PSA-based DPX showed a simple, rapid, and efficient clean-up effect for neonicotinoid pesticides in bee pollen. It was interesting to observe that the extract solution became colorless after the PSA-based DPX clean-up. Furthermore, a chromatogram of the cleaned-up solution (Figure 1a) indicated that the intensity of the matrix interferences was dramatically reduced, while a good resolution of analytes with negligible interference was achieved (Figure 1c). These results indicate the excellent clean-up performance of DPX. To reveal the effect of DPX conditions, parameters were systematically investigated.

Different types of sorbents, including C18 and silica gel, were compared with PSA for clean-up efficiency. In the case of C18, cloudy solutions were obtained after both the loading and the eluting steps. After standing for minutes, sediments were observed in the bottom of the solutions; therefore, the cloudy extract was not further investigated in HPLC. In the silica gel DPX, the final eluent was transparent and colorless; its HPLC chromatogram is shown in Figure 2a. The result indicated that matrix compounds were also dramatically removed by silica gel DPX. Compared with the PSA-based DPX (Figure 1c), the matrix peak located at RT 23.54 min was much higher in silica gel DPX. In addition, peaks emerged at RT 18.08 min and 38.73 min. More importantly, significant leakage of neonicotinoids was observed in the loading step of silica gel DPX, which led to the signal responses of analytes in Figure 2a that were only about 40% of that in PSA-based DPX (Figure 1c). These results indicated that PSA demonstrated better clean-up effect and analytes recovery than silica gel for the analysis of neonicotinoids in bee pollen.

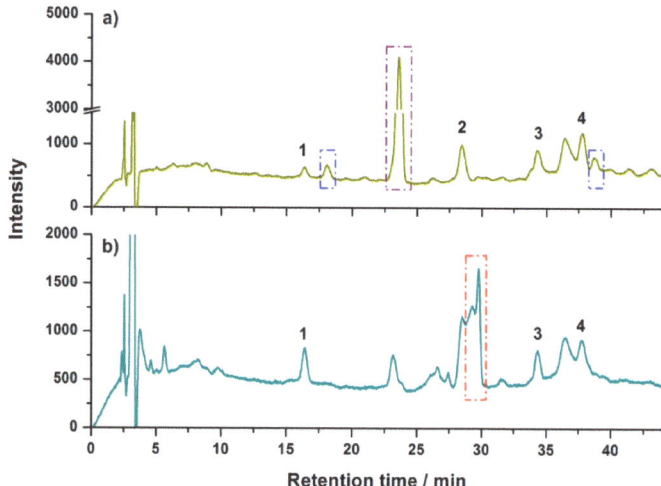

Figure 2. Representative chromatograms of extract cleaned up by (**a**) silica gel DPX, (**b**) dispersive solid-phase extraction (d-SPE). Peak 1: thiamethoxam, 2: acetamiprid, 3: thiacloprid, and 4: ethyl 6-chloropyridine-2-carboxylate (internal standard). The emerging peaks are marked in blue, the high-intensity peak from the matrix is marked in purple, and the matrix peaks overlapped with acetamiprid after d-SPE are marked in red.

The effect of the mass of PSA was studied in the range from 50 mg to 150 mg with increments of 25 mg. It was found that when the mass was from 50 mg to 100 mg, leakage of neonicotinoids was observed in the loading step, and the leaked content was increased with the decreasing of PSA mass. However, leaked analytes were not detected when the mass was further increased to 125 mg and 150 mg. This implies that not less than 125 mg of PSA would be required to prevent the leakage of neonicotinoids in the loading step. Additionally, the mass of PSA also affected the clean-up effect in the following elution step. As shown in Figure 3, in the DPX tips with 50 mg and 75 mg of PSA, peaks of matrix compounds were overlapped with acetamiprid. These interference compounds were eliminated when the mass of PSA was larger than 100 mg. However, as the mass increased to 150 mg, concentrations of neonicotinoids in the eluent were all slightly decreased, which might have resulted from the increasing retention of neonicotinoids in the larger mass of sorbent. Given the absence of analyte leaking in the loading step and its good performance in removing interferences, it was determined that 125 mg of PSA is a suitable choice for the DPX tip.

Sample loading was investigated in two modes: draw-in and top-loading. For the draw-in mode, sample solution was aspirated into the DPX column from the bottom of the tip. For the top-loading mode, sample solution was loaded on the top of the DPX tip. In the draw-in mode, increasing the draw-in and dispense-out repeat times (in-out cycles) could improve the clean-up effect. As can be observed in Figure 4, as the in–out cycles increased from 1 to 3 the intensity of matrix compounds decreased significantly. However, the peaks of target compounds were still indistinguishable. In the top-loading mode, the chromatogram of the filtrate solution which was collected by dispensing the loading solution through the DPX tip was very clear and had extremely low levels of matrix compounds (Figure 4). This means that target compounds are all retained on the DPX column; thus, an eluting step is required to wash out the neonicotinoids.

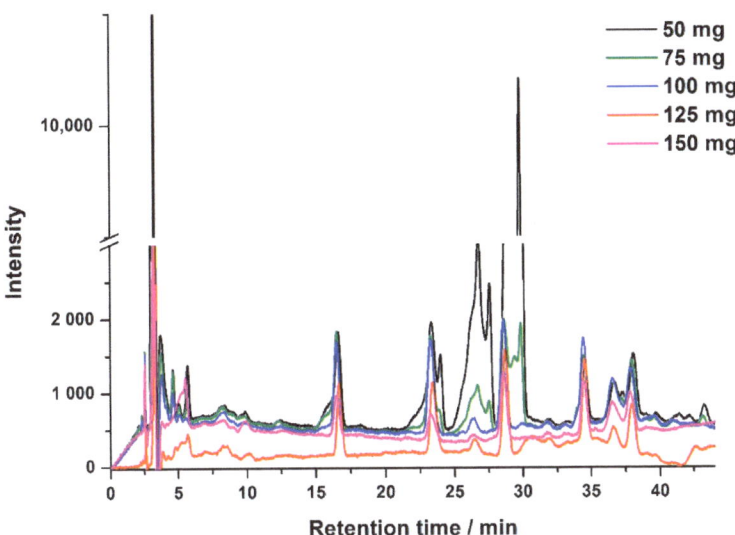

Figure 3. Representative chromatograms of extract cleaned up by DPX tips with different masses of PSA.

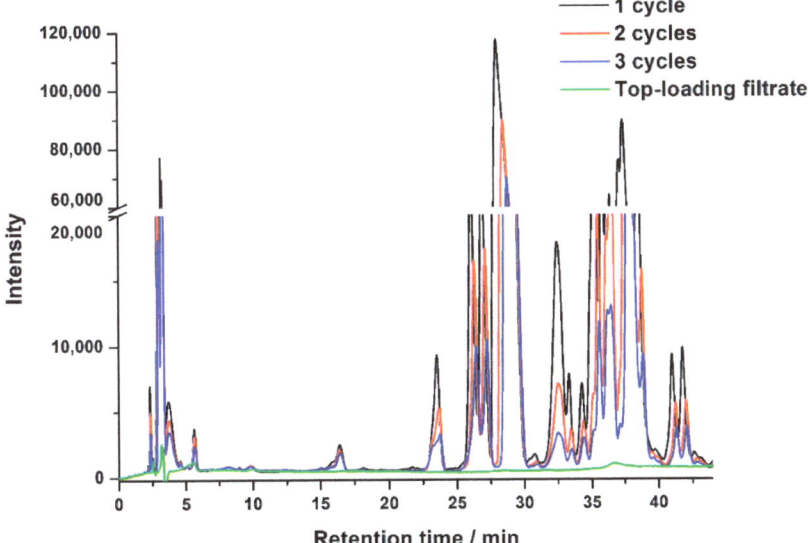

Figure 4. Representative chromatogram of the extract after different in–out cycles using draw-in sample loading, and the filtrate solution after top-loading.

After top-loading the extract, the target compounds were eluted by 100 µL of ACN without a washing step. This elution solvent was also loaded on top of the DPX tip and then dispensed out of the DPX column. Different types of solutions, including ACN, MeOH, H_2O, and ACN-H_2O mixture (25%, 50%, 75%, *v/v*), were investigated. The MeOH, H_2O, and ACN-H_2O mixture all showed the co-elution of matrix interferences. Furthermore, as the H_2O concentration increased, levels of interference compounds were significantly increased in the eluent. Different volumes of ACN were then compared, and 100 µL was selected as the suitable volume since the further increase in volume would reduce the concentration of neonicotinoids in the eluent, and also increase the risk of the co-elution of matrix interferences.

On the basis of these investigations, the DPX device and the optimal DPX procedure can be proposed, as shown schematically in Figure 5. The DPX column was simply assembled in a 1 mL pipette tip, in which 125 mg of PSA was used as the clean-up sorbent. After being conditioned with 300 µL of ACN solution, 100 µL of SALLE extract was loaded on top of the DPX tip. Then, 100 µL of ACN was applied as the elution solution, which was also added on the top of the pipette tip. The elution was carried out by dispensing the ACN through the DPX column and collecting the eluent for HPLC analysis. The proposed DPX procedure was very simple and rapid; the whole operation could be accomplished in 2 min, using a commercial pipette. Another advantage of the proposed method was the low consumption of sample and solvent, whose total volume was only 500 µL. This DPX method is eco-friendly and may have potential for use in green sample preparation.

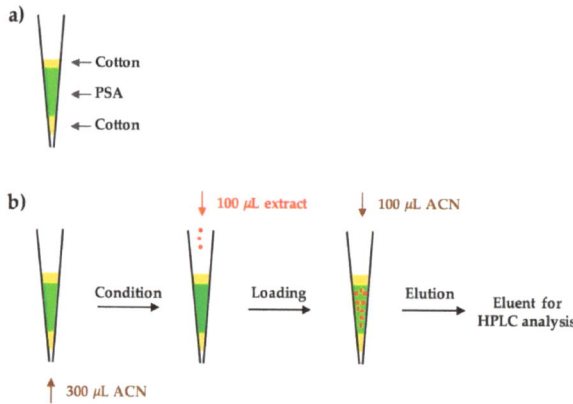

Figure 5. Schematic of (**a**) the DPX device, and (**b**) the optimal DPX procedure.

Finally, the developed DPX method was compared with the d-SPE procedure [41]. Due to the loss of solution in the d-SPE procedure, 100 µL of SALLE extract was diluted with 300 µL of ACN before being mixed with 125 mg of PSA. As shown in Figure 2b, although the d-SPE procedure also showed significant removal of matrix compounds, its clean-up ability was less efficient than the DPX method, especially for acetamiprid, whose peak was still overlapped with interferences after the d-SPE clean-up.

2.3. Analytical Performance

The final cleaned-up solution was separated in C18 reversed-phase HPLC with gradient elution. Target neonicotinoid pesticides were quantified by an inner standard based on our previous report [42]. The diode array detector (DAD) was used for the detection of pesticides, and a wavelength of 254 nm was applied for the quantification. Standards curves were in the linearity range from 0.05 to 3 µg/mL. The detection limit and quantification limit for the analytes were 100 µg/kg and 300 µg/kg with signal to noise of 3 and 10, respectively. The accuracy and precision at two spiked levels (1 × LOQ, 5 × LOQ) are shown in Table 1. Recoveries were between 89.63 and 94.56%, 96.41 and 100.85%, and 86.49 and 96.62% for thiamethoxam, acetamiprid, and thiacloprid, respectively. Intra-day and inter-day precisions were all less than 5%. These results were all in the acceptable range according to the Association of Official Analytical Chemist (AOAC) [43]. Finally, six commercial bee pollen samples were analyzed using the validated method. Results indicated that none of the target neonicotinoid pesticides was detected in these samples. Due to the DPX procedure's excellent clean-up performance in removing matrix interferences, detection sensitivity towards neonicotinoid pesticides could be significantly improved by combining this simple and rapid DPX method with mass spectrometry detection. These studies are now under way.

Table 1. Accuracy and precision of the proposed method at two spiked levels.

Analytes	Spiked Level (µg/kg)	Intra-Day						Inter-Day	
		Day 1		Day 2		Day 3		Mean ± SD (%, $n = 18$)	RSD (%, $n = 18$)
		Mean ± SD (%, $n = 6$)	RSD (%, $n = 6$)	Mean ± SD (%, $n = 6$)	RSD (%, $n = 6$)	Mean ± SD (%, $n = 6$)	RSD (%, $n = 6$)		
Thiamethoxam	300	94.56 ± 4.18	4.42	92.61 ± 3.01	3.25	89.63 ± 2.26	2.52	92.26 ± 3.70	4.01
	1500	92.40 ± 3.46	3.74	91.81 ± 3.68	4.01	89.70 ± 3.46	3.86	91.30 ± 3.53	3.87
Acetamiprid	300	99.13 ± 2.86	2.89	97.82 ± 2.25	2.30	98.18 ± 3.77	3.84	98.38 ± 2.89	2.94
	1500	97.24 ± 3.11	3.20	96.41 ± 2.63	2.73	100.85 ± 3.11	3.08	98.17 ± 3.42	3.48
Thiacloprid	300	86.49 ± 3.81	4.41	86.58 ± 3.36	3.88	87.68 ± 3.31	3.78	86.92 ± 3.49	4.02
	1500	93.29 ± 3.99	4.28	92.96 ± 3.38	3.64	96.62 ± 1.95	2.02	94.29 ± 3.47	3.68

3. Materials and Methods

3.1. Materials

Methanol and acetonitrile (ACN) at HPLC grade were obtained from Merck (Darmstadt, Germany). Acetic acid, anhydrous magnesium sulfate, and sodium chloride at analytical grade were obtained from Sinopharm Chemical Reagent Co., Ltd. (Shanghai, China). Standards of thiamethoxam, acetamiprid, thiacloprid, and ethyl 6-chloropyridine-2-carboxylate (internal standard) were supplied by Aladdin (Shanghai, China). C18 and PSA were from Sepax (Suzhou, China), and silica gel was from Sinopharm. Ultrapure water (18.2 MΩ) was used in all experiments. Rape (*Brassica campestris*) bee pollen used for method development was collected from an apiary in Hubei, China. Commercial rape bee pollen samples used in this application were purchased from local markets. Stock solutions of standards were prepared in ACN with a concentration of 0.2 mg/mL. Working solutions of standards were prepared by further dilution with ACN. All standard solutions were stored at 4 °C until used.

3.2. Salting-Out Assisted Liquid-Liquid Extraction

Bee pollen (1 g) was added into 4 mL of ACN-H_2O solution (1:1, *v/v*), then the mixture was homogenized for 30 s by a homogenizer (Fluko, China). After the addition of salts (0.4 g $MgSO_4$ and 0.2 g NaCl), the obtained mixture was vortexed for 30 s to dissolve the salts. Then, the final solution was centrifuged at 5000 rpm for 10 min to make phase separation clear. The upper phase was collected for DPX clean-up.

3.3. DPX Clean-Up

DPX was assembled in a commercial pipette tip (1 mL). Briefly, PSA (125 mg) was transferred into the pipette tip, with degreasing cotton preloaded at the tip exit for retaining the sorbent. Another pellet of degreasing cotton was then placed on top of the sorbent.

The obtained DPX tip was conditioned with 300 µL of ACN, which was aspirated into the tip from the bottom with an aspiration volume of 800 µL to draw in extra air bubbles for mixing. After bubbling was completed, the solution was dispensed to waste.

Then, 100 µL of SALLE extract was added to the top of the DPX tip and dispensed through the sorbent for sample loading. Finally, 100 µL of ACN was added to the top of the DPX tip and dispensed to elute target compounds. The eluent was collected and transferred into a vial for HPLC analysis.

3.4. Study of DPX Procedure

3.4.1. Comparing Matrix Solution without DPX Clean-Up

The SALLE extract (100 µL) was diluted 50 times with the ACN-H_2O mixture (50%, *v/v*) before HPLC analysis.

3.4.2. Effect of Different Sorbents

DPX tips were assembled with 125 mg of C18 or silica gel, and the clean-up was then performed as described in Section 3.3.

3.4.3. Effect of the Mass of PSA

DPX tips were assembled with different masses of PSA (50, 75, 100, 125, and 150 mg). Then, the filtrate solution after sample loading and the final eluent solution as described in Section 3.3 were collected for HPLC analysis.

3.4.4. Effect of Elution Solvent

DPX tips were assembled with 125 mg of PSA. Clean-up was then performed as described in Section 3.3 with different elution solvents, including MeOH, H_2O, and ACN-H_2O mixtures (25%, 50%, 75%, *v/v*).

3.4.5. Comparing Draw-In Sample Loading Mode

DPX tips assembled with 125 mg of PSA were conditioned as described in Section 3.3. Then, 100 µL of SALLE extract was aspirated into the DPX tip from the bottom with an aspiration volume of 800 µL. After bubbling was completed, the solution was dispensed through the sorbent and collected in another tube. This aspirating in–out operation was repeated three times, and solutions in different cycles were collected for HPLC analysis.

3.4.6. Comparing d-SPE

The d-SPE clean-up experiment was performed on the basis of the reported method [41]. Because of the loss of volume to the sorbent in the d-SPE method, 100 µL of the SALLE extract was diluted with 300 µL of ACN before being mixing with sorbent. The diluted solution was mixed with 125 mg of PSA and vortexed for 1 min. The mixture was then centrifuged at 5000 rpm for 10 min. After centrifugation, the supernatant was collected for HPLC analysis.

3.5. HPLC Analysis

HPLC analysis was performed on Shimadzu (Kyoto, Japan) LC-20AT, with a SIL-20AC autosampler, CTO-20AC column oven, and SPD-M20A photodiode array detector. A TSKgel (Tosoh, Japan) ODS-100V column (5 µm, 4.6 × 150 mm) was applied for the separation. The previously reported gradient elution [36,44] was modified for the separation of neonicotinoids. Solvent A was water with 0.1% acetic acid (*v/v*), and solvent B was methanol. Elution was as follows: 15% to 40% solvent B at 0–30 min, 40% to 46% solvent B at 30–44 min, post run with 46% to 100% solvent B for 6 min, then back to and maintained at 15% solvent B, each for 5 min. The column temperature was set at 35 °C, the injection volume was 10 µL, the flow rate was 0.8 mL/min, and the detection wavelength was 254 nm.

3.6. Method Validation

Seven levels of calibration curves were prepared by standard solutions containing neonicotinoid pesticides (0.05, 0.1, 0.2, 0.5, 1, 2, 3 µg/mL) and IS (2 µg/mL). The ratio of peak area (analyte/IS) versus the ratio of weight (analyte/IS) was used to construct the analytical curves. The y-intercept was set to zero and a linear fit was performed. Limit of detection (LOD) and limit of quantification (LOQ) were investigated in spiked bee pollen samples. Accuracy and precision were estimated by analyzing blank samples spiked at concentrations of 1 × LOQ and 5 × LOQ. Accuracy was expressed as recovery (%), and precision was measured as relative standard deviation (RSD) to the mean recovery of intra-day ($n = 6$) and inter-day ($n = 18$, three days) analyses.

4. Conclusions

In summary, a DPX sample preparation method was developed for the HPLC analysis of neonicotinoid pesticides in bee pollen. Effects of the DPX parameters on the clean-up efficiency were systematically investigated. Results revealed that using PSA as the DPX sorbent, combined with top-loading of the sample into the pipette tip and elution, resulted in a remarkable removal of matrix interferences. This PSA-based DPX method exhibited excellent clean-up ability with the merits of simple, rapid, and low solvent consumption. The present work might provide a new strategy for designing fast sample preparation procedures for the challenging and important bee pollen matrix.

Supplementary Materials: The following are available online, Figure S1: Structures and LogP values of analytes.

Author Contributions: Conceptualization, X.T. and W.C.; Investigation, X.T. and W.C.; Methodology, W.C.; Writing—original draft, W.C.; Writing—review & editing, X.T. and W.C. Both authors have read and agreed to the published version of the manuscript.

Funding: This work was funded by Natural Science Foundation of Fujian Province (NO.2020J01535, NO.2019J01409).

Conflicts of Interest: The authors declare no conflict of interest.

References

1. Neumann, P.; Carreck, N.L. Honey bee colony losses. *J. Apic. Res.* **2010**, *49*, 1–6. [CrossRef]
2. Goulson, D.; Nicholls, E.; Botías, C.; Rotheray, E.L. Bee declines driven by combined stress from parasites, pesticides, and lack of flowers. *Science* **2015**, *347*, 1255957. [CrossRef]
3. Campos, M.R.G.; Bogdanov, S.; de Almeida-Muradian, L.M.B.; Szczesna, T.; Mancebo, Y.; Frigerio, C.; Ferreira, F. Pollen composition and standardisation of analytical methods. *J. Apicult. Res.* **2008**, *47*, 156–163. [CrossRef]
4. Mao, W.; Schuler, M.A.; Berenbaum, M.R. Honey constituents up-regulate detoxification and immunity genes in the western honey bee Apis mellifera. *Proc. Natl. Acad. Sci. USA* **2013**, *110*, 8842–8846. [CrossRef] [PubMed]
5. Tu, X.; Chen, W. Overview of Analytical Methods for the Determination of Neonicotinoid Pesticides in Honeybee Products and Honeybee. *Crit. Rev. Anal. Chem.* **2020**, in press. [CrossRef] [PubMed]
6. Yáñez, K.P.; Martín, M.T.; Bernal, J.L.; Nozal, M.J. Trace Analysis of Seven Neonicotinoid Insecticides in Bee Pollen by Solid-Liquid Extraction and Liquid Chromatography Coupled to Electrospray Ionization Mass Spectrometry. *Food Anal. Methods* **2014**, *7*, 490–499. [CrossRef]
7. Zhang, L.; Wang, Y.; Sun, C.; Yang, S.; He, H. Simultaneous Determination of Organochlorine, Organophosphorus, and Pyrethroid Pesticides in Bee Pollens by Solid-Phase Extraction Cleanup Followed by Gas Chromatography Using Electron-Capture Detector. *Food Anal. Methods* **2012**, *6*, 1508–1514. [CrossRef]
8. López-Fernández, O.; Rial-Otero, R.; Simal-Gándara, J. High-throughput HPLC-MS/MS determination of the persistence of neonicotinoid insecticide residues of regulatory interest in dietary bee pollen. *Anal. Bioanal. Chem.* **2015**, *407*, 7101–7110. [CrossRef]
9. Vazquez-Quintal, P.E.; Rodríguez, D.M.; Medina-Peralta, S.; Moguel-Ordóñez, Y.B. Extraction of Organochlorine Pesticides from Bee Pollen by Matrix Solid-Phase Dispersion: Recovery Evaluation by GC-MS and Method Validation. *Chromatographia* **2012**, *75*, 923–930. [CrossRef]
10. Vázquez, P.P.; Lozano, A.; Uclés, S.; Ramos, M.G.; Fernández-Alba, A.R. A sensitive and efficient method for routine pesticide multiresidue analysis in bee pollen samples using gas and liquid chromatography coupled to tandem mass spectrometry. *J. Chromatogr. A* **2015**, *1426*, 161–173. [CrossRef]
11. Li, Y.; Kelley, R.A.; Anderson, T.D.; Lydy, M. Development and comparison of two multi-residue methods for the analysis of select pesticides in honey bees, pollen, and wax by gas chromatography-quadrupole mass spectrometry. *Talanta* **2015**, *140*, 81–87. [CrossRef] [PubMed]
12. Bernal, J.; Nozal, M.J.; Martín, M.T.; Bernal, J.L.; Ares, A.M. Trace analysis of flubendiamide in bee pollen using enhanced matrix removal-lipid sorbent clean-up and liquid chromatography-electrospray ionization mass spectrometry. *Microchem. J.* **2019**, *148*, 541–547. [CrossRef]

13. Valente, I.M.; Rodrigues, J.A. Recent advances in salt-assisted LLE for analyzing biological samples. *Bioanalysis* **2015**, *7*, 2187–2193. [CrossRef] [PubMed]

14. Schroeder, J.L.; Marinetti, L.J.; Smith, R.K.; Brewer, W.E.; Clelland, B.L.; Morgan, S.L. The analysis of delta9-tetrahydrocannabinol and metabolite in whole blood and 11-nor-delta9-tetrahydrocannabinol-9-carboxylic acid in urine using disposable pipette extraction with confirmation and quantification by gas chromatography-mass spectrometry. *J. Anal. Toxicol.* **2008**, *32*, 659–666. [CrossRef] [PubMed]

15. Bordin, D.C.M.; Alves, M.N.R.; De Campos, E.G.; De Martinis, B.S. Disposable pipette tips extraction: Fundamentals, applications and state of the art. *J. Sep. Sci.* **2016**, *39*, 1168–1172. [CrossRef]

16. Guan, H.; Brewer, W.E.; Garris, S.T.; Morgan, S.L. Disposable pipette extraction for the analysis of pesticides in fruit and vegetables using gas chromatography/mass spectrometry. *J. Chromatogr. A* **2010**, *1217*, 1867–1874. [CrossRef]

17. Guan, H.; Brewer, W.E.; Morgan, S.L. New Approach to Multiresidue Pesticide Determination in Foods with High Fat Content Using Disposable Pipette Extraction (DPX) and Gas Chromatography–Mass spectrometry (GC-MS). *J. Agric. Food Chem.* **2009**, *57*, 10531–10538. [CrossRef]

18. Scheidweiler, K.B.; Newmeyer, M.N.; Barnes, A.J.; Huestis, M.A. Quantification of cannabinoids and their free and glucuronide metabolites in whole blood by disposable pipette extraction and liquid chromatography-tandem mass spectrometry. *J. Chromatogr. A* **2016**, *1453*, 34–42. [CrossRef]

19. Mastrianni, K.R.; Metavarayuth, K.; Brewer, W.E.; Wang, Q. Analysis of 10 β-agonists in pork meat using automated dispersive pipette extraction and LC-MS/MS. *J. Chromatogr. B* **2018**, *1084*, 64–68. [CrossRef]

20. Bordin, D.C.M.; Alves, M.N.; Cabrices, O.G.; De Campos, E.G.; De Martinis, B.S. A Rapid Assay for the Simultaneous Determination of Nicotine, Cocaine and Metabolites in Meconium Using Disposable Pipette Extraction and Gas Chromatography-Mass Spectrometry (GC-MS). *J. Anal. Toxicol.* **2013**, *38*, 31–38. [CrossRef]

21. Pinto, M.A.L.; De Souza, I.D.; Queiroz, M.E.C. Determination of drugs in plasma samples by disposable pipette extraction with C18-BSA phase and liquid chromatography-tandem mass spectrometry. *J. Pharm. Biomed. Anal.* **2017**, *139*, 116–124. [CrossRef] [PubMed]

22. Samanidou, V.F.; Stathatos, C.; Njau, S.; Kovatsi, L. Disposable pipette extraction for the simultaneous determination of biperiden and three antipsychotic drugs in human urine by GC-nitrogen phosphorus detection. *Bioanalysis* **2013**, *5*, 21–29. [CrossRef] [PubMed]

23. Lehotay, S.J.; Mastovska, K.; Lightfield, A.R.; Nuñez, A.; Dutko, T.; Ng, C.; Bluhm, L. Rapid analysis of aminoglycoside antibiotics in bovine tissues using disposable pipette extraction and ultrahigh performance liquid chromatography-tandem mass spectrometry. *J. Chromatogr. A* **2013**, *1313*, 103–112. [CrossRef] [PubMed]

24. Chaves, A.R.; Moura, B.H.; Caris, J.A.; Rabelo, D.; Queiroz, M.E.C. The development of a new disposable pipette extraction phase based on polyaniline composites for the determination of levels of antidepressants in plasma samples. *J. Chromatogr. A* **2015**, *1399*, 1–7. [CrossRef] [PubMed]

25. Oenning, A.L.; Merib, J.; Carasek, E. An effective and high-throughput analytical methodology for pesticide screening in human urine by disposable pipette extraction and gas chromatography-mass spectrometry. *J. Chromatogr. B* **2018**, *1092*, 459–465. [CrossRef] [PubMed]

26. Zhang, H.; Li, Y.; Zhu, J.; Li, H.; Li, D.; Liu, Z.; Sun, X.; Wang, B.; Wang, Q.; Gao, Y. Disposable Pipette Extraction (DPX) Coupled with Liquid Chromatography-Tandem Mass Spectrometry for the Simultaneous Determination of Pesticide Residues in Wine Samples. *Food Anal. Methods* **2019**, *12*, 2262–2272. [CrossRef]

27. Júnior, C.A.S.A.; Dos Santos, A.L.R.; De Faria, A.M. Disposable pipette extraction using a selective sorbent for carbendazim residues in orange juice. *Food Chem.* **2020**, *309*, 125756. [CrossRef]

28. Tan, S.C.; Lee, H.K. Graphitic carbon nitride as sorbent for the emulsification-enhanced disposable pipette extraction of eight organochlorine pesticides prior to GC-MS analysis. *Microchim. Acta* **2020**, *187*, 1–10. [CrossRef]

29. Fernandes, V.C.; Domingues, V.F.; Mateus, N.; Delerue-Matos, C. Multiresidue pesticides analysis in soils using modified QuEChERS with disposable pipette extraction and dispersive solid-phase extraction. *J. Sep. Sci.* **2012**, *36*, 376–382. [CrossRef]

30. Lu, Z.; Fang, N.; Zhang, Z.; Hou, Z.; Lu, Z.; Li, Y. Residue analysis of fungicides fenpicoxamid, isofetamid, and mandestrobin in cereals using zirconium oxide disposable pipette extraction clean-up and ultrahigh-performance liquid chromatography-tandem mass spectrometry. *J. Chromatogr. A* **2020**, *1620*, 461004. [CrossRef]

31. Cadorim, H.R.; Schneider, M.; Hinz, J.; Luvizon, F.; Dias, A.N.; Carasek, E.; Welz, B. Effective and High-Throughput Analytical Methodology for the Determination of Lead and Cadmium in Water Samples by Disposable Pipette Extraction Coupled with High-Resolution Continuum Source Graphite Furnace Atomic Absorption Spectrometry (HR-CS GF AAS). *Anal. Lett.* **2019**, *52*, 2133–2149. [CrossRef]

32. Corazza, G.; Merib, J.; Magosso, H.A.; Bittencourt, O.R.; Carasek, E. A hybrid material as a sorbent phase for the disposable pipette extraction technique enhances efficiency in the determination of phenolic endocrine-disrupting compounds. *J. Chromatogr. A* **2017**, *1513*, 42–50. [CrossRef] [PubMed]

33. Morés, L.; Da Silva, A.C.; Merib, J.; Dias, A.N.; Carasek, E. A natural and renewable biosorbent phase as a low-cost approach in disposable pipette extraction technique for the determination of emerging contaminants in lake water samples. *J. Sep. Sci.* **2019**, *42*, 1404–1411. [CrossRef] [PubMed]

34. Turazzi, F.C.; Morés, L.; Carasek, E.; Merib, J.; Barra, G.M.D.O. A rapid and environmentally friendly analytical method based on conductive polymer as extraction phase for disposable pipette extraction for the determination of hormones and polycyclic aromatic hydrocarbons in river water samples using high-performance liquid chromatography/diode array detection. *J. Environ. Chem. Eng.* **2019**, *7*, 103156. [CrossRef]

35. Pena-Abaurrea, M.; De La Torre, V.G.; Ramos, L.; De La Torre, V.S.G. Ultrasound-assisted extraction followed by disposable pipette purification for the determination of polychlorinated biphenyls in small-size biological tissue samples. *J. Chromatogr. A* **2013**, *1317*, 223–229. [CrossRef] [PubMed]

36. Chen, W.; Tu, X.; Wu, D.; Gao, Z.; Wu, S.; Huang, S. Comparison of the Partition Efficiencies of Multiple Phenolic Compounds Contained in Propolis in Different Modes of Acetonitrile–Water-Based Homogenous Liquid-Liquid Extraction. *Molecules* **2019**, *24*, 442. [CrossRef] [PubMed]

37. Tu, X.; Sun, F.; Wu, S.; Liu, W.; Gao, Z.; Huang, S.; Chen, W. Comparison of salting-out and sugaring-out liquid-liquid extraction methods for the partition of 10-hydroxy-2-decenoic acid in royal jelly and their co-extracted protein content. *J. Chromatogr. B* **2018**, *1073*, 90–95. [CrossRef] [PubMed]

38. Chen, W.; Wu, S.; Zhang, J.; Yu, F.; Miao, X.; Tu, X. Salting-out-assisted liquid-liquid extraction of 5-hydroxymethylfurfural from honey and the determination of 5-hydroxymethylfurfural by high-performance liquid chromatography. *Anal. Methods* **2019**, *11*, 4835–4841. [CrossRef]

39. Zhou, J.; Qi, Y.; Ritho, J.; Zhang, Y.; Zheng, X.; Wu, L.; Li, Y.; Sun, L. Flavonoid glycosides as floral origin markers to discriminate of unifloral bee pollen by LC-MS/MS. *Food Control* **2015**, *57*, 54–61. [CrossRef]

40. Tu, X.; Ma, S.; Gao, Z.; Wang, J.; Huang, S.; Chen, W. One-Step Extraction and Hydrolysis of Flavonoid Glycosides in Rape Bee Pollen Based on Soxhlet-Assisted Matrix Solid Phase Dispersion. *Phytochem. Anal.* **2017**, *28*, 505–511. [CrossRef] [PubMed]

41. Anastassiades, M.; Lehotay, S.J.; Štajnbaher, D.; Schenck, F.J. Fast and Easy Multiresidue Method Employing Acetonitrile Extraction/Partitioning and "Dispersive Solid-Phase Extraction" for the Determination of Pesticide Residues in Produce. *J. AOAC Int.* **2003**, *86*, 412–431. [CrossRef] [PubMed]

42. Chen, W.; Wu, S.; Zhang, J.; Yu, F.; Hou, J.; Miao, X.; Tu, X. Matrix-Induced Sugaring-Out: A Simple and Rapid Sample Preparation Method for the Determination of Neonicotinoid Pesticides in Honey. *Molecules* **2019**, *24*, 2761. [CrossRef] [PubMed]

43. Taverniers, I.; De Loose, M.; Van Bockstaele, E. Trends in quality in the analytical laboratory. II. Analytical method validation and quality assurance. *TrAC Trends Anal. Chem.* **2004**, *23*, 535–552. [CrossRef]

44. Zhang, C.-P.; Huang, S.; Wei, W.-T.; Shun, P.; Shen, X.-G.; Li, Y.-J.; Hu, F.-L. Development of High-Performance Liquid Chromatographic for Quality and Authenticity Control of Chinese Propolis. *J. Food Sci.* **2014**, *79*, C1315–C1322. [CrossRef]

Sample Availability: Not available.

Publisher's Note: MDPI stays neutral with regard to jurisdictional claims in published maps and institutional affiliations.

Review

Sample Preparation Methods for Lipidomics Approaches Used in Studies of Obesity

Ivan Liakh [1,2], Tomasz Sledzinski [1], Lukasz Kaska [3], Paulina Mozolewska [1] and Adriana Mika [1,4,*]

[1] Department of Pharmaceutical Biochemistry, Medical University of Gdansk, Debinki 1, 80-211 Gdansk, Poland; liakh_ivan@mail.ru (I.L.); tomasz.sledzinski@gumed.edu.pl (T.S.); paulina.mozolewska@gumed.edu.pl (P.M.)

[2] Department of Toxicology, Medical University of Gdańsk, Al. Gen. Hallera 107, 80-416 Gdańsk, Poland

[3] Department of General, Endocrine and Transplant Surgery, Faculty of Medicine, Medical University of Gdansk, Smoluchowskiego 17, 80-214 Gdansk, Poland; lukasz.kaska@wp.pl

[4] Department of Environmental Analysis, Faculty of Chemistry, University of Gdansk, Wita Stwosza 63, 80-308 Gdansk, Poland

* Correspondence: adrianamika@tlen.pl; Tel.: +48-585235190

Received: 30 September 2020; Accepted: 12 November 2020; Published: 13 November 2020

Abstract: Obesity is associated with alterations in the composition and amounts of lipids. Lipids have over 1.7 million representatives. Most lipid groups differ in composition, properties and chemical structure. These small molecules control various metabolic pathways, determine the metabolism of other compounds and are substrates for the syntheses of different derivatives. Recently, lipidomics has become an important branch of medical/clinical sciences similar to proteomics and genomics. Due to the much higher lipid accumulation in obese patients and many alterations in the compositions of various groups of lipids, the methods used for sample preparations for lipidomic studies of samples from obese subjects sometimes have to be modified. Appropriate sample preparation methods allow for the identification of a wide range of analytes by advanced analytical methods, including mass spectrometry. This is especially the case in studies with obese subjects, as the amounts of some lipids are much higher, others are present in trace amounts, and obese subjects have some specific alterations of the lipid profile. As a result, it is best to use a method previously tested on samples from obese subjects. However, most of these methods can be also used in healthy, nonobese subjects or patients with other dyslipidemias. This review is an overview of sample preparation methods for analysis as one of the major critical steps in the overall analytical procedure.

Keywords: sample preparation; obesity; lipids; protein precipitation; liquid–liquid extraction; solid-phase extraction; biological samples

---

1. Introduction

Obesity remains one of the pressing problems of modern society, therefore, studies of the mechanisms underlying its occurrence and the therapies used to treat it continue to be relevant. Depending on the hypothesis, a wide range of research methods can be used, ranging from purely assessing biochemical parameters to deep psychological research. However, research usually involves standard procedures such as measuring body mass index (BMI) or fat content in human subjects. Among the biochemical parameters, those that undergo the greatest changes with obesity (lipid profile, fasting glucose, insulin, etc.) are first examined.

A large amount of accumulated data on obesity allows for their meta-analysis and underlies a large number of systematic and retrospective reviews [1]. In particular, studies related to bariatric surgery are a powerful source of data because these types of surgery involve altering the stomach, intestines, or both to induce weight loss [2–7]. In addition, a large number of studies of obesity are associated with cardiovascular diseases [8–11], diabetes and metabolic syndrome [12–14]; many studies are in the field of diets, psychology and neurology [15–23].

While the number of parameters in the study of obesity itself is limited only by the imagination of scientists and the equipment available in the laboratory [24–29], the study of obesity in connection with other diseases is strictly subordinate to the study area and is often limited to several parameters, such as BMI and total fat content. [30–33]. In addition, determining the diagnosis of obesity is always primary in that work; for this purpose, the most commonly used method is the calculation of BMI. The World Health Organization used BMI to categorize humans into underweight (< 18.5), normal weight (18.5–24.9), overweight (25–29.9) and obese (BMI ≥ 30) categories [34]. Since BMI may not be a good indicator of obesity for bodybuilders and other groups of athletes [35,36], body fat [37–39] and total body water [40,41] can be determined in these groups, as can concomitant states of lipid alterations in blood (dyslipidaemia) [42], hyperinsulinemia [43], etc. Since obesity is directly related to lipid metabolism, it is interesting to study not only standard plasma parameters but also alterations in specific lipid groups in serum [44]. However, due to the much higher lipid accumulation in obesity (Figure 1) and many alterations in the lipid composition, the methods used for sample preparations for lipidomic studies in samples from obese subjects sometimes have to be modified. Thus, the purpose of this review was to collect and systematize information on sample preparation for lipid analysis in the study of obesity. We focused on both standard procedures and new promising methods. However, most of these methods can be also used in healthy, nonobese subjects or patients with other dyslipidemias.

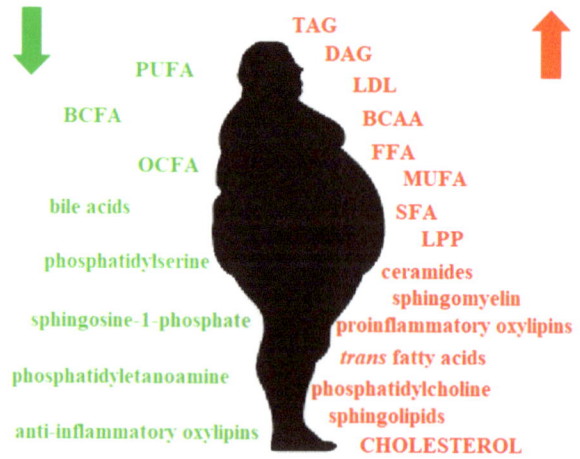

Figure 1. Lipid alterations in obesity. Lipids in red are elevated in obesity, and lipids in green are reduced. BCAA—branched chain amino acids; BCFA—branched chain fatty acids; DAG—diacylglycerols; FFA—free fatty acids; HDL—high density lipoproteins; LDL—low density lipoproteins; LPP—lipid peroxidation products; MUFA—monounsaturated fatty acids; OCFA—odd chain fatty acids; SFA—saturated fatty acids; PUFA—polyunsaturated fatty acids; TAG—triacyclglycerols.

In the study of obesity, determining triglyceride (TG) and cholesterol levels is of great clinical importance. Basic blood test results for total cholesterol (TC), TG and cholesterol in lipoprotein fractions (low density lipoproteins (LDL) and high density lipoproteins (HDL)) should be considered together. To study these indicators and related and often required for standard clinical practice indicators (C-reactive protein, glucose, insulin levels, etc.), there are many standard methods and their modifications that make these analyses routine in clinical practice, and they will not be presented in this review because of their routine nature. This work is focused more specifically on lipid groups, the study of which is carried out much less frequently because it requires the use of complex extraction methods and/or high-performance equipment. However, the alterations found in the extended lipidome may provide much more additional information about metabolic disorders in obese subjects than standard lipidograms.

2. Methods of Sample Preparation for Lipidomic Studies

2.1. Sample Collection and Storage

Fat metabolism disorders are detected by determining the lipid spectrum of the blood. Blood for a study is taken from a vein, always on an empty stomach (12–14 h after eating); otherwise, the results of the study are distorted, since 1-4 h after eating, alimentary hyperlipaemia occurs [45]. During blood sampling, adverse events such as haemolysis, coagulation, and platelet activation should be avoided, but the class of anticoagulants used should also be taken into account since calcium-chelating coagulants (ethylenediaminetetraacetic acid (EDTA) and citrate) can cause the calcium-dependent formation or degradation of certain classes of lipids ex vivo [46].

Different classes of lipids are subject to different changes during storage. Long-term storage of plasma at room temperature (RT) leads to an increase in lysophosphatidylethanolamines (LPE), lysophosphatidylcholines (LPC) and fatty acids (FAs), while phosphatidylethanolamines (PE) and phosphatidylcholines (PC) decrease, which suggests the breakdown of ester bonds in these phospholipids [47]. Avoiding freeze-thaw cycles is no less important because with their increase, the number of lipid metabolites decreases significantly [48].

2.2. Pre-Extraction Additives

Additives used during or before extraction serve a variety of purposes. Internal standards are a measure of extraction efficiency. In many cases, lipidomic studies of obesity are accompanied by the determination of obesity-associated hormone levels (such as ghrelin, obestatin, glucagon, leptin, and adiponectin); therefore, protease inhibitor cocktails are added to serum/plasma samples to increase hormone stability [49]. In addition, various detergents serve to facilitate cell destruction during homogenization, and buffers are used to maintain a stable pH. The most commonly added substances to prevent oxidative processes during extraction are antioxidants and radical scavengers such as butylated hydroxytoluene (BHT) [50,51]. This is especially important when studying unstable compounds such as oxylipins [52–54], which are the metabolites of polyunsaturated fatty acids.

2.3. Sample Stability

It is generally assumed that lipids are highly stable at RT, while it is advisable to not allow them to overheat during homogenization and to prevent oxidation by the addition of antioxidants [55]. Despite this, many studies consider the effects of storage conditions, the number of freeze/thaw cycles and the behaviour of organic compounds in experimental conditions. Jiang et al. validated a liquid chromatography-tandem mass spectrometry (LC-MS/MS) method for the determination of ceramides (Cer) in human plasma and determined the stability of each analyte at low- and high-quality control concentrations under long-term storage (39 days at -80 °C), freeze/thawing (five times), tabletop mode (14 h at RT before sample extraction) and autosampler conditions (3 days). The results showed that Cer (22:0) and Cer (24:0) were stable in human plasma under all conditions [56]. Ferreiro-Vera et al. assessed the stability of eicosanoids in serum under experimental conditions; every hour for 8 h, they analysed samples spiked with eicosanoids, and no significant differences in analyte concentrations were found [57]. Zeng and Cao also showed sufficient stability of short-chain FAs (SCFAs) and ketone body derivatives during autosampler storage (5 °C; 48 h), after 2 h at RT and after three freeze/thaw cycles [58]. Klawitter et al. showed that freeze/thaw cycles and long-term storage of plasma (6 h; RT) should be avoided to prevent changes in the composition of lipid classes of very low-density lipoprotein

(VLDL) (loss of cholesterol esters and phospholipids), while free fatty acid (FFA) concentrations did not change under the same conditions [59]. Oxylipins are especially unstable in this regard, and improper collection and storage of samples can lead both to a significant decrease in their level and to an increase in their content due to enzymatic and non-enzymatic oxidation [60]. Some oxylipins (resolvins and prostanoids) are unstable even at −20 °C [61], so the manufacturers of their standards recommend storing them at −80 °C, while the concentration of prostaglandins can significantly decrease with prolonged storage at even −80 °C [62].

2.4. Extraction Methods

2.4.1. Protein Precipitation

Protein precipitation (PPT) is used to remove protein from samples, therefore, when carrying out PPT during sample preparation, it is important that the chosen solvent causes protein denaturation and, at the same time, is a good solvent for lipids [55]. In addition, precipitation of proteins that make up a large volume of the analysed matrix is necessary since some groups of lipids are present in the matrix in trace amounts. This helps to minimize the risk of lack of detection or misidentification and to release protein-bound compounds prior to target lipid extraction [63]. Most often, PPT is preceded by subsequent solid-phase extraction (SPE) and liquid-liquid extraction (LLE). However, in the cases using high-performance equipment or shotgun lipidomics, PPT alone may be sufficient. Bellissimo et al. studied the metabolic profiles of obese individuals and used plasma acetonitrile (ACN) PPT before liquid chromatography-high-resolution mass spectrometry (LC-HRMS) analysis [64]. Jiang et al. when developing a method for validating Cer in human plasma, optimized the PPT method using 3 organic solvents (methanol (MeOH), ACN and isopropanol (IPA)) separately and in combination with chloroform. The mixture IPA:chloroform (9:1) was the most effective [56]. Drotleff et al. performed PPT during lipidomics of murine plasma by adding 55 μL of IPA and 20 μL of MeOH to 25 μL of plasma [65]. Söder et al. performed liquid chromatography-time-of-flight mass spectrometry (LC-TOF-MS)-based metabolomics analysis in overweight dogs using methanol extraction (5 μL of plasma:495 μL of MeOH), which allowed the determination of 317 phospholipids in the plasma samples [66].

2.4.2. Liquid–Liquid Extraction

The high solubility of the hydrocarbon chains of lipids in organic solvents allows the use of LLE for the separation of lipids in various immiscible liquids. Widely used methods such as those of Folch [67] and Bligh and Dyer [68] have the drawback of using toxic solvents [69]; in addition, some classes of lipids (for example, lysophospholipids (LPL)) can remain in the aqueous phase [70]; however, many proposed modifications of these methods can overcome the above disadvantages, and these methods are still widely used in lipidomics of obesity samples [70–76]. Methyl tert-butyl ether (MTBE) extraction, which has been popular recently, is undergoing various modifications and shows very good efficiency over classical methods [69]. In the study of obesity, MTBE extraction is used to isolate lipids from liver tissue [77], skeletal muscle [78], adipose tissue [79] and plasma [50,72].

2.4.3. Solid-Phase Extraction

SPE is more suitable than LLE for target lipidomics because it allows fractionation of specific lipid classes after LLE [80,81]. Therefore, in a lipidomics study, SPE is resorted to when it is necessary to isolate specific lipid groups or species that are present in the sample in a small amount, such as eicosanoids [57], LPL [70], oxidized phospholipids [75], serum sterols [82] oxysterols, endocannabinoids, and Cer [83], non-esterified FA and oxylipins [53,84–86]. Due to the wide variety of SPE protocols and commercially available SPE columns, there are studies in which these parameters are compared, for example, in studies of fatty acid esters of hydroxy fatty acids (FAHFAs) in serum [87] or oxylipins in human plasma [88,89]. Additionally, SPE helps to separate lipids in complex matrices with a large lipid abundance, such as adipose tissue [83,90] or brain tissue [80].

Since most of the extraction protocols described below are based on the SPE and LLE methods, the question arises—which of these protocols is more suitable for lipidomics studies. The advantages and disadvantages of PPT, LLE and SPE have been reviewed comprehensively elsewhere [91–96] and discussions on which of the methods is better are still ongoing. Among the main advantages of the LLE, one can note its low costs and the presence of a large number of well-established protocols. In turn, SPE is assumed to have a less pronounced matrix effect, there is less transfer of the aqueous phase, less toxic solvents are used, and it is less labour and faster. However, the last 2 statements are controversial [91]. On the other hand, the advantage of SPE, such as the high selectivity can be considered a disadvantage in the case when it is necessary to separate simultaneously several analytes with different physical and chemical properties which would require several different SPE columns. In addition, the complex structure of sorbents in SPE columns increases the risk of differences between individual bathes, in contrast to LLE, which uses highly purified simple organic solvents. It can be difficult to achieve high recovery efficiency with SPE since some analytes may partially elute in a different fraction during separation. However, despite the complication of the process and sample preparation time, the most comprehensive results can be obtained by combining LLE with SPE [70,79,80,83,87,97–101] or with TLC [74,86,102–104] approaches in one study.

2.4.4. Other Extraction Methods

In addition to the well-established routine extraction methods described above, such as LLE, SPE, and PPT, also more modern but at the same time rarer extraction methods are used in the studies of obesity, such as solid-phase microextraction (SPME), stir bar sorptive extraction (SBSE), dispersive liquid–liquid microextraction (DLLME) and their variants. The main disadvantage of the above-listed solvent extraction is the use of organic solvents that have such disadvantages as toxicity and harmfulness to the environment. In addition, they must be of high purity, which increases the cost of analysis [105]. However, SPME, SBSE, DLLME are a solvent-free sample preparation method that is easy to use, does not require preliminary sample preparation, and is easily automated [94].

The most widely used technique of above mentioned is SPME. In combination with gas chromatography-mass spectrometry (GC-MS) it can be used not only for analysis of volatile organic compounds, but also for the extraction of fatty acids and fatty acid esters from solid tissues and biofluids, which requires a very small sample volume and reduces the matrix effect [106]. Although SPME can be used for lipidomics studies, in obesity studies these methods are also used to study non-lipid compounds. SPME followed by GC/MS was used to analyze aroma compound headspace release from extra virgin olive oil after the interaction of saliva in obese and overweight individuals [107], to evaluate volatile organic compounds of gut microbiota of obese patients [108,109], and for urinary volatile organic compounds profiling in overweight children [110].

The SBSE method, like SPME, is a method of sample preparation without the use of solvents and with the use of a solid sorbent for preliminary concentration of the analyte before analysis. The surface area of the sorbing polymer is greater in SBSE than in SPME [111]. Eslami et al. used SBSE followed by HPLC for quantification of ghrelin in human plasma [112].

The DLLME method is based on the rapid mixing of dispersing and extraction solvents with an aqueous sample, resulting in the formation of an emulsion consisting of fine particles of the extraction solvent dispersed in the aqueous phase, then the solvent is separated from the sample by centrifugation [113]. Amin et al. used DLLME following GC/MS method for the evaluation of urinary Bisphenol A in obese subjects [114]. Krawczyńska et al. applied DLLME technique for the determination of vitamin D in obese patients plasma [115].

Thus, the relatively small number of studies in lipidomics using above methods is explained by their recent appearance, while such advantages as relative easiness of implementation, accuracy, small sample volume and lack of organic solvents make these extraction methods promising.

3. Preparation of Different Sample Types

Most methods of lipid extraction from biological samples are based on the dissolution of hydrocarbon chains in organic solvents mixed in various combinations (Folch [67] and Bligh and Dyer [68] and modifications of their method). However, there is no unified methodology, and the specific protocol should be selected depending on the lipid class being studied. Especially in studies in obese subjects, when the amounts of some lipids are elevated, others are present in trace amounts and obese subjects have some specific alterations of the lipid profile, it is best to use a method previously tested on samples from obese subjects. Table 1 summarizing sample preparation methods for lipidomics approaches described below is located at the end of this section.

3.1. Serum/Plasma Lipids

Plasma is the medium most responsive to changes in the body, which makes it one of the most important sources of information. However, at the same time, a large number of metabolites in the plasma and the fast rate of variation in their content can make it difficult to find specific markers of diseases. Most often, in the study of obesity, an important parameter such as the fasting lipid profile, which includes TC, HDL cholesterol, LDL cholesterol and TG, is determined for blood plasma [116]. A powerful indicator may be the content of FFAs in the blood, which reflects the influx of excess FAs from visceral fat into the liver, which is observed in obesity [117,118]. However, advanced approaches using lipidomics based on mass spectrometry can identify hundreds of lipid species belonging to dozens of lipid classes in plasma [119,120] (Figure 2).

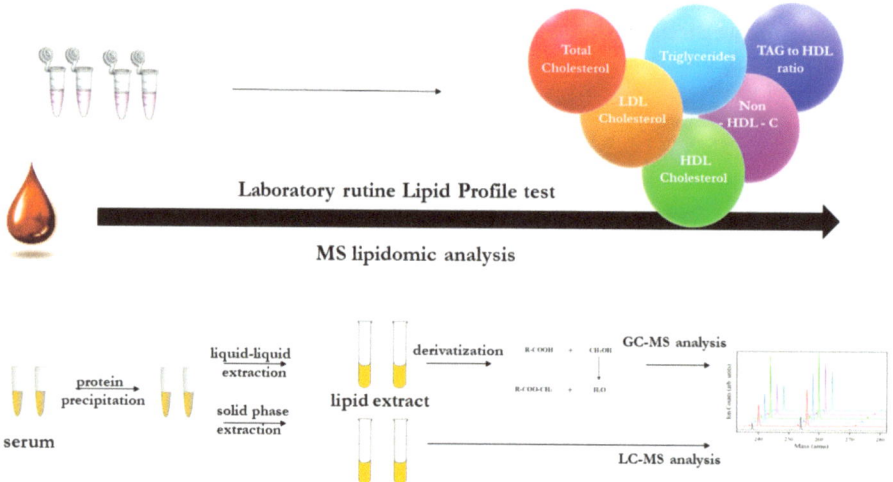

Figure 2. Lipid analysis in diagnostic laboratory vs. advanced lipidomic studies.

Blewett et al. used the modified Folch method to extract lipids from the plasma of obese rats, after which phospholipids were separated on silica G plates; for this, they were visualized using 8-anilino-1-naphthalenesulfonic acid under ultraviolet (UV) light and compared with the corresponding standards. The resulting phospholipid silica band was removed, and FA methyl esters (FAMEs) were prepared for further analysis by gas chromatography (GC) [121]. Additionally, Choromańska et al. used a combination of LLE and thin-layer chromatography (TLC) to extract lipid fractions (FFA and triacylglycerols (TAG)) from plasma in women with morbid obesity. Lipid fractions were extracted from 200 μL of plasma samples according to the method of Bligh and Dyer. Then, the samples were separated by TLC on silica gel plates (Silica Plate 60, 0.25 mm; Merck, Darmstadt, Germany). The separation was carried out in a solvent containing heptane, isopropyl and acetic acid (60:40:3;

v/v/v) after methylation with boron in trifluoride (BF3)-MeOH. Than samples were analysed by gas chromatography-mass spectrometry (GC-MS) [74].

Analysis of acyl-lysophosphatidic acids (LPAs) is of clinical importance in terms of prevention of obesity [122]. Yoon et al. established a method of LPA determination in human plasma. For this, extraction was performed with a MeOH/chloroform mixture (2:1) containing an internal standard (LPA C14:0). After back extraction with chloroform and water, the centrifuged lower phase was evaporated, redissolved in MeOH and analysed using ESI-MS-MS (directly injected into the ion source) [122].

Im et al. used a modified Matyash method [123] to isolate sphingomyelins (SM) from the plasma of men with abdominal obesity. Plasma was mixed with of 75% MeOH containing BHT and internal lipid standards. After MTBE (1 mL) was added to the mixture, it was shaken, and water was added to separate the phases. The mixture was then centrifuged, and the upper phase was dried and redissolved in of chloroform/MeOH (1:9; *v/v*) for liquid chromatography-tandem mass spectrometry (LC–MS/MS) analysis [50].

Wang et al. used a simple method of extraction in the study of the plasma lipidome in adults with obesity. Plasma was mixed, sonicated and incubated with a solution of chloroform/MeOH (2:1) containing internal standards after centrifugated and dried. The extracted lipids were resuspended in butanol and 10 mM NH_4CHOO in MeOH. After liquid chromatography–electrospay ionization-tandem mass spectrometry (LC–ESI-MS/MS) analysis, they quantified 328 lipid species from 24 lipid classes and subclasses [124].

Misra et al. used untargeted metabolomics analysis of serum in high-fat and high-cholesterol (HFHC) diet-fed baboons. Serum sample extraction was performed by sequential solvent extraction of serum samples initially with an ACN/IPA/water (3:3:2) mixture (1 mL) and later with ACN/water (1:1). The obtained extracts were mixed, dried and subjected to further chemical derivatization (N-methyl-trimethylsilyl-trifluoroacetamide (MSTFA) and N-(t-butyldimethylsilyl)-N-methyltrifluoro-acetamide (MTBSTFA) strategies), which allowed us to quantify 515 metabolites, many of which are involved in lipid metabolism [125].

Wang et al. developed a method to perform plasma lipidomics in spontaneously obese rhesus monkeys, based on cooling a plasma mixture with MeoH/ *n*-hexane with liquid nitrogen, followed by incubation with acetyl chloride for 24 h, adding a K_2CO_3 solution, and extracting the obtained methylated FAs with hexane for further analysis using GC-MS. This method allowed the identification, quantification and classification of 143 types of lipids [51].

3.1.1. Serum/Plasma Cholesterol

Dyslipidaemia that develops during obesity is reflected in the levels of both free cholesterol and its individual fractions [42]. To measure indicators such as TC, cholesterol, HDL and TAG in the blood, automated clinical systems based on enzymatic-calorimetric analysis are widely used, while the determination of LDL was not possible until recently. Its value was calculated according to the Friedewald formula [126]. At present, there are many methods for the direct measurement of LDL levels with more accurate results than those obtained by calculation [127].

In the study of cholesterol metabolism in obese people, isotope methods are often used that directly measure the fluxes of $[^2H_2]$- and $[U-^{13}C]$-labelled metabolites [128], and isotope enrichment can be estimated in lipid fractions extracted with organic solvents using the ^{13}C to ^{12}C ratio, which is measured by gas chromatography continuous-flow isotope ratio-mass spectrometry (which is also used for determining enrichment of $[^2H_2]$) (Finnigan Incos-XL GC-MS) [129]. Cho et al. decided that isotope-kinetic and sterile balance methods are not suitable for large-scale studies and developed their own simple and less-expensive approach for measuring GC-MS serum sterol signatures [82]. For this, serum samples mixed with MeOH and spiked with internal standards were used for hybrid solid-phase extraction-precipitation (H-PPT) and then eluted with MeOH three times. After drying, the pool of eluates was derivatized, and 12 sterol signatures were measured using GC-MS [82].

Recently, studies of fractions and subfractions of cholesterol have been carried out on the basis of size using specialized systems based on gel chromatography. In the work of Lindqvist et al. analysis of lipoprotein fractions during animal bariatric surgery modelling was performed using the Quantimetrix Lipoprint® system (Quantimetrix Corporation, Redondo Beach, CA, USA) [130]. Additionally, Kwon et al. in studies of low-density lipoprotein subfractions in overweight and obese women, also used this method to scan LDL subfractions. This system allowed them to divide LDL into seven subfractions [131]. Doğan et al. also used this system in LDL and HDL subfraction analysis in patients after laparoscopic sleeve gastrectomy [132].

3.1.2. Triglyceride Profiling

The study of blood TG levels is of great interest in the study of obesity because their level quickly responds to dietary changes [133,134] and is associated with insulin resistance [135,136], lipotoxicity [137] and dyslipidaemia [138]. Despite this, studies of the profile of individual FAs in plasma TG are quite rare; this parameter is usually tested in adipose tissue, where TG accumulates [71,139], or in breast milk [140]. Usually, the isolated plasma TAG fraction is examined as one of the components in the study of the whole lipid profile [74,128,141].

Perreault et al. in a study of the inflammatory state of metabolically healthy obese individuals, determined profiles of FAs from serum samples. They measured total FAs and fractionated FAs in PL and TAG fractions. In both cases, isolated lipid fractions were obtained by incubation of samples (45 min) on Silica-G TLC plates (Analtech, Newark, NJ, USA) with petroleum ether, ethyl ether and acetic acid (80:20:1; *v/v/v*). After that, both the collected PL and TAG lipid bands and the samples for the analysis of total FA were methylated for analysis on a GC with a flame ionization detector (FID) [102].

Much work has been done by Klawitter et al. who compared 4 different approaches for determining the plasma fatty acid desaturation index in response to a carbohydrate-rich diet. Using ultracentrifugation, TLC, SPE, and saponification followed by ultra-high-performance liquid chromatography (UHPLC)-MS for specific fatty acid analysis, they concluded that analysis of specific FAs in the VLDL fraction best reflects the activity of the stearoyl-CoA desaturase 1 (SCD1) metabolic pathway [59]. One of the approaches of UHPLC neutral loss MS was focused on analysing the fatty acid composition of TAG directly from plasma extracts or from plasma VLDL extracts without prior fractionation [59].

3.1.3. Fatty Acid Profiling

The FA profile is a rich source of information on dietary lipid intake and changes associated with obesity. Changes in the FA blood profile affect the production and secretion of cytokines, chemokines and eicosanoids, which can initiate inflammation of the whole body [102]. FFAs released into the blood from adipose tissue during TAG lipolysis may act as a marker of FA adipose tissue composition [142,143].

In our earlier study, we used high-performance liquid chromatography with a laser light scattering detector (HPLC-LLSD) to separate lipid extracts obtained by the Folch method during the identification of cyclopropane FA in the serum of obese people. The resulting FFA, TAG and PL were hydrolysed with 0.5 M KOH in MeOH. After neutralization with 6 M HCl and addition of water (1 mL), triple extraction of FA with *n*-hexane was performed, and the samples were dried again. Thereafter, FAMEs and picolinyl esters of FA were obtained from FFA and FA from the hydrolysis of complex lipids [144]. This method of extraction and hydrolysis followed by FAME synthesis can also be used to determine the levels of odd-chain fatty acids (OCFAs) and branched-chain fatty acids (BCFAs). The study of OCFAs and BCFAs is of interest since changes in these FAs in serum may be associated with insulin resistance in obese patients [145,146].

Kang et al. profiled FAs in the plasma of overweight subjects using a modified version of the Bondia-Pons method [147] to prepare samples for GC analysis. For this, *n*-hexane containing internal standards was added to plasma samples and then mixed with MeOH, and acetyl chloride was slowly added. After incubation 6% potassium carbonate was added to each tube, and after centrifugation, a transparent top layer of *n*-hexane with FAMEs was taken for GC-MS analysis [148]. Later, Lee et al. used this method to determine circulating fatty acid profiles in overweight individuals [149].

Wijayatunga et al. used improved direct synthesis of fatty acid methyl ester to measure serum FAs in patients after bariatric surgery. For this, serum was mixed with MeOH containing internal standard, water solution of 10 N KOH and MeOH. After incubation and shaking, samples were cooled, and 24 N H_2SO_4 in water was added and shaken. After that, the obtained FAMEs were extracted with *n*-hexane for subsequent analysis by GC-MS [7].

Aslan et al. studied changes in plasma polyunsaturated FAs (PUFAs) after bariatric surgery. For this, plasma was mixed with an internal standard solution (arachidonic acid (AA)-d8). After a mixture of ACN/37% hydrochloric acid (4:1; *v/v*) was added, the samples were hydrolysed, after which extraction was carried out with *n*-hexane. The upper phase, containing FFA, was dried, dissolved in MeOH-water (180:20; *v/v*) and filtered for LC-MS/MS analysis [150]. Later, Badoud et al. used this method to determine the profile of FAs in obese individuals. In both works, the authors were able to isolate, extract and quantify 28 FAs [151].

Ma et al. analysed serum FFA profiles to examine the relationship between FFA and the metabolic phenotype of obesity of two ethic groups in China. For this, a sample was mixed with ACN and incubated. After centrifugation, derivatives and 1-ethyl-3-[3-dimethylaminopropyl] carbodiimide were added to improve detection sensitivity, degree of separation, and binding efficiency, which allowed them to quantify 34 types of FFAs in serum [152].

Itariu et al. determined the profiles of FA in plasma phospholipids in severely obese nondiabetic patients. The main lipids were separated by TLC on silica gel plates (Merck) using *n*-hexane/diethyl ether/acetic acid (80:30:1; *v/v*) and BHT as the mobile phase and 11,2-dioleoyl-*sn*-glycero-3-phosphocholine (as the standard). After a solution (100 mg of berberine chloride in 100 mL of ethanol) was sprayed, lipid spots were visualized in ultraviolet light. Phospholipids were scraped into glass tubes, followed by methanolysis with a solution of MeOH/toluene (4:1). After 6% K_2CO_3 was added, the organic phase was collected and analysed by GC-MS [86].

Given the effect of high plasma NEFA levels on insulin resistance and insulin secretion, Nemati et al. studied changes in NEFA in patients with type 2 diabetes after bariatric surgery [153]. Dole extraction was performed to separate NEFAs. For this, serum was mixed with an internal standard (heptadecanoic acid in IPA and a modified Dole's mixture containing IPA/n-heptane/phosphoric acid (2 mol/L; 40/10/1; *v/v/v*). After incubation, n-heptane and water were added, and the samples were mixed and centrifuged. Then, the upper organic layer was dried and dissolved in IPA for analysis by LC-MS, which quantified 5 NEFAs (palmitic, stearic, oleic, palmitoleic and linoleic acid) [153].

Lin et al. described a procedure for FFA analysis in serum samples of patients after bariatric surgery in which a blood serum sample was evaporated in a nitrogen atmosphere and then addedd triheptadecanoin as an internal standard. After drying, the samples were subjected to direct esterification by dry 2.5 M HCl in methanol. The obtained FAMEs were extracted twice with isooctane, and a total of 16 FAs were determined using a GC instrument equipped with a FID [154].

Ramos-Molina et al. used two specific lipid extraction protocols to obtain the serum lipidome in obese subjects after bariatric surgery. Two separate ultrahigh-performance liquid chromatography (UPLC)-MS-based platforms analysing MeOH and chloroform/MeOH serum extracts were combined. In the first protocol, PPT was carried out by adding MeOH to serum. After vortexing and incubation, the supernatants of the samples were centrifuged, dried and restored in MeOH to determine FAs, bile acids (BA), steroids and LPL. A second protocol was used to determine glycerolipids, cholesterol esters, sphingolipids and phospholipids. For this, serum extracts were mixed with sodium chloride (50 mM) and chloroform/MeOH (2:1). After stirring and incubation, the samples were centrifuged, and the organic phase was collected, dried and restored in ACN/IPA (1: 1). Both extraction protocols used internal standards for each class of lipids. After that, two separate UPLC-MS-based platforms were used for analysis, allowing the identification of nearly 300 lipids present in human plasma [155].

Recently, new studies on FAHFAs that are related to diabetes and obesity have appeared [73,79,87,156]. Based on the fact that low concentrations of FAHFA are present in blood serum, López-Bascón et al. developed an automated online SPE LC-MS/MS method for sensitive and selective analysis of FAHFA [87]. Optimization of the SPE protocol included tests with four commercial sorbents (C8, C18, C18HD and Resin SH) packed with the same technology and based on nonpolar interactions. After testing the retention/elution ability of these sorbents using the generic reversed-phase protocol, the greatest retention/elution of FAHFAs was exhibited by C8. This sorbent was used in further optimization of the protocol, which involved varying the composition, volume and flow rate of the tested solvents; the gradient and the elution time [87].

Short-chain fatty acids (SCFAs) are saturated aliphatic FAs with fewer than six carbon atoms that can be produced by the anaerobic intestinal microbiota or by catabolism of branched-chain amino acids. The important role that SCFAs play in homeostasis and data indicating the association of SCFAs with multiple metabolic diseases make them an important object of research [58,157,158]. Other research methods targeting SCFAs are commonly based on extraction with a complex derivatization procedure for subsequent GC–MS analysis [158–160]. Zeng and Cao developed a novel LC-MS/MS method using fast derivatization with O-benzylhydroxylamine (O-BHA) and *N*-(3-dimethylaminopropyl)-*N*′-ethylcarbodiimide hydrochloride (EDC) in combination with LLE with dichloromethane for detecting SCFAs in the serum of obese and lean mice [58].

When describing FA profiling, it is important to mention such important metabolites of PUFAs such as oxylipins and endocannabinoids. It has long been known that endocannabinoids are key components of systems that regulate both nutrition and body mass, and oxylipins are also closely associated with obesity due to the wide range of their biological effects [53,161]. In our previous works, the features of sample preparation and methods for the analysis of oxylipins and endocannabinoids in biological samples are described in sufficient detail [60,162]. Compared to other classes of lipids, the analysis of oxylipins can be complicated by their low content in samples, susceptibility to oxidation, and ability to be synthesized de novo during sample preparation and extraction, all of which should be taken into account in the analysis of oxylipins.

Astarita et al. proposed an SPE method to fractionate lipid classes in which in one part of the organic phase after solvent extraction, high-abundance positively charged lipids were analysed by LC/MS. In the second part, the lipids were fractionated according to their relative polarities by serial elution with chloroform/MeOH (9:1 and 1:1; *v/v*) mixtures on open-bed silica gel columns (silica gel 60 230–400 mesh). Moreover, for further chromatographic separation, two different octadecyl (C18) columns were used [99]. Later, Argueta et al. used this protocol for the extraction of endocannabinoids from the plasma and jejunum of mice with western diet-induced obesity and their analysis by UPLC-MS/MS [100]. Additionally, Perez et al. used the method of Astarita et al. when studying the endocannabinoid system in plasma, pancreas and jejunum by UPLC-MS/MS in offspring obtained from obese mice fed a Western diet during pregnancy [101].

Ferreiro-Vera et al. developed a fast automatic method for quantitative analysis of oxylipins in serum samples from obese individuals. The approach is based on online SPE–LC–MS/MS method in which a sequence of automatic operations was performed on injected human serum. Further steps included rinsing the cartridge with MeOH, conditioning with water, loading sample into the cartridge with water and, after washing with 20% MeOH in water, switching a valve to elute through the cartridge into the chromatographic column using a mobile phase containing MeOH/water/ACN/acetic acid (76:22:2:0.02; *v/v*) [57].

Pickens et al. studied plasma lipid FA profiles in subjects with various BMIs and performed nonesterified plasma PUFA and oxylipins extraction and isolation. Oxilipids were isolated on SPE Phenomenex Strata-X columns (60 mg/3 mL, Phenomenex, Torrance, CA, USA) [163]. Later, Pickens et al. used this method for the extraction and isolation of nonesterified PUFAs and oxylipins [53].

Hernandez-Carretero et al. investigated changes in oxylipins, endocannabinoids, and Cer in mouse plasma after weight loss using a 96-well Ostro™ Pass Through Sample Preparation Plate (Waters Corp, Milford, MA, USA) to remove proteins and phospholipids. [52].

Azar et al. performed LC-MS/MS analysis of endocannabinoids in the serum of patients undergoing bariatric surgery. For this, after PPT with a mixture of acetones and Tris buffer, serum samples were homogenized in a mixture of MeOH and Tris buffer with the addition of an internal standard. The resulting homogenates were extracted with a mixture of chloroform/MeOH (2:1; *v/v*), washed three times with chloroform, dried and redissolved with MeOH [164].

Fan et al. investigated the production of unesterified oxylipins and endocannabinoids in mice fed a high-fat diet. To extract oxygenated PUFA metabolites, plasma samples were mixed with antioxidants (BHT/EDTA in MeOH:H_2O; 1:1) and deuterated surrogates in MeOH. Then, the proteins were precipitated by adding a mixture of MeOH/ACN (1:1). After centrifugation, the supernatants were filtered and quantified using UPLC-MS/MS] [54].

3.1.4. Ceramides and Sphingolipids

Sphingolipids are one of the most complex and structurally diverse families of compounds. Sphingolipids are not only components of biological structures such as membranes and lipoproteins but also highly biologically active compounds that affect dozens of biological processes [165]. Ceramides are sphingolipids that promote insulin resistance and are associated with the distribution of body fat, obesity, and type 2 diabetes [74]. Brozinick et al. used the method of single-phase extraction with MeOH-dichloromethane for the quantitative determination of seven types of Cer in rhesus macaque plasma on a Western-style diet [166].

León-Aguilar et al. used the method described by Croyal et al. [167] to quantify plasma Cer in offspring born to obese women. The method consisted of preparing seven standard dilutions (1, 5, 10, 50, 100, 250, and 500 nM) of 10 types of Cer in MeOH. Then, standard solutions and plasma samples (25 µL) were extracted using the Bligh and Dyer method with the addition of Cer (d18:1/17:0) as an internal standard for further UPLC-MS/MS analysis [76]. Neeland et al. when studying the role of Cer in the development of insulin resistance, used a method based on comparing total acyl carbon content and degrees of saturation between samples and a deuterated internal standard Cer (d18: 1) to identify 13 different types of Cer in plasma by UPLC-MS [168].

Özer et al. were able to quantify five Cer and three SM using successive chloroform/MeOH and chloroform/water extractions before ultrafast liquid chromatography (UFLC)-MS/MS analysis when studying changes in serum SM and Cer after bariatric surgery [169].

3.2. Adipose Tissue

In the study of obesity, adipose tissue is more useful than just its basic physiological functions (storage and release of fat); adipose tissue plays an important role in insulin sensitivity and is the site of the synthesis of many hormones and other signaling molecules associated with obesity [71,170,171]. Moreover, adipose tissue is relatively easily available as a research material and can be obtained during bariatric surgery. Most lipid extractions are carried out at RT, which is associated with poor lipid solubility in cold conditions; however, overheating can occur (during homogenization or sonication) and lead to the release of acyl FAs and the formation of lysolipid species [55]. For these reasons (in addition to reducing oxidation), antioxidants (BHT) and metal chelators (EDTA) are added to samples [52,83,103].

Roberts et al. described several approaches based on Folch extraction from white adipose tissue. The described methods, including GC–MS analysis of total fatty acid composition and intact lipid and acylcarnitine profiling by LC–MS, make it possible to obtain wide lipid profiles. At the same time, the authors suggest using SPE with LC–MS/MS analysis for eicosanoids from adipose tissue [90].

Kunešová et al. studied the fatty acid composition of TG in adipose tissue after weight loss. For this, the lipid fraction was extracted from a fat cake obtained during the extraction of RNA and then transmethylated to FAMEs (1 M sodium methoxide; RT; 60 min). After neutralization with 1 M acetic acid, the FAMEs were extracted twice with *n*-hexane, passed through a column (5 × 20 mm) of anhydrous sodium sulfate, mixed together and dried for further analysis on a GC instrument equipped with an FID [172]. Later, Montastier et al. also used this method (lipid fraction extraction from fat cakes produced during RNA extraction) for the analysis of FAs in the adipose tissue of obese women [173].

Hu et al. used a three-phase MTBE/MeOH/water extraction procedure for quantification of FAHFAs in a WAT sample of hamsters fed an HFD. In addition, the samples were subjected to additional removal neutral lipids by subsequent SPE, which allowed the identification of 64 FAHFAs [79].

Okada et al. measured the concentrations of lipid mediators in adipose tissue of obese mice. For this, the samples were extracted using an automated system (RapidTrace Biotage). Samples were pumped onto C18 cartridges, washed with water, *n*-hexane was added, and oxylipins were eluted with methyl formate [85].

Similar to plasma, when the levels of oxylipins are being examined in adipose tissue, additional purification with SPE is usually required. For example, Itariu et al. when studying chronic inflammation of adipose tissue in patients with severe obesity, used SPE (Oasis HLB Extraction Cartridge; Waters) to extract lipid mediators, which was followed by LC-MS [86].

Simple LLE extraction without SPE can also be quite effective in lipidomics. Therefore, Al-Sulaiti et al. successfully used the Bligh and Dyer method to extract lipids for further nontarget LC-MS determination of 76 TAG species from subcutaneous adipose tissue and omental depots from obese individuals [71]. Additionally, using a two-step chloroform/MeOH extraction, Grzybek et al. identified more than 300 lipids in different AT types of mice fed a high-fat diet (a rodent model of obesity) [174]. It is interesting that although all stages of fluid processing were performed using the Hamilton Robotics STARlet robotic platform with the Anti Droplet Control function, the authors encountered difficulties pipetting reproducible amounts of homogenized tissues due to the presence of large amounts of fat droplets when using an aqueous buffer. The problem was solved by homogenization of AT in 50 vol.% ethanol, followed by dilution in pure ethanol [174].

Tomášová et al. described their modification of the Folch lipid extraction method, which solved the problem of the high content of TAG in adipose tissue. They used TLC for separating lipid classes before LC-MS analysis, which allowed them to detect 37 lipids that were below the detection limit without TLC [103]. Similarly, Choromańska et al. applied the previously described method (Bligh and Dyer LLE with TLC on silica gel plates) for TAG, CER and DAG separation in adipose tissue of women with morbid obesity [74].

Depending on the studied lipid class and the selected extraction method, an appropriate organic solvent/mixture of solvents is added during or after homogenization. Mutemberezi et al. used dichloromethane/MeOH/water (8:4:2) extraction with subsequent SPE in LC-MS analysis of oxysterols, Cer, and endocannabinoids involved in obesity and metabolic syndrome [83]. The authors paid special attention to the elimination of the cholesterol oxidation during extraction. To minimize this process, they chose solvents with low prooxidant properties, and to remove cholesterol from the sample, they created an SPE procedure where three different solvent mixtures, *n*-hexane/IPA (99:1; *v/v*), *n*-hexane-diethylether (90:10; *v/v*) and *n*-hexane-dichloromethane (80:20; *v/v*), were tested. Based on their ability to selectively elute cholesterol and their tendency to oxidize cholesterol ($5\alpha,6\alpha$-epoxycholesterol,27-oxysterol), they chose the *n*-hexane-IPA mixture as the best for this purpose [83].

Serbulea et al. developed a targeted LC-MS approach to analyse oxidized phospholipids (OxPL) in lean and obese mice. OxPL was extracted by a modified Bligh and Dyer method. The method consisted in the fact that the aqueous phase remaining after the first extraction with a chloroform/MeOH mixture was extracted with chloroform, and the organic layer of the second extraction was combined with the first, dried, and resuspended in 300 μL of the mobile phase for further LC-MS analysis. In addition, phospholipids were separated on an EVO C18 column (Kinetex 5 μm; 100 × 4.6 mm; Phenomenex) using a binary gradient [75].

Hanzu et al. when studied visceral (VIS) and subcutaneous (SC) adipose tissue from obese subjects, examined not the tissue itself but the adipose tissue-conditioned medium samples, which were obtained by 24-h incubation of intact fat pads in serum-free medium, and further annotation and identification of metabolites was performed using GC-MS [175].

3.3. Liver

Although in the study of obesity, the liver is a pivotal tissue associated with lipid metabolism, similar to adipose tissue, there are few studies describing the extraction and identification of lipids in the liver. Undoubtedly, it is associated with difficulties in obtaining samples from this organ from humans. Most often, liver samples from rodent models of obesity are used. In various studies based on animal model liver tissue, lipids such as oxysterols, endocannabinoids, Cer [83], FAHFAs [79], SM, Cer [176], and glucosylceramides [177] have been studied. As mentioned above, Fan et al. used simple MeOH extraction for the analysis of oxylipins and endocannabinoids from the livers of mice fed a HFD [54].

Pakiet et al. studied the effect of a western diet on mouse brain lipid composition, but they also examined the liver; after a Folch extraction, they used two SPE procedures to separate lipid species [80]. In the first procedure, various eluents were used during SPE with chloroform/IPA (2:1; *v*/*v*) that allowed neutral lipids (NL) to be obtained; diethyl ether with 2% acetic acid (*v*/*v*) was used to elute FFAs, and MeOH was used to extract PLs. After that, neutral lipids were dissolved in *n*-hexane, fractionated on a new SPE cartridge and eluted, and cholesterol esters (with 6 mL *n*-hexane) were discarded. TAG (elution with methylene chloride/diethyl ether/*n*-hexane; 10:1:89; *v*/*v*/*v*), cholesterol (elution with 5% ethyl acetate in *n*-hexane (*v*/*v*)), diacylglycerols (DAG) (elution with 15% ethyl acetate in *n*-hexane (*v*/*v*) and monoacylglycerols (MAG) (elution with chloroform-MeOH (2:1; *v*/*v*)) were obtained. After that, the MAG, DAG and TAG fractions were mixed together into a mixture of acylglycerols [80]. Using the second method, from the lipid extracts recovered in chloroform, the following were obtained by extraction: NL (15% ethyl acetate in *n*-hexane (*v*/*v*) was eluted), Cer (chloroform/MeOH; 23:1, *v*/*v*), FFA and α-hydroxy-FFA (5% acetic acid in diisopropyl ether (*v*/*v*)), glycosphingolipids (GSPL) (acetone-MeOH (9:1.35; *v*/*v*), and SM (chloroform/MeOH (2:1; *v*/*v*)). After hydrolysis of all collected lipid fractions and derivatization, the obtained FAMEs were determined using GC-MS [80].

Lytle et al. used chloroform/MeOH (2:1) extraction followed by TLC to fractionate total lipids in a study of steatohepatitis in obese mice; after that, the lipids were saponified, and the FAs were extracted with *n*-hexane and identified using RP-HPLC with a UV detector. Additionally, for some other types of lipids from saponified FAs, FAMEs were prepared, which were extracted in *n*-hexane for subsequent quantification by GC with an FID [104].

Du et al. studied the hepatic expression of sirtuin 5 (SIRT5) in obese mice and determined the liver TAG content. For this, Soxhlet extraction of liver lipids using diethyl ether as a solvent was performed; in addition, to study changes in the size and number of lipid droplets (LD) in hepatocytes, neutral lipids were stained with probe-lipid TOX Green, and haematoxylin and eosin staining was performed [178].

Yetukuri et al. performed lipid profiling in the liver in obese mice with hepatic steatosis. For this, 20–30 mg tissue sample was subjected to chloroform/MeOH extraction with an internal standards mixture (diacylglycerophosphocholine (GPCho) (17:0/17:0), diacylglyceroethanolamine (17:0/17:0), GPCho (17:0/0:0), Cer (d18:1/17:0), and TG (17:0/17:0/17:0)) after which to the separated lower phase, 10 μL of a labelled standard mixture was added, and the sample was analysed by UPLC-MS [179].

Wang et al. when analysing obese mouse liver by shotgun lipidomics, developed an approach that solves the problem of analysing lipids such as LPL that remain in the aqueous phase and are usually discarded after the classical extraction procedure of Bligh and Dyer (because they cannot be directly used for MS analysis due to a high salt content). The method is based on the purification of the aqueous phase solution remaining after LLE extraction. For this purpose, the aqueous phase was loaded onto a Hybrid SPE cartridge that had been twice washed with MeOH; next, lysophosphatidylinositol (LPI), lysophosphatidylserine (LPS) and lysophosphatidylglycerol (LPG) species were eluted with a 10% solution of MeOH in ammonia, and the LPAs were eluted with a solution of 20% ammonia in MeOH. Furthermore, SPE eluents were dried and reduced in MeOH prior to analysis by MS [70].

One of the most comprehensive approaches to liver lipidomics was shown by Garcia-Yaramillo et al. when studying Western diet-induced nonalcoholic steatohepatitis in mice. They used GC analysis of free and saponified FAs converted to FAMES to quantify SFA, MUFA, and ω3, and ω6 PUFA. For quantification of DAG, TAG, PC, phosphatidylserines (PS), phosphatidylinositols (PI), phosphatidylglycerols, PE, LPL and SM, they used an untargeted LC/MS approach. For this, extraction was performed with methylene chloride/IPA/MeOH (25:10:65; $v/v/v$; -20 °C). In addition, they quantified ω3 and ω6 PUFA and PUFA-derived oxylipins using targeted LC/MS. For this, MeOH extraction with additional SPE on Strata-X columns was used. Both targeted and nontargeted analyses were performed on the same UHPLC system (a Shimadzu Nexera system coupled to a triple time-of-flight (TOF) 5600 mass spectrometer) and the same column (Waters Acquity (UPLC); CSH C18) [98].

In their studies of obesity, Preuss et al. drew attention to the composition of LDs in various tissues, including the liver. To separate LDs from the cytosolic and membrane fractions in homogenates, the authors used 1 h of centrifugation (100,000× g; 4 °C), after which LDs formed a clear top layer. Then, lipids from LDs were extracted according to the Folch method. The authors focused on the use of a Centri-Tube slicer (Beckman Coulter, Brea, CA, USA) that increases the purity of the collected LD fraction. Subsequent SPE (Sep Pak Diol Cartridges; Waters, MA, USA) followed by LC-MS/MS analysis allows the determination of diacylglycerols and Cer in all cell fractions [180]. The authors noted that compared with other tissues, the liver had a high level of lipids; therefore, 10 mL of buffer was used to homogenize 20 mg of liver tissue, whereas in the case of muscle and cardiac tissue, 50 mg was homogenized in 500 µL of buffer [180].

3.4. Brain

Due to the difficulties of collecting material for research in humans, the study of lipid metabolism in the brain tissue most often occurs with material obtained from animals with obesity induced by a high-energy diet. In addition to difficulties with biopsy, in cases of animals with large brains as in humans, additional difficulties arise due to the heterogeneity of cell populations in grey and white matter, which can lead to different ratios of cell populations in samples [181].

Yang et al. in a study of circulating Cer in HFD-fed mice, used the methodology of Bielawski et al. [182]. After homogenization of frozen tissues with a buffer (0.25 M sucrose, 25 mM KCl, 50 mM Tris, and 0.5 mM EDTA, pH 7.4) in a 1:10 (w/v) ratio, homogenate was filtered through layers of gauze. After that, tissue homogenates were spiked with IS solution and extracted twice with IPA/water/ethyl acetate (30:10:60; $v/v/v$) mixture with subsequent vortexing, sonication and centrifugation. The combined upper layers were dried and reconstituted in the mobile phase for LC-MS analysis [183]. Later, Gao et al. used this approach when exploring ceramide metabolism in the hypothalamus of mice in a model of leptin hypothalamic control of feeding [184].

Rutkowsky et al. performed metabolic analysis of complex lipids in the brains of mice fed a western diet. For this, they used the Matyash protocol [123] based on methyl tert-butyl ether (MTBE) extraction. [185]. Additionally, LLE protocol with MeOH/ethyl acetate in this work was used, for analysis of non-esterified oxylipins and endocannabinoids in the brain was carried out [185].

Rawish et al. performed hypothalamic lipid analyses in mice fed a high-fat diet. For this, they combined several approaches, including using fluorescence microscopy to quantify neutral lipids in the LD hypothalamus (staining LD with lipophilic dye (LD540)) and studying the metabolism of FAs in slices of the hypothalamus using the tracer alkyne oleate; lipidomics analysis of the hypothalamus (analysis of neutral glycerolipids, phosphoglycerol, sphingolipids and acyl-carnitine lipids) was also performed by LC-MS [186].

Kirkham et al. studied the effects of diet on endocannabinoid levels in the rat forebrain and hypothalamus using 5 v of chloroform/MeOH/50 mM Tris HCl (2:1:1) for endocannabinoid extraction from tissue homogenates. After consequent double extraction with 1 v of chloroform, organic phases were pooled, dried and resuspended in chloroform/MeOH (99:1; *v/v*) for GC-MS analysis, but before this, the obtained solutions were purified by open bed chromatography on silica and further fractionated by normal-phase high-pressure liquid chromatography (NP-HPLC) on a silica column (Spherisorb S5W; Phase Sep, Queensferry, Clwyd, UK) using a 40 min linear gradient [161].

3.5. Skeletal Muscle

Gudbrandsen et al. analysed lipid metabolism in rats after bariatric surgery and extracted FAs from skeletal muscle. They used the Bligh and Dyer method with the addition of heneicosanoic acid as an internal standard. Then, the extracts were methylated in anhydrous MeOH containing 2.5 M HCl (100 °C; 2 h) and extracted twice with isooctane, and methyl esters were quantified using GC with an FID [187].

The most commonly studied classes of muscle lipids in obesity are TAG and CER since the accumulation of these lipids in muscle tissue is associated with the development of IR [188]. The aforementioned Preuss et al. also examined muscles when studying the composition of LDs and determined DAG and Cer in cell fractions [180].

Van Hees et al. when checking the "lipid overflow" hypothesis, did not limit their analysis to TAG content in muscles during the study of IR because they also divided the total amount of lipids obtained after extraction by chloroform-MeOH into FFA, DAG, TAG and PL by TLC before further GC-MS analysis [129].

Laurentius et al. developed a new method to identify and classify FAMEs by GC-MS in HFD-fed rats. At the initial stage, CE, TAG and GPL fractions were extracted from muscle homogenates using tert-butyl methyl ether (90%, tert-BME) and MeOH, and the FFA fraction was extracted using a chloroform/MeOH (2:1) mixture. In the further separation of lipid classes, they used SPE during which they sequentially eluted the CE fraction (elution 1% methyl acetate/*n*-hexane; *v/v*) and TAG fraction (elution of 2.5% methyl acetate/*n*-hexane; *v/v*), and after the column was washed with acetone, the GPL fraction was eluted with MeOH. FFA were extracted with 80% *n*-hexane/diethyl ether (*v/v*) on separate SPE columns, followed by incubation with 1% sulfuric acid in MeOH, addition of 5% sodium chloride, and three washes of the *n*-hexane layer with water [97].

One of the latest approaches developed by Eum et al. allows the identification of hundreds of individual lipid species in the muscle tissue of mice with HFD [78]. A two-stage extraction method was performed using MTBE for the first extraction and subsequent secondary extraction of lipids from the lower aqueous layer with MeOH. After mixing, the organic layers were dried and reduced with a mixture of chloroform/MeOH (1:9; *v/v*) for subsequent nanoflow ultrahigh-performance liquid chromatography with tandem mass spectrometry (nUHPLC-ESI-MS/MS) analysis [78].

3.6. Heart

The study of lipid metabolism in heart tissue is of interest primarily due to the strong influences of obesity and a high-fat diet, particularly the influences on the phospholipid composition of mitochondrial membranes of cardiomyocytes [189]. Due to the unique lipid composition of mitochondrial membranes, they are very sensitive to high levels of FAs and their oxidation products in the blood [189].

Harmancey et al. when studying changes in the composition of cardiac acyl-CoA in obese rats, used the Bligh and Dyer method to extract cardiac lipids from heart tissue; then, extraction was repeated three times to complete lipid recovery for further quantification of Cer and diacylglycerols by HPLC-UV [190]. When the cardiac ceramide content in rats fed a high-fat diet was studied, the Bligh and Dyer method was also used to extract total cardiac lipids [191]. Later, they used the extraction method described by Merrill et al. [165], which is based on extraction with a mixture of MeOH and chloroform, followed by sonication (48 °C; overnight), which is necessary for the extraction of sphingolipids, followed by incubation (37 °C; 1 h) to remove interfering glycerolipids (in particular PC). After another extraction with chloroform, ceramide species were detected by LC-MS [165].

Table 1. Sample preparation methods for lipidomics approaches used in the studies of obesity.

Lipid Class(es)	Matrix	Sample Preparation Method			Analysis Method	References
		Pre-Preparation	Extraction Method	Derivatization Step		
PL	Plasma	-	modified Folch method	sodium methoxide—FAME	GC	Blewett et al. [121]
FFA, TAG	Plasma	-	Bligh and Dyer's method	14% BF_3 - MeOH - FAME	GC-MS	Choromańska et al. [74]
LPAs	Plasma	-	hydrochloric acid + MeOH/chloroform (2:1)	-	ESI-MS–MS	Yoon et al. [122]
SM	Plasma	MeOH	modified Matyash method	-	LC-MS/MS	Im et al. [50]
lipidomic profile (328 lipid species from 24 lipid classes: dhCer, Cer. MHC, DHC, THC, GM3, SM, PC, PC(0), PC(P), LPC, PE, PE(0), PE(P), LPE, PI, LPI, PS, PG, CE, COH, DG, TG)	Plasma	MeOH	chloroform/MeOH (2:1)	-	LC ESI-MS/MS	Wang et al. [124]
untargeted metabolomics analysis /lipidomic profile? (515 metabolites)	Serum	-	ACN: isopropanol: water (3:3:2)	MSTFA + MTBSTFA	2D GC-ToF-MS	Misra et al. [125]
lipidomic profile (143 lipid species from lipid classes: FA, FFA, PC, PE, PI, PS, PG, LPC, LPA, SM)	Plasma	-	MeOH/n-hexane (4:1)	acetyl chloride + 6% K_2CO_3	GC- FID/MS	Wang et al. [51]
sterols	Serum	MeOH	solid-phase extraction (hybrid solid-phase extraction-precipitation (H-PPT) cartridge)	MSTFA/ammonium iodide (NH 4 I)/dithioerythritol (DTE) (500:4:2)	GC-MS	Cho et al. [82]
total FAs + circulating PL, TG	Serum	-	chloroform/MeOH (2:1)	methylation (100 °C; 1.5 h)	GC- FID	Perreault et al. [102]
FFA, TAG, PL	Serum	-	Folch method	10% BF_3 - MeOH	GC-MS	Śledziński et al. [144]
SFA, MUFAs, PUFAs	Plasma	-	MeOH	acetyl chloride + 6% K_2CO_3	GC-MS	Kang et al. [148]

Table 1. *Cont.*

Lipid Class(es)	Matrix	Sample Preparation Method			Analysis Method	References
		Pre-Preparation	Extraction Method	Derivatization Step		
MCFAs, NEFAs	Serum	MeOH	-	10M KOH in MeOH + 24 N H2SO4	GC-MS	Wijayatunga et al. [7]
PUFAs	Plasma	ACN/37% hydrochloric acid (4:1)	n-hexane	-	LC-MS/MS	Aslan et al. [150]
MUFAs, PUFAs, OCFAs	Serum	ACN	chloroform/MeOH (2:1)	-	UHPLC-MS	Ma et al. [152]
PUFAs	Plasma	MeOH	-	acetyl chloride + 6% K_2CO_3	LC-MS	Itariu et al. [86]
NEFAs	Plasma	-	Dole extraction	-	LC-MS	Nemati et al. [153]
MUFAs, PUFAs, SFA	Serum	-	chloroform/MeOH (2:1)	HCl in MeOH	GLC-FID	Lin et al. [154]
FAs, bile acids (BA), steroids, LPL, glycerolipids, cholesterol esters, SPL, PL	Serum	For FAs, bile acids (BA), steroids and LPL - MeOH	For glycerolipids, cholesterol esters, sphingolipids and phospholipids - NaCl + chloroform/MeOH (2:1)	-	UPLC-MS	Ramos-Molina et al. [155]
	Serum	-	chloroform/MeOH (2:1)	-	UHPLC-MS	Ramos-Molina et al. [155]
FAHFAs	Serum	MeOH	solid-phase extraction (hysphere C8 cartridges)	-	LC-MS/MS	López-Bascón et al. [87]
SCFAs	Plasma	MeOH	dichloromethane	0.1 M O-benzylhydroxylamine (O-BHA) in MeOH and 0.25 M N-(3-dimethylaminopropyl)-N′-ethylcarbodiimide hydrochloride (EDC) in MeOH	LC-MS/MS	Zeng and Cao [58]

Table 1. *Cont.*

Lipid Class(es)	Matrix	Sample Preparation Method			Analysis Method	References
		Pre-Preparation	Extraction Method	Derivatization Step		
MAG, FAE, oxFAE, oxMAG, FA, oxFA, DAG, TAG, PE, PI, NAPE, LNAPE, PC	Plasma	-	Chloroform	-	LC/MS	Astarita et al. [99]
endocannabinoids	Plasma	-	Chloroform	-	UPLC-MS/MS	Argueta et al. [100]
endocannabinoids	Plasma	-	Chloroform	-	UPLC-MS/MS	Perez et al. [101]
oxylipins	Serum	-	solid-phase extraction (C18 cartridges)	-	online SPE–LC–MS/MS	Ferreiro-Vera et al. [57]
nonesterified PUFAs and oxylipins	Plasma	MeOH + formic acid	solid-phase extraction (Strata-X)	-	UHPLC-MS/MS	Pickens et al. [163]
oxylipins, endocannabinoid, Cer	Serum	IPA with 10 mM ammonium formate + 1% formic acid	-	-	UPLC-MS/MS	Hernandez-Carretero et al. [52]
endocannabinoids	Serum	acetones + Tris buffer (50 mM, pH 8.0)	chloroform/MeOH (2:1)	-	LC-MS/MS	Azar et al. [164]
unesterified oxylipins, endocannabinoids	Plasma	MeOH/ACN (1:1)	solid-phase extraction (BEH C18 colum)	-	UPLC-MS/MS	Fan et al. [54]
Cer	Plasma	-	Bligh and Dyer method	-	UPLC-MS/MS	León-Aguilar et al. [76]
Cer, SM	Serum	-	chloroform/MeOH (2:1)	-	(UFLC)-MS/MS	Özer et al. were [169]
FA	Adipose tissue	-	MeOH/chloroform (2:1)	10 % BF3 - MeOH	GC-MS	Roberts et al. [90]
acylcarnitines	Adipose tissue	-	MeOH/chloroform (2:1)	-	LC-MS	Roberts et al. [90]
SFA, MUFAs, TFA, PUFAs	Adipose tissue	-	*n*-hexane	sodium methoxide	GC-FID/MS analysis	Kunešová et al. [172]

Table 1. *Cont.*

Lipid Class(es)	Matrix	Sample Preparation Method			Analysis Method	References
		Pre-Preparation	Extraction Method	Derivatization Step		
FAHFAs	Adipose tissue	-	MTBE/MeOH/water	-	UPLC-MS/MS	Hu et al. [79]
oxylipins	Adipose tissue	MeOH	RapidTrace Biotage	-	LC-MS-MS	Okada et al. [85]
TAG	Adipose tissue	-	Bligh and Dyer method	-	LC-MS	Al-Sulaiti et al. [71]
more than 300 lipid species from lipid classes: CL, Cer, ST. HexCer, LPA, LPC, LPE, LPG, LPI, LPS, SM, TAG, CE, DAG, PA, PC, PE, PG, PI, PS	Adipose tissue	-	two-step chloroform/MeOH extraction	-	MS	Grzybek et al. [174]
TAG, MAG, DAG, LysoPC, PC, LysoPE, PE, Cer, SM, PI, PS, FA	Adipose tissue	-	modified Folch method	-	LC-MS	Tomášová et al. [103]
oxysterols, Cer, endocannabinoids	Adipose tissue	-	dichloromethane/MeOH/water (8:4:2) + solid-phase extraction (C18 colum)	-	LC-MS	Mutemberezi et al. [83]
OxPL	Adipose tissue	-	chloroform/MeOH (3:1) + BHT	-	LC-MS	Serbulea et al. [75]
MAG, DAG, TAG, NL, Cer, FFA, GSPL, SM	Liver	-	Folch extraction	10% BF3 - MeOH	GC-MS	Pakiet et al. [80]
STA, MUFA, PUFA	HepG2 cells	-	chloroform/MeOH (2:1) + BHT	hexane + 0.05% BHT	GC- FID	Lytle et al. [104]
Cer, SM, GPCho, GPEtn, GPSer, GPA, GPGro, DG, TG	Liver	-	chloroform/MeOH (2:1)	-	UPLC-MS	Yetukuri et al. [179]
LPL (LPS, LPA, LPI, LPG, LPC, LPE)	Liver	4% formic acids in MeOH	modified method of Bligh and Dyer + solid-phase extraction (HybridSPE cartridge)	-	ESI-MS	Wang et al. [70]

Table 1. Cont.

Lipid Class(es)	Matrix	Sample Preparation Method			Analysis Method	References
		Pre-Preparation	Extraction Method	Derivatization Step		
SFA, MUFA, PUFA	Liver	-	chloroform:MeOH (2:1) plus 1 mM BHT	1% H2SO4 in MeOH	GC-FID	Garcia-Yaramillo et al. [98]
PUFA, PUFA-derived oxylipins	Liver	MeOH	solid-phase extraction (Strata-X)	-	targeted UPLC-TOF-MS/MS	Garcia-Yaramillo et al. [98]
DAG, TAG, PC, PS, PI, PG, PE, LPL, SM	Liver	-	methylene chloride/IPA/MeOH (25:10:65)	-	untargeted UPLC-TOF-MS/MS	Garcia-Yaramillo et al. [98]
DAG, Cer	Liver	-	Folch method + solid-phase extraction (Sep Pak Diol Cartridges)	-	LC-MS/MS	Preuss et al. [180]
	Brain	-	IPA/water/ethyl acetate (30:10:60)	-	LC-MS	Yang et al. [184]
non-esterified oxylipins, endocannabinoids, PUFAs	Brain	MeOH	MTBE	-	UHPLC-QTOF-MS	Rutkowsky et al. [185]
Cer, DG, ClcCer, LPC, PC, PE, FA, PI, SM	Brain	-	MeOH/ethyl acetate	-	CSH-ESI QTOF MS/MS	Rutkowsky et al. [185]
endocannabinoid	Brain	-	chloroform/MeOH/50 mM Tris HCl (2:1:1)	MSTFA + 1% trimethylchlorosylane	GC-MS	Kirkham et al. [161]
SFA, MUFA, PUFA	Skeletal muscle	-	Bligh and Dyer method	anhydrous MeOH containing 2.5 M HCl (100 °C; 2 h)	GC-FID	Gudbrandsen et al. [187]
FFA, DAG, TAG, PL	Skeletal muscle	-	chloroform/MeOH (2:1)	14% BF3 - MeOH	GC-MS	Van Hees et al. [129]
CE, TAG, GPL	Skeletal muscle	-	for CE, TAG and GPL fractions – tert-butyl methyl ether (90%, tert-BME) and MeOH; for FFA fraction chloroform/MeOH (2:1) + solid phase extraction	2 M sodium methoxide solution	GC-MS	Laurentius et al. [97]

Table 1. *Cont.*

Lipid Class(es)	Matrix	Sample Preparation Method			Analysis Method	References
		Pre-Preparation	Extraction Method	Derivatization Step		
LPC, LPE, PI, PG, Cer, PC, PE, PS, TG, HexCer, SM	Skeletal muscle	-	two-stage extraction method using MTBE/MeOH	-	nUHPLC-ESI-MS/MS	Eum et al. [78]
DAG, Cer, acyl-CoA	Heart	-	Bligh and Dyer method	-	HPLC-UV	Harmancey et al. [190]
sphingolipids	Heart	-	MeOH/chloroform	-	LC-MS/MS	Merrill et al. [165]
SFA, MUFA, PUFA	Heart	-	chloroform/MeOH (2:1) + solid-phase extraction (Strata)	10% BF3—MeOH	GC-MS	Pakiet et al. [192]
SM, PC, PE, PG, PI, PS	Urine	-	ACQUITY UPLC HSS-T3 C_{18} column	-	UPLC-QTOF-MS/MS	Feng et al. [193]
Cholesterol, 7-ketocholesterol	Saliva	-	for cholesterol and 7-ketocholesterol chloroform/MeOH (2:1); for 25-hydroxyvitamins D2 and D3 MeOH/IPA	-	HPLC-DAD	Araujo and Santos [194]
FFA, PI, PC, LPC, PS, PE, TG	Follicular fluid and serum	-	IPA/acetonitrile/water (3:3:2) or MTBE	MSTFA + 1% TMCS	GC-MS	Ruebel et al. [195]
FFA, PI, PC, LPC, PE, PS	Follicular fluid and serum	-	IPA/acetonitrile/water (3:3:2) or MTBE	-	untargeted CSH-ESI QTOF MS/MS	Ruebel et al. [195]

Table 1. *Cont.*

Lipid Class(es)	Matrix	Sample Preparation Method			Analysis Method	References
		Pre-Preparation	Extraction Method	Derivatization Step		
SCFAs	Faecal samples	-	-	CTC Combipal 3 autosampler in HS/SPME mode equipped with a gray fibe	GC-MS	Cuesta-Zuluaga et al. [196]

2D GC-ToF-MS: two-dimensional gas chromatography time-of-flight mass spectrometry; ESI-MS-MS: turbo electrospray ionization tandem mass spectrometry; GC- FID/MS: gas chromatography-flame ionization detector-mass spectrometer; GC-FID: gas chromatography-flame ionization detection; GLC-FID: gas liquid chromatography-flame ionization detector; HPLC-UV: high-performance liquid chromatography with UV-detection; LC-HRMS/MS: liquid-chromatography high-resolution tandem mass spectrometry; LC-MS-MS liquid chromatography with tandem mass spectrometry; nUHPLC-ESI-MS/MS: nanoflow ultrahigh performance liquid chromatography with tandem mass spectrometry; UHPLC/Q-TOF-MS: ultra-high performance liquid chromatography-quadrupole time-of-flight mass spectrometry; UPLC-QTOF-MS: ultra-performance liquid chromatography coupled to a Triple time-of-flight mass spectrometer; PL: phospholipids; FFA: free fatty acids; TAG: triacylglycerol; LPA: lysophosphatidic acid; SM: sphingomyelin; dhCer: dihydroceramide; Cer: ceramides; MHC: monohexosylceramide; DHC: dihexosylceramide; THC: trihexosylceramide; GM3:ganglioside; PC: phosphatidylcholine; PC(0): alkylphosphatidylcholine; PC(P): phosphatidylcholine plasmalogen; LPC: lysophosphatidylcholine; PE: phosphatidylethanolamine, PE(0): akylphosphatidylethanolamine, PE(P): phosphatidylethanolamine plasmalogen; LPE: lysophosphatidylethanolamine; PI: phosphatidylinositol; LPI: lysophosphatidylinositol, PS: phosphatidylserine; PG: phosphatidylglycerol; CE: chofesterol ester; COH: free cholesterol; DG: diacylglycerol; TG: triacylglycerol; FA: fatty acid; SFA: saturated fatty acid; MUFAs: monounsaturated fatty acids; PUFAs: polyunsaturated fatty acids; NEFAs: non-esterified fatty acids; OCFAs: odd-carbon fatty acids; SPL: sphingolipids; FAHFAs: fatty acid esters of hydroxy fatty acids; SCFAs: short-chain fatty acids; MAG: monoacylglycerol; FAE: fatty-acid ethanolamides; oxFA: oxygenated FAE; oxMAG: oxygenated MAG; oxFA: oxygenated fatty acids; NAPE: N-acyl-phosphatidylethanolamine; LNAPE: lyso-NAPE; TFA: trans fatty acids; CL: cardiolipin; ST: cholesterol; HexCer: hexosylceramide; LPG: lysophosphatidylglycerol; LPI: lysophosphatidylinositol; LPS: lysophosphatidylserine; PA: Phosphate; LysoPC: Lysophosphatidylcholines; LysoPE: Lysophosphatidylethanolamines; OxPL: oxidized phospholipids; NL: neutral lipids; GSPL: glycosphingolipids; GPcho: diacylglycerophosphocholine; GPEtn: glycerophosphoethanolamines; GPSer: glycerophosphoserine; GPGro: Glycerophosphoglycerol; GlcCer: Glucosylceramides; Acyl:CoA:dihydroxyacetone phosphate acyltransferase; CSH-ESI QTOF MS/MS: charged-surface hybrid column-electrospray ionization quadrupole time-of-flight tandem mass spectrometry.

The abovementioned protocol of Pakiet et al. with the use of LLE and two SPE methods for extracting lipids from the brains of mice was also successfully used to extract lipids from the hearts of mice fed a high-fat diet [192].

3.7. Other Biological Materials

3.7.1. Urine

Feng et al. used an approach based on UPLC-MS to study the effects of HFD on glycolipid metabolism in young obese men. Metabolomic profiling was performed on urine samples that were centrifuged (14,000× g; 4 °C; 10 min) and then analysed using UPLC-QTOF-MS/MS [193]. However, to obtain a more complete urinary lipid profile, extraction with organic solvents was necessary [197].

3.7.2. Saliva

Araujo and Santos used high performance liquid chromatography–diode array detector (HPLC–DAD) Shimadzu analytical system to show the possibility of using saliva as a biomarker. The authors used chloroform/MeOH extraction for measurements of cholesterol and 7-ketocholesterol, and MeOH/IPA extraction was used for measurements of 25-hydroxyvitamins D2 and D3. The authors claim that saliva, with its simplicity and non-invasiveness of collection, can be an important source of markers of nutritional status [194]. This may indeed be true since the study of changes in the metabolism of vitamin D associated with obesity in adipose tissue, in addition to requiring invasive intervention, requires complex MTBE extraction followed by SPE for further liquid-chromatography high-resolution tandem mass spectrometry (LC-HRMS/MS) determination of 25-hydroxyvitamin D and 1,25-dihydroxyvitamin D in adipose tissue [198].

3.7.3. Follicular Fluid

Ruebel et al. used two untargeted metabolomics approaches for primary metabolism analyses and metabolomics assessment of complex lipids in the study of follicular fluid (FF) in obese women. The first approach included GC-MS analysis of FF samples after extraction and derivatization by silylation/methyloximation [195]. During the second approach after MTBE extraction, FF samples were analysed using CSH-ESI QTOF MS/MS; negative ion MS was used for FFA and PI analysis, while PC, LPC, PE and PS were analysed in positive ion mode [195].

3.7.4. Faecal samples

Regarding nonstandard approaches, it is worth mentioning that de la Cuesta-Zuluaga et al. developed a method for determining SCFAs in faecal samples from obese subjects. They determined volatiles using a CTC Combipal 3 autosampler in HS/SPME mode equipped with a grey fibre (Carboxen/DVB/PDMS–ref. SU57329U; Supelco) in water extracts heated to 80 °C, which was followed by desorption of the fibre (at 250 °C) and GC-MS analysis [196].

4. Conclusions

Using lipidomics in the study of obesity is very important to understanding the regulatory and diagnostic value of changes in the lipidome associated with obesity. Depending on the studied tissue, the role of changes in specific lipids may have different levels of importance (CER and TAG in muscles, fatty acid composition of TG in adipose tissues, etc.). Despite the fact that new modifications of extraction methods such as those of Bligh and Dyer and new methods (e.g., MTBE) are constantly emerging, none of the extraction methods allows for quantitative extraction of all types of lipids due to the huge differences in their chemical characteristics. This may explain the great popularity of the targeted approach and the SPE method, which can reduce lipid degradation during extraction and can also be automated. At the same time, LLE using phase separation is a more suitable method for nontarget lipidomics, where fractionation is not required. Thus, the study of changes in lipid metabolism in obesity contributed to the development of new methods for sample preparation, extraction and quantification, not only to improve the accuracy and sensitivity of existing methods but also to develop methods for detecting new specific markers among a huge variety of lipid compounds.

Author Contributions: Conceptualization, I.L. and A.M.; resources, I.L.; writing—original draft preparation, I.L.; writing—review and editing, A.M., L.K., and T.S.; visualization, P.M. and A.M.; supervision, A.M. and T.S.; funding acquisition, T.S., L.K. All authors have read and agreed to the published version of the manuscript.

Funding: This research was funded by the Medical University of Gdansk, grant number ST40 and S89; and by the National Science Centre of Poland, grant number NCN 2016/21/D/NZ5/00219.

Conflicts of Interest: The authors declare no conflict of interest.

References

1. Champion, J.D.; Collins, J.L. Retrospective Chart Review for Obesity and Associated Interventions Among Rural Mexican-American Adolescents Accessing Healthcare Services. *J. Am. Assoc. Nurse Pract.* **2013**, *25*, 604–610. [CrossRef] [PubMed]
2. Dogan, U.; Ellidag, H.Y.; Aslaner, A.; Cakir, T.; Oruc, M.T.; Koc, U.; Mayir, B.; Gomceli, I.; Bulbuller, N.; Yılmaz, N. The Impact of Laparoscopic Sleeve Gastrectomy on Plasma Obestatin and Ghrelin Levels. *Eur. Rev. Med. Pharmacol. Sci.* **2016**, *20*, 2113–2122. [PubMed]
3. Blüher, S.; Raschpichler, M.; Hirsch, W.; Till, H. A Case Report and Review of the Literature of Laparoscopic Sleeve Gastrectomy in Morbidly Obese Adolescents: Beyond Metabolic Surgery and Visceral Fat Reduction. *Metabolism* **2013**, *62*, 761–767. [CrossRef] [PubMed]
4. Carswell, K.A.; Belgaumkar, A.P.; Amiel, S.A.; Patel, A.G. A Systematic Review and Meta-Analysis of the Effect of Gastric Bypass Surgery on Plasma Lipid Levels. *Obes. Surg.* **2016**, *26*, 843–855. [CrossRef]
5. Franco, J.V.A.; Ruiz, P.A.; Palermo, M.; Gagner, M. A Review of Studies Comparing Three Laparoscopic Procedures in Bariatric Surgery: Sleeve Gastrectomy, Roux-en-Y Gastric Bypass and Adjustable Gastric Banding. *Obes. Surg.* **2011**, *21*, 1458–1468. [CrossRef]
6. Howard, M.L.; Steuber, T.D.; Nisly, S.A. Glycemic Management in the Bariatric Surgery Population: A Review of the Literature. *Pharmacotherapy* **2018**, *38*, 663–673. [CrossRef]
7. Wijayatunga, N.N.; Sams, V.G.; Dawson, J.A.; Mancini, M.L.; Mancini, G.J.; Moustaid-Moussa, N. Roux-en-Y Gastric Bypass Surgery Alters Serum Metabolites and Fatty Acids in Patients with Morbid Obesity. *Diabetes Metab. Res. Rev.* **2018**, *34*, e3045. [CrossRef]
8. Hansen, D.; Marinus, N.; Remans, M.; Courtois, I.; Cools, F.; Calsius, J.; Massa, G.; Takken, T. Exercise Tolerance in Obese vs. Lean Adolescents: A Systematic Review and Meta-Analysis. *Obes. Rev.* **2014**, *15*, 894–904. [CrossRef]
9. Supariwala, A.; Makani, H.; Kahan, J.; Pierce, M.; Bajwa, F.; Dukkipati, S.S.; Teixeira, J.; Chaudhry, F.A. Feasibility and Prognostic Value of Stress Echocardiography in Obese, Morbidly Obese, and Super Obese Patients Referred for Bariatric Surgery. *Echocardiography* **2013**, *31*, 879–885. [CrossRef]
10. Sommer, A.; Twig, G. The Impact of Childhood and Adolescent Obesity on Cardiovascular Risk in Adulthood: A Systematic Review. *Curr. Diab. Rep.* **2018**, *18*, 91. [CrossRef]
11. Kiliçaslan, B.; Tigen, M.K.; Tekin, A.S.; Çiftçi, H. Cardiac Changes with Subclinical Hypothyroidism in Obese Women. *Turk Kardiyol Dern Ars.* **2013**, *41*, 471–477. [PubMed]
12. Hajian-Tilaki, K.; Heidari, B. Variations in the Pattern and Distribution of Non-Obese Components of Metabolic Syndrome across Different Obesity Phenotypes among Iranian Adults' Population. *Diabetes Metab. Syndr. Clin. Res. Rev.* **2019**, *13*, 2419–2424. [CrossRef] [PubMed]
13. Rao, R.S.; Yanagisawa, R.; Kini, S. Insulin Resistance and Bariatric Surgery. *Obes. Rev.* **2012**, *13*, 316–328. [CrossRef] [PubMed]
14. Sejooti, S.S.; Naher, S.; Hoque, M.M.; Zaman, M.S.; Aminur Rashid, H.M. Frequency of Insulin Resistance in Nondiabetic Adult Bangladeshi Individuals of Different Obesity Phenotypes. *Diabetes Metab. Syndr. Clin. Res. Rev.* **2019**, *13*, 62–67. [CrossRef] [PubMed]
15. Di Vincenzo, A.; Beghetto, M.; Vettor, R.; Rossato, M.; Bond, D.; Pagano, C. SAT-108 Effects of Bariatric and Non-Bariatric Weight Loss on Migraine Headache in Obesity. A Systematic Review and Meta-Analysis. *J. Endocr. Soc.* **2019**, *3*. [CrossRef]

16. Großschädl, F.; Freidl, W.; Rásky, É.; Burkert, N.; Muckenhuber, J.; Stronegger, W.J. A 35-Year Trend Analysis for Back Pain in Austria: The Role of Obesity. *PLoS ONE* **2014**, *9*, e107436. [CrossRef]
17. Chao, H.-L. Body Image Change in Obese and Overweight Persons Enrolled in Weight Loss Intervention Programs: A Systematic Review and Meta-Analysis. *PLoS ONE* **2015**, *10*, e0124036. [CrossRef]
18. Li, W.; Rukavina, P. A Review on Coping Mechanisms against Obesity Bias in Physical Activity/Education Settings. *Obes. Rev.* **2009**, *10*, 87–95. [CrossRef]
19. Budd, G.M.; Mariotti, M.; Graff, D.; Falkenstein, K. Health Care Professionals' Attitudes about Obesity: An Integrative Review. *Appl. Nurs. Res.* **2011**, *24*, 127–137. [CrossRef]
20. Devoto, F.; Zapparoli, L.; Bonandrini, R.; Berlingeri, M.; Ferrulli, A.; Luzi, L.; Banfi, G.; Paulesu, E. Hungry Brains: A Meta-Analytical Review of Brain Activation Imaging Studies on Food Perception and Appetite in Obese Individuals. *Neurosci. Biobehav. Rev.* **2018**, *94*, 271–285. [CrossRef]
21. Amiri, S.; Behnezhad, S. Obesity and Anxiety Symptoms: A Systematic Review and Meta-Analysis. *Neuropsychiatrie* **2019**, *33*, 72–89. [CrossRef] [PubMed]
22. Kim, M.K.; Kim, W.; Kwon, H.-S.; Baek, K.-H.; Kim, E.K.; Song, K.-H. Effects of Bariatric Surgery on Metabolic and Nutritional Parameters in Severely Obese Korean Patients with Type 2 Diabetes: A Prospective 2-Year Follow Up. *J. Diabetes Investig.* **2014**, *5*, 221–227. [CrossRef] [PubMed]
23. Rashad, N.M.; Sayed, S.E.; Sherif, M.H.; Sitohy, M.Z. Effect of a 24-Week Weight Management Program on Serum Leptin Level in Correlation to Anthropometric Measures in Obese Female: A Randomized Controlled Clinical Trial. *Diabetes Metab. Syndr. Clin. Res. Rev.* **2019**, *13*, 2230–2235. [CrossRef] [PubMed]
24. Morais, L.C.; Rocha, A.P.R.; Turi-Lynch, B.C.; Ferro, I.S.; Koyama, K.A.K.; Araújo, M.Y.C.; Codogno, J.S. Health Indicators and Costs among Outpatients According to Physical Activity Level and Obesity. *Diabetes Metab. Syndr. Clin. Res. Rev.* **2019**, *13*, 1375–1379. [CrossRef] [PubMed]
25. Chen, C.M. Overview of Obesity in Mainland China. *Obes. Rev.* **2008**, *9*, 14–21. [CrossRef]
26. Nishide, R.; Ando, M.; Funabashi, H.; Yoda, Y.; Nakano, M.; Shima, M. Association of Serum Hs-CRP and Lipids with Obesity in School Children in a 12-Month Follow-Up Study in Japan. *Environ. Health Prev. Med.* **2015**, *20*, 116–122. [CrossRef]
27. Fernandes, L.A.; Braz, L.G.; Koga, F.A.; Kakuda, C.M.; Mōdolo, N.S.P.; De Carvalho, L.R.; Vianna, P.T.G.; Braz, J.R.C. Comparison of Peri-Operative Core Temperature in Obese and Non-Obese Patients. *Anaesthesia* **2012**, *67*, 1364–1369. [CrossRef]
28. Pérez-Pérez, R.; García-Santos, E.; Ortega-Delgado, F.J.; López, J.A.; Camafeita, E.; Ricart, W.; Fernández-Real, J.-M.; Peral, B. Attenuated Metabolism is a Hallmark of Obesity as Revealed by Comparative Proteomic Analysis of Human Omental Adipose Tissue. *J. Proteomics* **2012**, *75*, 783–795. [CrossRef]
29. Dolinková, M.; Dostálová, I.; Lacinová, Z.; Michalský, D.; Haluzíková, D.; Mráz, M.; Kasalický, M.; Haluzík, M. The Endocrine Profile of Subcutaneous and Visceral Adipose Tissue of Obese Patients. *Mol. Cell. Endocrinol.* **2008**, *291*, 63–70. [CrossRef]
30. Myung, Y.; Heo, C.-Y. Relationship between Obesity and Surgical Complications after Reduction Mammaplasty: A Systematic Literature Review and Meta-Analysis. *Aesthetic Surg. J.* **2017**, *37*, 308–315. [CrossRef]
31. Faucher, M.; Hastings-Tolsma, M.; Song, J.; Willoughby, D.; Bader, S.G. Gestational Weight Gain and Preterm Birth in Obese Women: A Systematic Review and Meta-Analysis. *BJOG* **2016**, *123*, 199–206. [CrossRef] [PubMed]
32. Freeman, C.M.; Woodle, E.S.; Shi, J.; Alexander, J.W.; Leggett, P.L.; Shah, S.A.; Paterno, F.; Cuffy, M.C.; Govil, A.; Mogilishetty, G.; et al. Addressing morbid obesity as a barrier to renal transplantation with laparoscopic sleeve gastrectomy. *Am. J. Transplant.* **2015**, *15*, 1360–1368. [CrossRef] [PubMed]
33. Rashad, N.M.; Al-sayed, R.M.; Yousef, M.S.; Saraya, Y.S. Kisspeptin and Body Weight Homeostasis in Relation to Phenotypic Features of Polycystic Ovary Syndrome; Metabolic Regulation of Reproduction. *Diabetes Metab. Syndr. Clin. Res. Rev.* **2019**, *13*, 2086–2092. [CrossRef] [PubMed]
34. Nuttall, F.Q. Body Mass Index: Obesity, BMI, and Health: A Critical Review. *Nutr. Today* **2015**, *50*, 117–128. [CrossRef] [PubMed]
35. Grier, T.; Canham-Chervak, M.; Sharp, M.; Jones, B.H. Does Body Mass Index Misclassify Physically Active Young Men. *Prev. Med. Reports* **2015**, *2*, 483–487. [CrossRef]
36. Van Marken Lichtenbelt, W.D.; Hartgens, F.; Vollaard, N.B.J.; Ebbing, S.; Kuipers, H. Body Composition Changes in Bodybuilders: A Method Comparison. *Med. Sci. Sports Exerc.* **2004**, *36*, 490–497. [CrossRef]

37. Sweeting, H.N. Measurement and Definitions of Obesity in Childhood and Adolescence: A Field Guide for the Uninitiated. *Nutr. J.* **2007**, *6*, 1–8. [CrossRef]

38. Deurenberg, P.; Yap, M. The Assessment of Obesity: Methods for Measuring Body Fat and Global Prevalence of Obesity. *Baillieres Best Pract. Clin. Endocrinol. Metab.* **1999**, *13*, 1–11. [CrossRef]

39. Peltz, G.; Aguirre, M.T.; Sanderson, M.; Fadden, M.K. The Role of Fat Mass Index in Determining Obesity. *Am. J. Hum. Biol.* **2010**, *22*, 639–647. [CrossRef]

40. Butte, N.F.; Brandt, M.L.; Wong, W.W.; Liu, Y.; Mehta, N.R.; Wilson, T.A.; Adolph, A.L.; Puyau, M.R.; Vohra, F.A.; Shypailo, R.J.; et al. Energetic Adaptations Persist after Bariatric Surgery in Severely Obese Adolescents. *Obesity* **2015**, *23*, 591–601. [CrossRef]

41. Sartorio, A.; Malavolti, M.; Agosti, F.; Marinone, P.G.; Caiti, O.; Battistini, N.; Bedogni, G. Body Water Distribution in Severe Obesity and Its Assessment from Eight-Polar Bioelectrical Impedance Analysis. *Eur. J. Clin. Nutr.* **2005**, *59*, 155–160. [CrossRef] [PubMed]

42. Vekic, J.; Zeljkovic, A.; Stefanovic, A.; Jelic-Ivanovic, Z.; Spasojevic-Kalimanovska, V. Obesity and Dyslipidemia. *Metabolism* **2019**, *92*, 71–81. [CrossRef] [PubMed]

43. Erion, K.A.; Corkey, B.E. Hyperinsulinemia: A Cause of Obesity? *Curr. Obes. Rep.* **2017**, *6*, 178–186. [CrossRef]

44. Mika, A.; Sledzinski, T. Alterations of Specific Lipid Groups in Serum of Obese Humans: A Review. *Obes. Rev.* **2017**, *18*, 247–272. [CrossRef] [PubMed]

45. Apryatin, S.A.; Sidorova, Y.S.; Shipelin, V.A.; Balakina, A.; Trusov, N.V.; Mazo, V.K. Neuromotor Activity, Anxiety and Cognitive Function in the In Vivo Model of Alimentary Hyperlipidemia and Obesity. *Bull. Exp. Biol. Med.* **2017**, *163*, 37–41. [CrossRef]

46. Burla, B.; Arita, M.; Arita, M.; Bendt, A.K.; Cazenave-Gassiot, A.; Dennis, E.A.; Ekroos, K.; Han, X.; Ikeda, K.; Liebisch, G.; et al. MS-Based Lipidomics of Human Blood Plasma: A Community-Initiated Position Paper to Develop Accepted Guidelines. *J. Lipid Res.* **2018**, *59*, 2001–2017. [CrossRef]

47. Jørgenrud, B.; Jäntti, S.; Mattila, I.; Pöhö, P.; Rønningen, K.S.; Yki-Järvinen, H.; Orešič, M.; Hyötyläinen, T. The Influence of Sample Collection Methodology and Sample Preprocessing on the Blood Metabolic Profile. *Bioanalysis* **2015**, *7*, 991–1006. [CrossRef]

48. Ishikawa, M.; Maekawa, K.; Saito, K.; Senoo, Y.; Urata, M.; Murayama, M.; Tajima, Y.; Kumagai, Y.; Saito, Y. Plasma and Serum Lipidomics of Healthy White Adults Shows Characteristic Profiles by Subjects' Gender and Age. *PLoS ONE* **2014**, *9*, e91806. [CrossRef]

49. Kawano, Y.; Ohta, M.; Hirashita, T.; Masuda, T.; Inomata, M.; Kitano, S. Effects of Sleeve Gastrectomy on Lipid Metabolism in an Obese Diabetic Rat Model. *Obes. Surg.* **2013**, *23*, 1947–1956. [CrossRef]

50. Im, S.-S.S.; Park, H.Y.; Shon, J.C.; Chung, I.-S.S.; Cho, H.C.; Liu, K.-H.H.; Song, D.-K.K. Plasma Sphingomyelins Increase in Pre-Diabetic Korean Men with Abdominal Obesity. *PLoS ONE* **2019**, *14*, e0213285. [CrossRef]

51. Wang, J.; Zhang, L.; Xiao, R.; Li, Y.; Liao, S.; Zhang, Z.; Yang, W.; Liang, B. Plasma Lipidomic Signatures of Spontaneous Obese Rhesus Monkeys. *Lipids Health Dis.* **2019**, *18*, 8. [CrossRef] [PubMed]

52. Hernandez-Carretero, A.; Weber, N.; La Frano, M.R.; Ying, W.; Lantero Rodriguez, J.; Sears, D.D.; Wallenius, V.; Börgeson, E.; Newman, J.W.; Osborn, O. Obesity-Induced Changes in Lipid Mediators Persist after Weight Loss. *Int. J. Obes.* **2018**, *42*, 728–736. [CrossRef] [PubMed]

53. Pickens, C.A.; Sordillo, L.M.; Zhang, C.; Fenton, J.I.; Austin, C.; Sordillo, L.M.; Zhang, C.; Fenton, J.I. Obesity is Positively Associated with Arachidonic Acid-Derived 5-and 11-Hydroxyeicosatetraenoic Acid (HETE). *Metabolism* **2017**, *70*, 177–191. [CrossRef] [PubMed]

54. Fan, R.; Kim, J.; You, M.; Giraud, D.; Toney, A.M.; Shin, S.H.; Kim, S.Y.; Borkowski, K.; Newman, J.W.; Chung, S. α-Linolenic Acid-Enriched Butter Attenuated High Fat Diet-Induced Insulin Resistance and Inflammation by Promoting Bioconversion of n-3 PUFA and Subsequent Oxylipin Formation. *J. Nutr. Biochem.* **2020**, *76*, 108285. [CrossRef] [PubMed]

55. Rupasinghe, T.W.T. *Lipidomics: Extraction Protocols for Biological Matrices*; Humana Press Inc: Totowa, NJ, USA, 2013; Volume 1055, pp. 71–80, ISBN 9781627035767.

56. Jiang, H.; Hsu, F.F.; Farmer, M.S.; Peterson, L.R.; Schaffer, J.E.; Ory, D.S.; Jiang, X. Development and Validation of LC-MS/MS Method for Determination of Very Long Acyl Chain (C22:0 and C24:0) Ceramides in Human Plasma. *Anal. Bioanal. Chem.* **2013**, *405*, 7357–7365. [CrossRef] [PubMed]

57. Ferreiro-Vera, C.; Priego-Capote, F.; Mata-Granados, J.M.; Luque De Castro, M.D. Short-Term Comparative Study of the Influence of Fried Edible Oils Intake on the Metabolism of Essential Fatty Acids in Obese Individuals. *Food Chem.* **2013**, *136*, 576–584. [CrossRef]

58. Zeng, M.; Cao, H. Fast Quantification of Short Chain Fatty Acids and Ketone Bodies by Liquid Chromatography-Tandem Mass Spectrometry after Facile Derivatization Coupled with Liquid-Liquid Extraction. *J. Chromatogr. B* **2018**, *1083*, 137–145. [CrossRef]

59. Klawitter, J.J.; Bek, S.; Zakaria, M.; Zeng, C.; Hornberger, A.; Gilbert, R.; Shokati, T.; Klawitter, J.J.; Christians, U.; Boernsen, K.O. Fatty Acid Desaturation Index in Human Plasma: Comparison of Different Analytical Methodologies for the Evaluation of Diet Effects. *Anal. Bioanal. Chem.* **2014**, *406*, 6399–6408. [CrossRef]

60. Liakh, I.; Pakiet, A.; Sledzinski, T.; Mika, A. Modern Methods of Sample Preparation for the Analysis of Oxylipins in Biological Samples. *Molecules* **2019**, *24*, 1639. [CrossRef]

61. Colas, R.A.; Shinohara, M.; Dalli, J.; Chiang, N.; Serhan, C.N. Identification and Signature Profiles for Pro-Resolving and Inflammatory Lipid Mediators in Human Tissue. *Am. J. Physiol. Cell Physiol.* **2014**, *307*, C39–C54. [CrossRef]

62. Golovko, M.Y.; Murphy, E.J. An Improved LC-MS/MS Procedure for Brain Prostanoid Analysis Using Brain Fixation with Head-Focused Microwave Irradiation and Liquid-Liquid Extraction. *J. Lipid Res.* **2008**, *49*, 893–902. [CrossRef] [PubMed]

63. Vuckovic, D. Current Trends and Challenges in Sample Preparation for Global Metabolomics Using Liquid Chromatography-Mass Spectrometry. *Anal. Bioanal. Chem.* **2012**, *403*, 1523–1548. [CrossRef] [PubMed]

64. Bellissimo, M.P.; Cai, Q.; Ziegler, T.R.; Liu, K.H.; Tran, P.H.; Vos, M.B.; Martin, G.S.; Jones, D.P.; Yu, T.; Alvarez, J.A. Plasma High-Resolution Metabolomics Differentiates Adults with Normal Weight Obesity from Lean Individuals. *Obesity* **2019**, *27*, 1729–1737. [CrossRef] [PubMed]

65. Drotleff, B.; Illison, J.; Schlotterbeck, J.; Lukowski, R.; Lämmerhofer, M. Comprehensive Lipidomics of Mouse Plasma Using Class-Specific Surrogate Calibrants and SWATH Acquisition for Large-Scale Lipid Quantification in Untargeted Analysis. *Anal. Chim. Acta* **2019**, *1086*, 90–102. [CrossRef] [PubMed]

66. Söder, J.; Wernersson, S.; Dicksved, J.; Hagman, R.; Östman, J.R.; Moazzami, A.A.; Höglund, K. Indication of Metabolic Inflexibility to Food Intake in Spontaneously Overweight Labrador Retriever Dogs. *BMC Vet. Res.* **2019**, *15*, 96. [CrossRef] [PubMed]

67. Folch, J.; Lees, M.; Sloane Stanley, G.H. A Simple Method for the Isolation and Purification of Total Lipides from Animal Tissues. *J. Biol. Chem.* **1957**, *226*, 497–509. [PubMed]

68. Bligh, E.G.; Dyer, W.J. A Rapid Method of Total Lipid Extraction and Purification. *Can. J. Biochem. Physiol.* **1959**, *37*, 911–917. [CrossRef]

69. Pizarro, C.; Arenzana-Rámila, I.; Pérez-del-Notario, N.; Pérez-Matute, P.; González-Sáiz, J.-M. Plasma Lipidomic Profiling Method Based on Ultrasound Extraction and Liquid Chromatography Mass Spectrometry. *Anal. Chem.* **2013**, *85*, 12085–12092. [CrossRef]

70. Wang, C.; Wang, M.; Han, X. Comprehensive and Quantitative Analysis of Lysophospholipid Molecular Species Present in Obese Mouse Liver by Shotgun Lipidomics. *Anal. Chem.* **2015**, *87*, 4879–4887. [CrossRef]

71. Al-Sulaiti, H.; Diboun, I.; Banu, S.; Al-Emadi, M.; Amani, P.; Harvey, T.M.; Dömling, A.S.; Latiff, A.; Elrayess, M.A. Triglyceride Profiling in Adipose Tissues from Obese Insulin Sensitive, Insulin Resistant and Type 2 Diabetes Mellitus Individuals. *J. Transl. Med.* **2018**, *16*, 175. [CrossRef]

72. Fernández-Arroyo, S.; Hernández-Aguilera, A.; de Vries, M.A.; Burggraaf, B.; van der Zwan, E.; Pouw, N.; Joven, J.; Castro Cabezas, M. Effect of Vitamin D3 on the Postprandial Lipid Profile in Obese Patients: A Non-Targeted Lipidomics Study. *Nutrients* **2019**, *11*, 1194. [CrossRef] [PubMed]

73. Yore, M.M.; Syed, I.; Moraes-Vieira, P.M.; Zhang, T.; Herman, M.A.; Homan, E.A.; Patel, R.T.; Lee, J.; Chen, S.; Peroni, O.D.; et al. Discovery of a Class of Endogenous Mammalian Lipids with Anti-Diabetic and Anti-Inflammatory Effects. *Cell* **2014**, *159*, 318–332. [CrossRef] [PubMed]

74. Choromańska, B.; Myśliwiec, P.; Razak Hady, H.; Dadan, J.; Myśliwiec, H.; Chabowski, A.; Mikłosz, A. Metabolic Syndrome is Associated with Ceramide Accumulation in Visceral Adipose Tissue of Women with Morbid Obesity. *Obesity* **2019**, *27*, 444–453. [CrossRef] [PubMed]

75. Serbulea, V.; Upchurch, C.M.; Schappe, M.S.; Voigt, P.; DeWeese, D.E.; Desai, B.N.; Meher, A.K.; Leitinger, N. Macrophage Phenotype and Bioenergetics are Controlled by Oxidized Phospholipids Identified in Lean and Obese Adipose Tissue. *Proc. Natl. Acad. Sci. USA* **2018**, *115*, E6254–E6263. [CrossRef] [PubMed]

76. León-Aguilar, L.F.; Croyal, M.; Ferchaud-Roucher, V.; Huang, F.; Marchat, L.A.; Barraza-Villarreal, A.; Romieu, I.; Ramakrishnan, U.; Krempf, M.; Ouguerram, K.; et al. Maternal Obesity Leads to Long-Term Altered Levels of Plasma Ceramides in the Offspring as Revealed by a Longitudinal Lipidomic Study in Children. *Int. J. Obes.* **2019**, *43*, 1231–1243. [CrossRef] [PubMed]

77. Jaramillo, M.G.; Lytle, K.A.; Spooner, M.H.; Jump, D.B.; García-Jaramillo, M.; Lytle, K.A.; Spooner, M.H.; Jump, D.B. A Lipidomic Analysis of Docosahexaenoic Acid (22:6, ω3) Mediated Attenuation of Western Diet Induced Nonalcoholic Steatohepatitis in Male Ldlr-/-Mice. *Metabolites* **2019**, *9*, 252. [CrossRef] [PubMed]

78. Eum, J.Y.; Lee, G.B.; Yi, S.S.; Kim, I.Y.; Seong, J.K.; Moon, M.H. Lipid Alterations in the Skeletal Muscle Tissues of Mice after Weight Regain by Feeding a High-Fat Diet Using Nanoflow Ultrahigh Performance Liquid Chromatography-Tandem Mass Spectrometry. *J. Chromatogr. B* **2020**, *1141*, 122022. [CrossRef] [PubMed]

79. Hu, T.; Lin, M.; Zhang, D.; Li, M.; Zhang, J. A UPLC/MS/MS Method for Comprehensive Profiling and Quantification of Fatty Acid Esters of Hydroxy Fatty Acids in White Adipose Tissue. *Anal. Bioanal. Chem.* **2018**, *410*, 7415–7428. [CrossRef]

80. Pakiet, A.; Jakubiak, A.; Czumaj, A.; Sledzinski, T.; Mika, A. The Effect of Western Diet on Mice Brain Lipid Composition. *Nutr. Metab.* **2019**, *16*, 81. [CrossRef]

81. Kim, H.; Salem, N. Separation of Lipid Classes by Solid Phase Extraction. *J. Lipid Res.* **1990**, *31*, 2285–2289.

82. Cho, A.R.; Moon, J.Y.; Kim, S.; An, K.Y.; Oh, M.; Jeon, J.Y.; Jung, D.H.; Choi, M.H.; Lee, J.W. Effects of Alternate Day Fasting and Exercise on Cholesterol Metabolism in Overweight or Obese Adults: A Pilot Randomized Controlled Trial. *Metabolism* **2019**, *93*, 52–60. [CrossRef] [PubMed]

83. Mutemberezi, V.; Masquelier, J.; Guillemot-Legris, O.; Muccioli, G.G. Development and Validation of an HPLC-MS Method for the Simultaneous Quantification of Key Oxysterols, Endocannabinoids, and Ceramides: Variations in Metabolic Syndrome. *Anal. Bioanal. Chem.* **2016**, *408*, 733–745. [CrossRef] [PubMed]

84. Ramsden, C.E.; Hennebelle, M.; Schuster, S.; Keyes, G.S.; Johnson, C.D.; Kirpich, I.A.; Dahlen, J.E.; Horowitz, M.S.; Zamora, D.; Feldstein, A.E.; et al. Effects of Diets Enriched in Linoleic Acid and Its Peroxidation Products on Brain Fatty Acids, Oxylipins, and Aldehydes in Mice. *Biochim. Biophys. Acta Mol. Cell Biol. Lipids* **2018**, *1863*, 1206–1213. [CrossRef] [PubMed]

85. Okada, K.; Hosooka, T.; Shinohara, M.; Ogawa, W. Modulation of Lipid Mediator Profile May Contribute to Amelioration of Chronic Inflammation in Adipose Tissue of Obese Mice by Pioglitazone. *Biochem. Biophys. Res. Commun.* **2018**, *505*, 29–35. [CrossRef]

86. Itariu, B.K.; Zeyda, M.; Hochbrugger, E.E.; Neuhofer, A.; Prager, G.; Schindler, K.; Bohdjalian, A.; Mascher, D.; Vangala, S.; Schranz, M.; et al. Long-Chain n−3 PUFAs Reduce Adipose Tissue and Systemic Inflammation in Severely Obese Nondiabetic Patients: A Randomized Controlled Trial. *Am. J. Clin. Nutr.* **2012**, *96*, 1137–1149. [CrossRef] [PubMed]

87. López-Bascón, M.A.; Calderón-Santiago, M.; Priego-Capote, F. Confirmatory and Quantitative Analysis of fatty acid esters of hydroxy fatty acids in serum by solid phase extraction coupled to liquid Chromatography Tandem Mass Spectrometry. *Anal. Chim. Acta* **2016**, *943*, 82–88. [CrossRef]

88. Ostermann, A.I.; Willenberg, I.; Schebb, N.H. Comparison of Sample Preparation Methods for the Quantitative Analysis of Eicosanoids and other Oxylipins in Plasma by Means of LC-MS/MS. *Anal. Bioanal. Chem.* **2015**, *407*, 1403–1414. [CrossRef]

89. Galvão, A.F.; Petta, T.; Flamand, N.; Bollela, V.R.; Silva, C.L.; Jarduli, L.R.; Malmegrim, K.C.R.; Simões, B.P.; de Moraes, L.A.B.; Faccioli, L.H. Plasma Eicosanoid Profiles Determined by High-Performance Liquid Chromatography Coupled with Tandem Mass Spectrometry in Stimulated Peripheral Blood from Healthy Individuals and Sickle Cell Anemia Patients in Treatment. *Anal. Bioanal. Chem.* **2016**, *408*, 3613–3623. [CrossRef]

90. Roberts, L.D.; West, J.A.; Vidal-Puig, A.; Griffin, J.L. Methods for Performing Lipidomics in White Adipose Tissue. In Methods in Enzymology. 2014; 538, 211–231. *538*.

91. Andrade-Eiroa, A.; Canle, M.; Leroy-Cancellieri, V.; Cerdà, V. Solid-Phase Extraction of Organic Compounds: A Critical Review. Part II. *TrAC-Trends Anal. Chem.* **2016**, *80*, 655–667. [CrossRef]

92. Bhattacharya, S.K. Lipidomics. In *Methods in Molecular Biology*; Bhattacharya, S.K., Ed.; Springer New York: New York, NY, USA, 2017; Volume 1609, ISBN 978-1-4939-6995-1.

93. Benedusi, V. Lipidomics. *Mater. Methods* **2018**, *8*, 5139. [CrossRef]

94. Ruiz-Rodriguez, A.; Reglero, G.; Ibañez, E. Recent Trends in the Advanced Analysis of Bioactive Fatty Acids. *J. Pharm. Biomed. Anal.* **2010**, *51*, 305–326. [CrossRef] [PubMed]

95. Jurowski, K.; Kochan, K.; Walczak, J.; Barańska, M.; Piekoszewski, W.; Buszewski, B. Comprehensive Review of Trends and Analytical Strategies Applied for Biological Samples Preparation and Storage in Modern Medical Lipidomics: State of the Art. *TrAC Trends Anal. Chem.* **2017**, *86*, 276–289. [CrossRef]

96. Campíns-Falcó, P.; Sevillano-Cabeza, A.; Herráez-Hernández, R.; Molins-Legua, C.; Moliner-Martínez, Y.; Verdú-Andrés, J. Solid-Phase Extraction and Clean-Up Procedures in Pharmaceutical Analysis. *Encycl. Anal. Chem.* **2012**. [CrossRef]

97. Laurentius, T.; Kob, R.; Fellner, C.; Nourbakhsh, M.; Bertsch, T.; Sieber, C.C.; Bollheimer, L.C. Long-Chain Fatty Acids and Inflammatory Markers Coaccumulate in the Skeletal Muscle of Sarcopenic Old Rats. *Dis. Markers* **2019**, *2019*, 1–11. [CrossRef] [PubMed]

98. Garcia-Jaramillo, M.; Spooner, M.H.; Löhr, C.V.; Wong, C.P.; Zhang, W.; Jump, D.B. Lipidomic and Transcriptomic Analysis of Western Diet-Induced Nonalcoholic Steatohepatitis (NASH) in Female Ldlr-/-mice. *PLoS ONE* **2019**, *14*, e0214387. [CrossRef] [PubMed]

99. Astarita, G.; Piomelli, D. Lipidomic Analysis of Endocannabinoid Metabolism in Biological Samples. *J. Chromatogr. B* **2009**, *877*, 2755–2767. [CrossRef] [PubMed]

100. Argueta, D.A.; DiPatrizio, N.V. Peripheral Endocannabinoid Signaling Controls Hyperphagia in Western Diet-Induced Obesity. *Physiol. Behav.* **2017**, *171*, 32–39. [CrossRef] [PubMed]

101. Perez, P.A.; DiPatrizio, N.V. Impact of Maternal Western Diet-Induced Obesity on Offspring Mortality and Peripheral Endocannabinoid System in Mice. *PLoS ONE* **2018**, *13*, e0205021. [CrossRef]

102. Perreault, M.; Zulyniak, M.A.; Badoud, F.; Stephenson, S.; Badawi, A.; Buchholz, A.; Mutch, D.M. A Distinct Fatty Acid Profile Underlies the Reduced Inflammatory State of Metabolically Healthy Obese Individuals. *PLoS ONE* **2014**, *9*, e88539. [CrossRef]

103. Tomášová, P.; Čermáková, M.; Pelantová, H.; Vecka, M.; Kratochvílová, H.; Lipš, M.; Lindner, J.; Šedivá, B.; Haluzík, M.; Kuzma, M. Minor Lipids Profiling in Subcutaneous and Epicardial Fat Tissue Using LC/MS with an Optimized Preanalytical Phase. *J. Chromatogr. B* **2019**, *1113*, 50–59. [CrossRef]

104. Lytle, K.A.; Wong, C.P.; Jump, D.B. Docosahexaenoic Acid Blocks Progression of Western Diet-Induced Nonalcoholic Steatohepatitis in Obese Ldlr-/-Mice. *PLoS ONE* **2017**, *12*, 1–26. [CrossRef] [PubMed]

105. Prosen, H. Applications of Liquid-Phase Microextraction in the Sample Preparation of Environmental Solid Samples. *Molecules* **2014**, *19*, 6776–6808. [CrossRef] [PubMed]

106. Teo, C.C.; Chong, W.P.K.; Tan, E.; Basri, N.B.; Low, Z.J.; Ho, Y.S. Advances in Sample Preparation and Analytical Techniques for Lipidomics Study of Clinical Samples. *TrAC Trends Anal. Chem.* **2015**, *66*, 1–18. [CrossRef]

107. Genovese, A.; Rispoli, T.; Sacchi, R. Extra Virgin Olive Oil Aroma Release after Interaction with Human Saliva from Individuals with Different Body Mass Index. *J. Sci. Food Agric.* **2018**, *98*, 3376–3383. [CrossRef] [PubMed]

108. Del Chierico, F.; Nobili, V.; Vernocchi, P.; Russo, A.; De Stefanis, C.; Gnani, D.; Furlanello, C.; Zandonà, A.; Paci, P.; Capuani, G.; et al. Gut Microbiota Profiling of Pediatric Nonalcoholic Fatty Liver Disease and Obese Patients Unveiled by an Integrated Meta-Omics-Based Approach. *Hepatology* **2017**, *65*, 451–464. [CrossRef] [PubMed]

109. Zamora-Gasga, V.M.; Montalvo-González, E.; Loarca-Piña, G.; Vázquez-Landaverde, P.A.; Tovar, J.; Sáyago-Ayerdi, S.G. Microbial Metabolites Profile during In Vitro Human Colonic Fermentation of Breakfast Menus Consumed by Mexican School Children. *Food Res. Int.* **2017**, *97*, 7–14. [CrossRef]

110. Cozzolino, R.; De Giulio, B.; Marena, P.; Martignetti, A.; Günther, K.; Lauria, F.; Russo, P.; Stocchero, M.; Siani, A. Urinary Volatile Organic Compounds in Overweight Compared to Normal-Weight Children: Results from the Italian I.Family Cohort. *Sci. Rep.* **2017**, *7*, 1–13. [CrossRef]

111. Marisol Encerrado Manriquez, A. Method Development For The Analysis Of Fatty Acids In Adipose Tissue Using Stir Bar Sorptive Extraction Coupled With Gas Chromatography-Mass Spectrometry. Masters's Thesis, Biochemistry. The University of Texas at el El Paso, El Paso, TX, USA, August 2020.

112. Eslami, Z.; Torabizadeh, M.; Talebpour, Z.; Talebpour, M.; Ghassempour, A.; Aboul-Enein, H.Y. Simple and Sensitive Quantification of Ghrelin Hormone in Human Plasma Using SBSE-HPLC/DAD-MS. *J. Chromatogr. Sci.* **2016**, *54*, 1652–1660. [CrossRef]

113. Rezaee, M.; Assadi, Y.; Milani Hosseini, M.-R.; Aghaee, E.; Ahmadi, F.; Berijani, S. Determination of Organic Compounds in Water Using Dispersive Liquid–Liquid Microextraction. *J. Chromatogr. A* **2006**, *1116*, 1–9. [CrossRef]

114. Amin, M.M.; Ebrahim, K.; Hashemi, M.; Shoshtari-Yeganeh, B.; Rafiei, N.; Mansourian, M.; Kelishadi, R. Association of Exposure to Bisphenol A with Obesity and Cardiometabolic Risk Factors in Children and Adolescents. *Int. J. Environ. Health Res.* **2019**, *29*, 94–106. [CrossRef]

115. Krawczyńska, A.; Konieczna, L.; Skrzypkowska, M.; Siebert, J.; Reiwer-Gostomska, M.; Gutknecht, P.; Kaska, Ł.; Bigda, J.; Proczko-Stepaniak, M.; Bączek, T. Decreased Level of Vitamin D in Obesity Patients Measured by the LC-MS/MS Method. In Proceedings of the CECE 2019—16th International Interdisciplinary Meeting on Bioanalysis, Gdańsk, Poland, 24–26 September 2019; p. 33.

116. Nigam, P.K. Serum Lipid Profile: Fasting or Non-Fasting? *Indian J. Clin. Biochem.* **2011**, *26*, 96–97. [CrossRef] [PubMed]

117. Tchernof, A.; Després, J.-P. Pathophysiology of Human Visceral Obesity: An Update. *Physiol. Rev.* **2013**, *93*, 359–404. [CrossRef] [PubMed]

118. Nielsen, S.; Guo, Z.; Johnson, C.M.; Hensrud, D.D.; Jensen, M.D. Splanchnic Lipolysis in Human Obesity. *J. Clin. Invest.* **2004**, *113*, 1582–1588. [CrossRef] [PubMed]

119. Koelmel, J.P.; Ulmer, C.Z.; Fogelson, S.; Jones, C.M.; Botha, H.; Bangma, J.T.; Guillette, T.C.; Luus-Powell, W.J.; Sara, J.R.; Smit, W.J.; et al. Lipidomics for Wildlife Disease Etiology and Biomarker Discovery: A Case Study of Pansteatitis Outbreak in South Africa. *Metabolomics* **2019**, *15*, 38. [CrossRef]

120. Mousa, A.; Naderpoor, N.; Mellett, N.; Wilson, K.; Plebanski, M.; Meikle, P.J.; de Courten, B. Lipidomic Profiling Reveals Early-Stage Metabolic Dysfunction in Overweight or Obese Humans. *Biochim. Biophys. Acta-Mol. Cell Biol. Lipids* **2019**, *1864*, 335–343. [CrossRef] [PubMed]

121. Blewett, H.J.; Gerdung, C.A.; Ruth, M.R.; Proctor, S.D.; Field, C.J. Vaccenic Acid Favourably Alters Immune Function in Obese JCR:LA-Cp Rats. *Br. J. Nutr.* **2009**, *102*, 526–536. [CrossRef]

122. Yoon, H.-R.; Kim, H.; Cho, S.-H. Quantitative Analysis of Acyl-Lysophosphatidic Acid in Plasma Using Negative Ionization Tandem Mass Spectrometry. *J. Chromatogr. B* **2003**, *788*, 85–92. [CrossRef]

123. Matyash, V.; Liebisch, G.; Kurzchalia, T.V.; Shevchenko, A.; Schwudke, D. Lipid Extraction by Methyl-Tert-Butyl ETHER for high-Throughput Lipidomics. *J. Lipid Res.* **2008**, *49*, 1137–1146. [CrossRef]

124. Wang, Y.; Jiang, C.-T.; Song, J.-Y.; Song, Q.-Y.; Ma, J.; Wang, H.-J. Lipidomic Profile Revealed the Association of Plasma Lysophosphatidylcholines with Adolescent Obesity. *Biomed Res. Int.* **2019**, *2019*, 1–9. [CrossRef]

125. Misra, B.B.; Puppala, S.R.; Comuzzie, A.G.; Mahaney, M.C.; VandeBerg, J.L.; Olivier, M.; Cox, L.A. Analysis of Serum Changes in Response to a High Fat High Cholesterol Diet Challenge Reveals Metabolic Biomarkers of Atherosclerosis. *PLoS ONE* **2019**, *14*, e0214487. [CrossRef]

126. Friedewald, W.T.; Levy, R.I.; Fredrickson, D.S. Estimation of the Concentration of Low-Density Lipoprotein Cholesterol in Plasma, without Use of the Preparative Ultracentrifuge. *Clin. Chem.* **1972**, *18*, 499–502. [CrossRef] [PubMed]

127. Nauck, M.; Warnick, G.R.; Rifai, N. Methods for Measurement of LDL-Cholesterol: A Critical Assessment of Direct Measurement by Homogeneous Assays versus Calculation. *Clin. Chem.* **2002**, *48*, 236–254. [CrossRef] [PubMed]

128. Van Der Kolk, B.W.; Goossens, G.H.; Jocken, J.W.; Kersten, S.; Blaak, E.E. Angiopoietin-Like Protein 4 and Postprandial Skeletal Muscle Lipid Metabolism in Overweight and Obese Prediabetics. *J. Clin. Endocrinol. Metab.* **2016**, *101*, 2332–2339. [CrossRef] [PubMed]

129. Van Hees, A.M.J.; Jans, A.; Hul, G.B.; Roche, H.M.; Saris, W.H.M.; Blaak, E.E. Skeletal Muscle Fatty Acid Handling in Insulin Resistant Men. *Obesity* **2011**, *19*, 1350–1359. [CrossRef] [PubMed]

130. Lindqvist, A.; Ekelund, M.; Garcia-Vaz, E.; Ståhlman, M.; Pierzynowski, S.; Gomez, M.F.; Rehfeld, J.F.; Groop, L.; Hedenbro, J.; Wierup, N.; et al. The Impact of Roux-en-Y Gastric Bypass Surgery on Normal Metabolism in a Porcine Model. *PLoS ONE* **2017**, *12*, e0173137. [CrossRef] [PubMed]

131. Kwon, Y.J.; Lee, H.S.; Lee, J.W. Direct Bilirubin is Associated with Low-Density Lipoprotein Subfractions and Particle Size in Overweight and Centrally Obese Women. *Nutr. Metab. Cardiovasc. Dis.* **2018**, *28*, 1021–1028. [CrossRef] [PubMed]

132. Doğan, S.; Aslan, I.; Eryılmaz, R.; Ensari, C.O.; Bilecik, T.; Aslan, M. Early Postoperative Changes of HDL Subfraction Profile and HDL-Associated Enzymes after Laparoscopic Sleeve Gastrectomy. *Obes. Surg.* **2013**, *23*, 1973–1980. [CrossRef] [PubMed]

133. Sofi, F.; Dinu, M.; Pagliai, G.; Cesari, F.; Gori, A.M.; Sereni, A.; Becatti, M.; Fiorillo, C.; Marcucci, R.; Casini, A. Low-Calorie Vegetarian Versus Mediterranean Diets for Reducing Body Weight and Improving Cardiovascular Risk Profile. *Circulation* **2018**, *137*, 1103–1113. [CrossRef]

134. Parks, E.J. Effect of Dietary Carbohydrate on Triglyceride Metabolism in Humans. *J. Nutr.* **2001**, *131*, 2772S–2774S. [CrossRef]

135. Vogelzangs, N.; van der Kallen, C.J.H.; van Greevenbroek, M.M.J.; van der Kolk, B.W.; Jocken, J.W.E.; Goossens, G.H.; Schaper, N.C.; Henry, R.M.A.; Eussen, S.J.P.M.; Valsesia, A.; et al. Metabolic Profiling of Tissue-Specific Insulin Resistance in Human Obesity: Results from the Diogenes Study and the Maastricht Study. *Int. J. Obes.* **2020**, *44*, 1376–1386. [CrossRef]

136. Sparks, J.D.; Sparks, C.E.; Adeli, K. Selective Hepatic Insulin Resistance, VLDL Overproduction, and Hypertriglyceridemia. *Arterioscler. Thromb. Vasc. Biol.* **2012**, *32*, 2104–2112. [CrossRef] [PubMed]

137. Engin, A.B.; Engin, A. *Obesity and Lipotoxicity*; Engin, A.B., Engin, A., Eds.; Advances in Experimental Medicine and Biology; Springer International Publishing: Cham, Switzerland, 2017; Volume 960, ISBN 978-3-319-48380-1.

138. Björnson, E.; Adiels, M.; Taskinen, M.R.; Borén, J. Kinetics of Plasma Triglycerides in Abdominal Obesity. *Curr. Opin. Lipidol.* **2017**, *28*, 11–18. [CrossRef] [PubMed]

139. Alsulaiti, H. Triglycerides Analysis in Adipose Tissue from Insulin Sensitive and Insulin Resistance Obese Patients. *Atherosclerosis* **2017**, *263*, e267–e268. [CrossRef]

140. Koulman, A.; Furse, S.; Baumert, M.; Goldberg, G.; Bluck, L. Rapid Profiling of Triglycerides in Human Breast Milk Using Liquid Extraction Surface Analysis Fourier Transform Mass Spectrometry Reveals New Very Long Chain Fatty Acids and Differences within Individuals. *Rapid Commun. Mass Spectrom.* **2019**, *33*, 1267–1276. [CrossRef] [PubMed]

141. Crowe, F.L.; Skeaff, C.M.; Green, T.J.; Gray, A.R. Serum Fatty Acids as Biomarkers of Fat Intake Predict Serum Cholesterol Concentrations in a Population-Based Survey of New Zealand Adolescents and Adults. *Am. J. Clin. Nutr.* **2006**, *83*, 887–894. [CrossRef] [PubMed]

142. Hodge, A.M.; English, D.R.; O'Dea, K.; Sinclair, A.J.; Makrides, M.; Gibson, R.A.; Giles, G.G. Plasma Phospholipid and Dietary Fatty Acids as Predictors of Type 2 Diabetes: Interpreting the Role of Linoleic Acid. *Am. J. Clin. Nutr.* **2007**, *86*, 189–197. [CrossRef] [PubMed]

143. Wolk, A.; Furuheim, M.; Vessby, B. Fatty Acid Composition of Adipose Tissue and Serum Lipids Are Valid Biological Markers Of Dairy Fat Intake in Men. *J. Nutr.* **2001**, *131*, 828–833. [CrossRef]

144. Sledzinski, T.; Mika, A.; Stepnowski, P.; Proczko-Markuszewska, M.; Kaska, L.; Stefaniak, T.; Swierczynski, J. Identification of Cyclopropaneoctanoic Acid 2-Hexyl in Human Adipose Tissue and Serum. *Lipids* **2013**, *48*, 839–848. [CrossRef]

145. Su, X.; Magkos, F.; Zhou, D.; Eagon, J.C.; Fabbrini, E.; Okunade, A.L.; Klein, S. Adipose Tissue Monomethyl Branched-Chain Fatty Acids and Insulin Sensitivity: Effects of Obesity and Weight Loss. *Obesity* **2015**, *23*, 329–334. [CrossRef]

146. Mika, A.; Stepnowski, P.; Kaska, L.; Proczko, M.; Wisniewski, P.; Sledzinski, M.; Sledzinski, T. A Comprehensive Study of Serum Odd- and Branched-Chain Fatty Acids in Patients with Excess Weight. *Obesity* **2016**, *24*, 1669–1676. [CrossRef]

147. Bondia-Pons, I.; Castellote, A.I.; López-Sabater, M.C. Comparison of Conventional and Fast Gas Chromatography in Human Plasma Fatty Acid Determination. *J. Chromatogr. B Anal. Technol. Biomed. Life Sci.* **2004**, *809*, 339–344. [CrossRef] [PubMed]

148. Kang, M.; Lee, A.; Yoo, H.J.; Kim, M.; Kim, M.; Shin, D.Y.; Lee, J.H. Association between Increased Visceral Fat Area and Alterations in Plasma Fatty acid Profile in Overweight Subjects: A Cross-Sectional Study. *Lipids Health Dis.* **2017**, *16*, 248. [CrossRef] [PubMed]

149. Lee, Y.J.; Lee, A.; Yoo, H.J.; Kim, M.; Kim, M.; Jee, S.H.; Shin, D.Y.; Lee, J.H. Effect of Weight Loss on Circulating Fatty Acid Profiles in Overweight Subjects with High Visceral Fat Area: A 12-Week Randomized Controlled Trial. *Nutr. J.* **2018**, *17*, 28. [CrossRef] [PubMed]

150. Aslan, M.; Aslan, I.; Özcan, F.; Eryılmaz, R.; Ensari, C.O.; Bilecik, T. A Pilot Study Investigating Early Postoperative Changes of Plasma Polyunsaturated Fatty Acids after Laparoscopic Sleeve Gastrectomy. *Lipids Health Dis.* **2014**, *13*, 62. [CrossRef] [PubMed]

151. Badoud, F.; Lam, K.P.; Perreault, M.; Zulyniak, M.A.; Britz-McKibbin, P.; Mutch, D.M. Metabolomics Reveals Metabolically Healthy and Unhealthy Obese Individuals Differ in Their Response to a Caloric Challenge. *PLoS ONE* **2015**, *10*, e0134613. [CrossRef] [PubMed]

152. Ma, Y.; Qiu, T.; Zhu, J.; Wang, J.; Li, X.; Deng, Y.; Zhang, X.; Feng, J.; Chen, K.; Wang, C.; et al. Serum FFAs Profile Analysis of Normal Weight and Obesity Individuals of Han and Uygur Nationalities in China. *Lipids Health Dis.* **2020**, *19*, 13. [CrossRef]

153. Nemati, R.; Lu, J.; Tura, A.; Smith, G.; Murphy, R. Acute Changes in Non-Esterified Fatty Acids in Patients with Type 2 Diabetes Receiving Bariatric Surgery. *Obes. Surg.* **2017**, *27*, 649–656. [CrossRef]

154. Lin, C.; Våge, V.; Mjøs, S.A.; Kvalheim, O.M. Changes in Serum Fatty Acid Levels During the First Year After Bariatric Surgery. *Obes. Surg.* **2016**, *26*, 1735–1742. [CrossRef]

155. Ramos-Molina, B.; Castellano-Castillo, D.; Alcaide-Torres, J.; Pastor, Ó.; de Luna Díaz, R.; Salas-Salvadó, J.; López-Moreno, J.; Fernández-García, J.C.; Macías-González, M.; Cardona, F.; et al. Differential Effects of Restrictive and Malabsorptive Bariatric Surgery Procedures on the Serum Lipidome in Obese Subjects. *J. Clin. Lipidol.* **2018**, *12*, 1502–1512. [CrossRef]

156. Muoio, D.M.; Newgard, C.B. The Good in Fat. *Nature* **2014**, *516*, 49–50. [CrossRef]

157. Pingitore, A.; Chambers, E.S.; Hill, T.; Maldonado, I.R.; Liu, B.; Bewick, G.; Morrison, D.J.; Preston, T.; Wallis, G.A.; Tedford, C.; et al. The Diet-Derived Short Chain Fatty Acid Propionate Improves Beta-Cell Function in Humans and Stimulates Insulin Secretion from Human Islets In Vitro. *Diabetes Obes. Metab.* **2017**, *19*, 257–265. [CrossRef] [PubMed]

158. He, L.; Prodhan, M.A.I.; Yuan, F.; Yin, X.; Lorkiewicz, P.K.; Wei, X.; Feng, W.; McClain, C.; Zhang, X. Simultaneous Quantification of Straight-Chain and Branched-Chain Short Chain Fatty Acids by Gas Chromatography Mass Spectrometry. *J. Chromatogr. B* **2018**, *1092*, 359–367. [CrossRef] [PubMed]

159. Amer, B.; Nebel, C.; Bertram, H.C.; Mortensen, G.; Dalsgaard, T.K. Direct Derivatization vs Aqueous Extraction Methods of Fecal Free Fatty Acids for GC–MS Analysis. *Lipids* **2015**, *50*, 681–689. [CrossRef] [PubMed]

160. García-Villalba, R.; Giménez-Bastida, J.A.; García-Conesa, M.T.; Tomás-Barberán, F.A.; Carlos Espín, J.; Larrosa, M. Alternative Method for Gas Chromatography-Mass Spectrometry Analysis of Short-Chain Fatty Acids in Faecal Samples. *J. Sep. Sci.* **2012**, *35*, 1906–1913. [CrossRef]

161. Kirkham, T.C.; Williams, C.M.; Fezza, F.; Marzo, V. Di Endocannabinoid Levels in Rat Limbic Forebrain and Hypothalamus in Relation to Fasting, Feeding and Satiation: Stimulation of Eating by 2-Arachidonoyl Glycerol. *Br. J. Pharmacol.* **2002**, *136*, 550–557. [CrossRef]

162. Liakh, I.; Pakiet, A.; Sledzinski, T.; Mika, A. Methods of the Analysis of Oxylipins in Biological Samples. *Molecules* **2020**, *25*, 349. [CrossRef]

163. Pickens, C.A.; Sordillo, L.M.; Comstock, S.S.; Harris, W.S.; Hortos, K.; Kovan, B.; Fenton, J.I. Plasma Phospholipids, Non-Esterified Plasma Polyunsaturated Fatty Acids and Oxylipids are Associated with BMI. *Prostaglandins Leukot. Essent. Fat. Acids* **2015**, *95*, 31–40. [CrossRef]

164. Azar, S.; Sherf-Dagan, S.; Nemirovski, A.; Webb, M.; Raziel, A.; Keidar, A.; Goitein, D.; Sakran, N.; Shibolet, O.; Tam, J.; et al. Circulating Endocannabinoids Are Reduced Following Bariatric Surgery and Associated with Improved Metabolic Homeostasis in Humans. *Obes. Surg.* **2019**, *29*, 268–276. [CrossRef]

165. Merrill, A.H.; Sullards, M.C.; Allegood, J.C.; Kelly, S.; Wang, E. Sphingolipidomics: High-Throughput, Structure-Specific, and Quantitative Analysis of Sphingolipids by Liquid Chromatography Tandem Mass Spectrometry. *Methods* **2005**, *36*, 207–224. [CrossRef]

166. Brozinick, J.T.; Hawkins, E.; Hoang Bui, H.; Kuo, M.S.; Tan, B.; Kievit, P.; Grove, K. Plasma Sphingolipids Are Biomarkers of Metabolic Syndrome in Non-Human Primates Maintained on a Western-Style Diet. *Int. J. Obes.* **2013**, *37*, 1064–1070. [CrossRef]

167. Croyal, M.; Kaabia, Z.; León, L.; Ramin-Mangata, S.; Baty, T.; Fall, F.; Billon-Crossouard, S.; Aguesse, A.; Hollstein, T.; Sullivan, D.R.; et al. Fenofibrate Decreases Plasma Ceramide in Type 2 Diabetes Patients: A Novel Marker of CVD? *Diabetes Metab.* **2018**, *44*, 143–149. [CrossRef] [PubMed]

168. Neeland, I.J.; Singh, S.; McGuire, D.K.; Vega, G.L.; Roddy, T.; Reilly, D.F.; Castro-Perez, J.; Kozlitina, J.; Scherer, P.E. Relation of Plasma Ceramides to Visceral Adiposity, Insulin Resistance and the Development of Type 2 Diabetes Mellitus: The Dallas Heart Study. *Diabetologia* **2018**, *61*, 2570–2579. [CrossRef] [PubMed]

169. Özer, H.; Aslan, İ.; Oruç, M.T.; Çöpelci, Y.; Afşar, E.; Kaya, S.; Aslan, M. Early Postoperative Changes of Sphingomyelins and Ceramides after Laparoscopic Sleeve Gastrectomy. *Lipids Health Dis.* **2018**, *17*, 269. [CrossRef] [PubMed]

170. Roth, C.L.; Kratz, M.; Ralston, M.M.; Reinehr, T. Changes in Adipose-Derived Inflammatory Cytokines and Chemokines after Successful Lifestyle Intervention in Obese Children. *Metabolism* **2011**, *60*, 445–452. [CrossRef]

171. Van der Kolk, B.W.; Vogelzangs, N.; Jocken, J.W.E.; Valsesia, A.; Hankemeier, T.; Astrup, A.; Saris, W.H.M.; Arts, I.C.W.; van Greevenbroek, M.M.J.; Blaak, E.E. Plasma Lipid Profiling of Tissue-Specific Insulin Resistance in Human Obesity. *Int. J. Obes.* **2019**, *43*, 989–998. [CrossRef]

172. Kunešová, M.; Hlavatý, P.; Tvrzická, E.; Staňková, B.; Kalousková, P.; Viguerie, N.; Larsen, T.M.; Van Baak, M.A.; Jebb, S.A.; Martinez, J.A.; et al. Fatty Acid Composition of Adipose Tissue Triafter Weight Loss and Weight Maintenance: The Diogenes Study. *Physiol. Res.* **2012**, *61*, 597–607. [CrossRef]

173. Montastier, E.; Villa-Vialaneix, N.; Caspar-Bauguil, S.; Hlavaty, P.; Tvrzicka, E.; Gonzalez, I.; Saris, W.H.M.; Langin, D.; Kunesova, M.; Viguerie, N. System Model Network for Adipose Tissue Signatures Related to Weight Changes in Response to Calorie Restriction and Subsequent Weight Maintenance. *PLOS Comput. Biol.* **2015**, *11*, e1004047. [CrossRef]

174. Grzybek, M.; Palladini, A.; Alexaki, V.I.; Surma, M.A.; Simons, K.; Chavakis, T.; Klose, C.; Coskun, Ü. Comprehensive and Quantitative Analysis of White and Brown Adipose Tissue by Shotgun Lipidomics. *Mol. Metab.* **2019**, *22*, 12–20. [CrossRef]

175. Hanzu, F.A.; Vinaixa, M.; Papageorgiou, A.; Párrizas, M.; Correig, X.; Delgado, S.; Carmona, F.; Samino, S.; Vidal, J.; Gomis, R. Obesity Rather Than Regional Fat Depots Marks the Metabolomic Pattern of Adipose Tissue: An Untargeted Metabolomic Approach. *Obesity* **2014**, *22*, 698–704. [CrossRef]

176. Mitsutake, S.; Zama, K.; Yokota, H.; Yoshida, T.; Tanaka, M.; Mitsui, M.; Ikawa, M.; Okabe, M.; Tanaka, Y.; Yamashita, T.; et al. Dynamic Modification of Sphingomyelin in Lipid Microdomains Controls Development of Obesity, Fatty Liver, and Type 2 Diabetes. *J. Biol. Chem.* **2011**, *286*, 28544–28555. [CrossRef]

177. Zhao, H.; Przybylska, M.; Wu, I.H.; Zhang, J.; Siegel, C.; Komarnitsky, S.; Yew, N.S.; Cheng, S.H. Inhibiting Glycosphingolipid Synthesis Improves Glycemic Control and Insulin Sensitivity in Animal Models of Type 2 Diabetes. *Diabetes* **2007**, *56*, 1210–1218. [CrossRef] [PubMed]

178. Du, Y.; Hu, H.; Qu, S.; Wang, J.; Hua, C.; Zhang, J.; Wei, P.; He, X.; Hao, J.; Liu, P.; et al. SIRT5 Deacylates Metabolism-Related Proteins and Attenuates Hepatic Steatosis in Ob/Ob Mice. *EBioMedicine* **2018**, *36*, 347–357. [CrossRef] [PubMed]

179. Yetukuri, L.; Katajamaa, M.; Medina-Gomez, G.; Seppänen-Laakso, T.; Vidal-Puig, A.; Orešič, M. Bioinformatics Strategies for Lipidomics Analysis: Characterization of Obesity Related Hepatic Steatosis. *BMC Syst. Biol.* **2007**, *1*, 1–15. [CrossRef] [PubMed]

180. Preuss, C.; Jelenik, T.; Bódis, K.; Müssig, K.; Burkart, V.; Szendroedi, J.; Roden, M.; Markgraf, D.F. A New Targeted Lipidomics Approach Reveals Lipid Droplets in Liver, Muscle and Heart as a Repository for Diacylglycerol and Ceramide Species in Non-Alcoholic Fatty Liver. *Cells* **2019**, *8*, 277. [CrossRef] [PubMed]

181. Wang, M.; Han, X. Advanced Shotgun Lipidomics for Characterization of Altered Lipid Patterns in Neurodegenerative Diseases and Brain Injury. *Methods Mol. Biol.* **2016**, *1303*, 405–422.

182. Bielawski, J.; Szulc, Z.M.; Hannun, Y.A.; Bielawska, A. Simultaneous Quantitative Analysis of Bioactive Sphingolipids by High-Performance Liquid Chromatography-Tandem Mass Spectrometry. *Methods* **2006**, *39*, 82–91. [CrossRef]

183. Yang, G.; Badeanlou, L.; Bielawski, J.; Roberts, A.J.; Hannun, Y.A.; Samad, F. Central Role of Ceramide Biosynthesis in Body Weight Regulation, Energy Metabolism, and the Metabolic Syndrome. *Am. J. Physiol.-Endocrinol. Metab.* **2009**, *297*. [CrossRef]

184. Gao, S.; Zhu, G.; Gao, X.; Wu, D.; Carrasco, P.; Casals, N.; Hegardt, F.G.; Moran, T.H.; Lopaschuk, G.D. Important Roles of Brain-Specific Carnitine Palmitoyltransferase and Ceramide Metabolism in Leptin Hypothalamic Control of Feeding. *Proc. Natl. Acad. Sci.USA* **2011**, *108*, 9691–9696. [CrossRef]

185. Rutkowsky, J.M.; Lee, L.L.; Puchowicz, M.; Golub, M.S.; Befroy, D.E.; Wilson, D.W.; Anderson, S.; Cline, G.; Bini, J.; Borkowski, K.; et al. Reduced Cognitive Function, Increased Bloodbrain-Barrier Transport and Inflammatory Responses, and Altered Brain Metabolites in LDLr-/-and C57BL/6 Mice Fed a Western Diet. *PLoS ONE* **2018**, *13*, e0191909. [CrossRef]

186. Rawish, E.; Nickel, L.; Schuster, F.; Stölting, I.; Frydrychowicz, A.; Saar, K.; Hübner, N.; Othman, A.; Kuerschner, L.; Raasch, W. Telmisartan Prevents Development of Obesity and Normalizes Hypothalamic Lipid Droplets. *J. Endocrinol.* **2020**, *244*, 95–110. [CrossRef]

187. Gudbrandsen, O.A.; Kodama, Y.; Mjøs, S.A.; Zhao, C.-M.; Johannessen, H.; Brattbakk, H.-R.; Haugen, C.; Kulseng, B.; Mellgren, G.; Chen, D. Effects of Duodenal Switch Alone or in Combination with Sleeve Gastrectomy on Body Weight and Lipid Metabolism in Rats. *Nutr. Diabetes* **2014**, *4*, e124-e124. [CrossRef] [PubMed]

188. De la Maza, M.P.; Rodriguez, J.M.; Hirsch, S.; Leiva, L.; Barrera, G.; Bunout, D. Skeletal Muscle Ceramide Species in Men with Abdominal Obesity. *J. Nutr. Health Aging* **2015**, *19*, 389–396. [CrossRef] [PubMed]

189. Stanley, W.C.; Dabkowski, E.R.; Ribeiro, R.F.; O'Connell, K.A. Dietary Fat and Heart Failure: Moving From Lipotoxicity to Lipoprotection. *Circ. Res.* **2012**, *110*, 764–776. [CrossRef] [PubMed]

190. Harmancey, R.; Wilson, C.R.; Wright, N.R.; Taegtmeyer, H. Western Diet Changes Cardiac Acyl-CoA Composition in Obese Rats: A Potential Role for Hepatic Lipogenesis. *J. Lipid Res.* **2010**, *51*, 1380–1393. [CrossRef]

191. Butler, T.J.; Ashford, D.; Seymour, A.-M. Western Diet Increases Cardiac Ceramide Content in Healthy and Hypertrophied Hearts. *Nutr. Metab. Cardiovasc. Dis.* **2017**, *27*, 991–998. [CrossRef]

192. Pakiet, A.; Jakubiak, A.; Mierzejewska, P.; Zwara, A.; Liakh, I.; Sledzinski, T.; Mika, A. The Effect of a High-Fat Diet on the Fatty Acid Composition in the Hearts of Mice. *Nutrients* **2020**, *12*, 824. [CrossRef]

193. Feng, R.; Sun, G.; Zhang, Y.; Sun, Q.; Ju, L.; Sun, C.; Wang, C. Short-Term High-Fat Diet Exacerbates Insulin Resistance and Glycolipid Metabolism Disorders in Young Obese Men with Hyperlipidemia, as Determined by Metabolomics Analysis Using Ultra-HPLC-Quadrupole Time-of-Flight Mass Spectrometry. *J. Diabetes* **2019**, *11*, 148–160. [CrossRef]

194. Araujo, D.S.; Guedes de Oliveira Scudine, K.; Pedroni-Pereira, A.; Gavião, M.B.D.; Pereira, E.C.; Fonseca, F.L.A.; Castelo, P.M. Salivary Uric Acid is a Predictive Marker of Body Fat Percentage in Adolescents. *Nutr. Res.* **2020**, *74*, 62–70. [CrossRef]

195. Ruebel, M.L.; Piccolo, B.D.; Mercer, K.E.; Pack, L.; Moutos, D.; Shankar, K.; Andres, A. Obesity Leads to Distinct Metabolomic Signatures in Follicular Fluid of Women Undergoing In Vitro Fertilization. *Am. J. Physiol. Metab.* **2019**, *316*, E383–E396. [CrossRef]

196. De la Cuesta-Zuluaga, J.; Mueller, N.; Álvarez-Quintero, R.; Velásquez-Mejía, E.; Sierra, J.; Corrales-Agudelo, V.; Carmona, J.; Abad, J.; Escobar, J. Higher Fecal Short-Chain Fatty Acid Levels Are Associated with Gut Microbiome Dysbiosis, Obesity, Hypertension and Cardiometabolic Disease Risk Factors. *Nutrients* **2018**, *11*, 51. [CrossRef]

197. Tipthara, P.; Thongboonkerd, V. Differential Human Urinary Lipid Profiles Using Various Lipid-Extraction Protocols: MALDI-TOF and LIFT-TOF/TOF Analyses. *Sci. Rep.* **2016**, *6*, 1–9. [CrossRef] [PubMed]

198. Bonnet, L.; Margier, M.; Svilar, L.; Couturier, C.; Reboul, E.; Martin, J.C.; Landrier, J.F.; Defoort, C. Simple Fast Quantification of Cholecalciferol, 25-Hydroxyvitamin D and 1,25-Dihydroxyvitamin D in Adipose Tissue Using LC-HRMS/MS. *Nutrients* **2019**, *11*, 1977. [CrossRef] [PubMed]

Publisher's Note: MDPI stays neutral with regard to jurisdictional claims in published maps and institutional affiliations.

Review

Sample Preparation to Determine Pharmaceutical and Personal Care Products in an All-Water Matrix: Solid Phase Extraction

Daniele Sadutto * and Yolanda Picó *

Food and Environmental Safety Research Group, Desertification Research Centre—CIDE (CSIC-UV-GV), University of Valencia (SAMA-UV), Moncada-Naquera Road, Km 4.5, 46113 Moncada, Spain
* Correspondence: sadutto@uv.es (D.S.); yolanda.pico@uv.es (Y.P.); Tel.: +34-96342135528 (D.S.)

Academic Editors: Victoria Samanidou and Irene Panderi
Received: 12 October 2020; Accepted: 5 November 2020; Published: 9 November 2020

Abstract: Pharmaceuticals and personal care products (PPCPs) are abundantly used by people, and some of them are excreted unaltered or as metabolites through urine, with the sewage being the most important source to their release to the environment. These compounds are in almost all types of water (wastewater, surface water, groundwater, etc.) at concentrations ranging from ng/L to µg/L. The isolation and concentration of the PPCPs from water achieves the appropriate sensitivity. This step is mostly based on solid-phase extraction (SPE) but also includes other approaches (dispersive liquid-liquid microextraction (DLLME), buckypaper, SPE using multicartridges, etc.). In this review article, we aim to discuss the procedures employed to extract PPCPs from any type of water sample prior to their determination via an instrumental analytical technique. Furthermore, we put forward not only the merits of the different methods available but also a number of inconsistencies, divergences, weaknesses and disadvantages of the procedures found in literature, as well as the systems proposed to overcome them and to improve the methodology. Environmental applications of the developed techniques are also discussed. The pressing need for new analytical innovations, emerging trends and future prospects was also considered.

Keywords: pharmaceuticals and personal care products; isolation; concentration; solid-phase extraction; cartridges; disks; online; dispersive liquid-liquid microextraction; water samples

1. Introduction

The production and consumption of pharmaceutical and personal care products (PPCPs) is considered an important environmental risk [1–4]. In the last decades, the occurrence of these compounds in nature increased, as described in a number of studies [5–7]. PPCPs can be detected as the active substance, with an unaltered chemical structure, or as a metabolite or a degradation product produced by human and environmental enzymatic activity [8], weather conditions, wastewater treatments [9] and by chemical-physical properties of matrices. There are many reasons for the increasing occurrence of these compounds in different environmental compartments, e.g., their intensive use in farms and aquaculture [10–12] or their inefficient removal from wastewater treatment plants [13,14]. The latter explains why PPCPs used and excreted at home or in hospital can ultimately be released into the environment. Another important source of contamination by PPCPs is industrial waste [15], which is not always processed in the correct form. Furthermore, treated wastewaters are reused for agriculture activity, especially in arid regions [16], contributing to the spread of PPCPs in more matrices, such as soil, wild animals, vegetation, and even food crops.

These considerations on the sources of PPCPs shows the key role that the analysis of water plays to fight against contamination. In fact, water—the most affected environmental compartment—may

Molecules **2020**, *25*, 5204

be considered as a mirror of the pollution status of an area, and also a scarce resource that must be preserved with optimal quality and zero pollution. In addition, water contaminants, depending on their physicochemical properties, may also (bio)accumulate in sediments and biota, consequently harming human health [17]. Therefore, it was considered a relevant vehicle to different environmental compartments.

Determining and quantifying PPCPs in different types of water provides considerable interesting information not only related to pollution status. For example, the analysis of wastewater samples, divided into influent and effluent waters, could offer information on the PPCPs consumption of a community, estimate the wastewater treatment plant efficiency and establish the most recalcitrant compounds difficult to eliminate. River, lake, and seawater samples could give us an idea of the more persistence substances. The detection in irrigation channels could identify food quality issues. In addition, drinking water is certainly another matrix that should be monitored to assess the potential risks on human and animal health for long-term use [18,19].

For an accurate analysis of PPCPs in different types of water, it is fundamental to consider all water characteristics that could influence recovery of the contaminants. The pH could affect the structure of molecule, promote ionization according to pKa, or activate a prodrug with a change of structure. Many substances were thermolabile and photosensitive, for this reason, sample temperature must always be considered. The salinity of water could increase or decrease extraction efficiency due to different ionic strengths of the media, or the formation of molecular complexes between PPCPs and multivalent metal cations present in the samples that are soluble in water [20]. Other water components have also a strong influence on the PPCPs recovery because are responsible for degradation and/or metabolism. A large range of different metabolites can be formed depending on the specific enzymatic activities, presence of fulvic and humic acids, microorganisms, etc. [21–23].

In addition to the matrix, it is also important to consider the structural variability of PPCPs, designed to interact with specific targets. The presence of distinct functional groups (such as esters, carboxylic acid, ketones, amides, etc.) or the existence of nucleophile/electrophile substituents contribute to all chemical-physical characteristics of each active substance. Influencing stability, reactivity, and solubility in water are all parameters that need to be considered before a sample's preparation for analysis. Despite the variability in the PPCPs' chemical structures, for most laboratories specialized in the analysis of these compounds, the use of multiresidue methods is very attractive because not only attains a reduction of cost and time, but also offers global patterns of contamination with only one analysis. Moreover, these methods easily facilitate an eco-friendly analysis with decreases in waste. In these multiresidue methods, the sample preparation becomes the heart of analysis that influences any other procedural steps from sample collection and storage to the specific instruments selected for final quantification (high performance liquid chromatography-mass spectrometry (HPLC-MS), gas chromatography-mass spectrometry (GC-MS), etc.). The choice of a liquid or gas chromatography (LC or GC), to analyze the final extract, is guided by the analytes' polarity. Generally, compounds with polar characteristics are more suitable for LC, and those with non-polar properties are more amenable to GC; most PPCPs are polar or moderately polar [24]. Surprisingly, some contemporary review articles either cover broader aspects of environmental analysis [25–27], or focus on a particular type of extraction process (e.g., microextraction, use of nanomaterials, magnetic, ultrasonic, etc.) [28–30] but do not cover the entire sample preparation.

Therefore, the goal of this review was to critically analyze the status of sample preparation to determine PPCPs in an all-water matrix. Each step of sample preparation, including all analytical variants (conventional and innovative methods) used to detect these contaminants were considered. Furthermore, each sample preparation method was critically analyzed, highlighting advantages and disadvantages. This review performs an examination of all studies published from January 2018 to May 2020. The search was conducted on the database Scopus (Elsevier), with two different inputs: "extraction pharmaceutical environmental"/"extraction personal care products environmental" and "extraction pharmaceutical water"/"extraction personal care products water". More than one thousand one hundred works have been viewed. The selection criteria to choose the studies were based on (i) the presence at least of 10 PPCPs to include attractive multiresidue methods; and (ii) water compartmentation, in all variants (wastewater, rivers, irrigation channels, lakes, drinking water, seas, urban storms, swimming pools and thermal water), was chosen. In addition, some reviews outside the interval of time were also chosen.

2. Extraction and Clean-Up of PPCPs in Water

PPCPs are organic compounds, and traditionally this type of compound has been extracted by solid-phase extraction (SPE). This technique was commercialized in the late 1970s and rapidly replaced the liquid-liquid extraction that was previously used [31]. Table 1 and Figure 1 show the analytical methods applied to extract PPCPs in water. The most common were still based on SPE in all possible variants (cartridge, disk, offline, online, etc.). The classic SPE process (cartridges offline) was used in 71.3% of the studies, the online version was employed in the determination in 9.2%, and disks were utilized instead of cartridges in 3.2%.

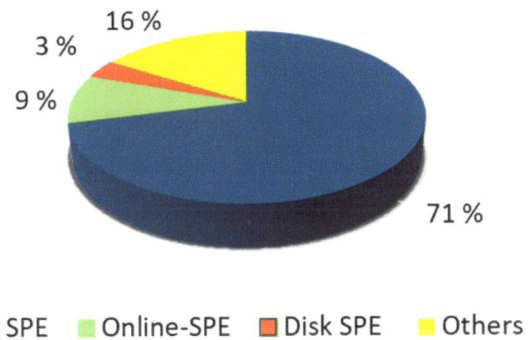

Figure 1. Pharmaceuticals and personal care products (PPCPs) extraction procedures according to the percentage of studies that applied them. SPE: solid-phase extraction.

Only 16% of the studies use other types of methods, such as direct injection, dispersive liquid-liquid extraction (DLLME) (based on liquid-liquid extraction), polyether sulfone microextraction (PES) or buckypaper devices. It is important to note here that many of the methods classified as "other" are based on the basic principles of SPE, but using new phases or formats.

2.1. SPE

This technique involved the use of a small amount of sorbent (commonly hundreds of mg) in a cartridge or syringe barrel. After activation of the sorbent, a water sample of hundreds of mL was passed through the sorbent, which retained the analytes of interest (in this case PPCPs) whereas the water was discarded. Then, the analytes retained in the sorbent were eluted using a few mL of organic solvent. This technique has some advantages, such as the minor investment in reagent and materials, and rapidity.

Table 1. Selected applications extraction approaches to determine PPCPs in water samples.

Matrix *	No. of PPCPs	Preservation	Volume (mL)	Extraction Method	Sorbent or Cartridge	Detection	Recovery %	Reference
WW, SF	168	Na$_4$EDTA	50	SPE	Cleanert PEP-2	HPLC-MS/MS	0.05–127	[32]
WW, SF	168	Na$_4$EDTA	-	Direct injection	-	HPLC-MS/MS	0.05–127	[32]
SF	59	Na$_2$EDTA	1000	SPE	Oasis HLB	HPLC-MS/MS	52–137	[33]
WW, SF, DW	27	-	-	SPE	Cleanert PEP	HPLC-MS/MS	74–120	[34]
WW	55	Na$_2$EDTA	150	SPE	Oasis HLB	HPLC-MS/MS	9–119	[35]
SW	91	-	1000	SPE	Oasis HLB	HPLC-MS/MS	70–110	[36]
WW	12	-	7.9	DLLME	-	GC-MS/MS	91–115	[37]
WW	12	-	1000	SPE	Oasis HLB	GC-MS/MS	65–115	[38]
WW, SF, DW	58	-	1.8	Online-SPE	PLRP-s	HPLC-MS/MS	70–120 (82% of total)	[39]
SW	62	-	≤20	Online-SPE	Oasis HLB	HPLC-MS/MS	81–120	[40]
SW	62	-	200	SPE	Oasis HLB	HPLC-MS/MS	81–121	[40]
WW, SW	44	Na$_2$EDTA	200	SPE	Strata-X	HPLC-MS/MS	8–239	[41]
SW	11	-	200	SPE	Oasis HLB	HPLC-MS/MS	40–120	[42]
SW	34	Na$_2$EDTA	400	SPE	Oasis HLB	HPLC-MS/MS	41–125	[43]
WW, SW	30	Na$_2$EDTA	250	SPE	Oasis MCX	HPLC-MS/MS	78–106	[44]
SW	10	-	500	SPE	Oasis HLB	HPLC-MS/MS	69–88	[45]
WW	11	-	-	Online-SPE	TurboFlow™ column	HPLC-MS/MS	45–150	[46]
SW	16	-	10	DLLME	-	HPLC-MS/MS	70–120	[47]
WW, SW	27	Na$_2$EDTA	125–500	SPE	Oasis MCX	HPLC-MS/MS	73–116	[48]
WW, SW	25 (of 41)	Na$_2$EDTA	120	PES microextraction	-	HPLC-MS/MS	80–119	[49]
WW, SW	25 (of 41)	Na$_2$EDTA	100–250	SPE	Oasis HLB	HPLC-MS/MS	71–131	[49]
WW, SW	10	-	20 uL	Online-SPE	Oasis HLB	HPLC-MS/MS	-	[50]
SW	12	-	500	SPE	Oasis HLB	HPLC-MS/MS	55–120	[51]
WW, SW	44	-	500	SPE innovative	GCHM, Oasis HLB	HPLC-MS/MS	76	[52]
WW	190	-	100	SPE innovative	Oasis HLB, Isolute ENV+, Strata-X-AW, Strata-X-CV	UPLC-Q-TOF-MS/MS	57–120	[53]
WW	52	-	100	Disk SPE	BAKERBOND C18 Polar Plus	GC-TOF-MS	-	[54]
SW	24	Na$_2$EDTA	1000	SPE	Chromabond HR-X	HPLC-MS/MS	52–117	[55]
SW	13	Na$_2$EDTA	250	SPE	Strata-X	HPLC-MS/MS	51–102	[56]
SW	32	-	200	SPE	Strata-X	HPLC-MS/MS	36–119	[57]
SW	32	-	200	SPE	Strata-X-CW	HPLC-MS/MS	25–110	[57]
SP	111	-	150	SPE	Strata-X-CW	SFC-MS/MS	77 (average)	[58]

Table 1. Cont.

Matrix *	No. of PPCPs	Preservation	Volume (mL)	Extraction Method	Sorbent or Cartridge	Detection	Recovery %	Reference
WW, SW	40	-	250	SPE	Oasis HLB	HPLC-MS/MS	17–146	[59]
WW	11	-	250	SPE	Oasis HLB	HPLC-MS/MS	53–124	[60]
SW	39	-	1000	SPE	Oasis HLB	HPLC-MS/MS	1–125	[61]
WW	15	-	250	SPE innovative	Strata-X, PSA, Alumina	GC-MS	19–103	[62]
SW	69	-	100	SPE	Strata X-CW	SFC-MS/MS	76	[63]
WW, SW	31	-	100–500	SPE	Chromabond HR-X	HPLC-MS/MS	32–97	[64]
SW	130	Na$_2$EDTA	2000	SPE innovative	Oasis WAX, Oasis HLB, Sep-Pak Plus AC 2	HPLC-MS/MS	50–150	[65]
WW, DW	28	-	1000	SPE	C18 Cartridges	HPLC-MS/MS	n.r.–293	[66]
WW, SW	10	Na$_2$EDTA	500	SPE	Oasis HLB	UPLC-Q-TOF-MS/MS	n.r.–128	[20]
WW, SW	23	-	500	SPE	Oasis MCX	HPLC-MS/MS	54–117	[67]
WW	52	Na$_2$EDTA	10	Online-SPE	Shim-pack MAYI-ODS	HPLC-MS/MS	74–104	[68]
SW	20	Na$_2$EDTA	100	SPE	Strata-X	HPLC-MS/MS	70–119	[69]
WW, SW	20	-	200	SPE	Strata-X-Drug B	HPLC-MS/MS	39–102	[70]
SW	61	Na$_2$EDTA	1000	Disk SPE	Speedisk®	HPLC-MS/MS	-	[71]
SW	61	Na$_2$EDTA	200	SPE	Oasis HLB	HPLC-MS/MS	-	[72]
WW	26	Na$_2$EDTA	500	SPE	Oasis HLB	HPLC-MS/MS	-	[72]
WW	10	-	-	SPE	Oasis HLB	HPLC-MS/MS	85–94	[73]
SW	35	Na$_2$EDTA	1000	SPE	Oasis HLB	HPLC-MS/MS	58–194	[74]
WW, SW	20	-	300–400	SPE	Oasis HLB Prime	GC-MS	≥40%	[75]
WW	83	Na$_2$EDTA	50–100	SPE	Strata-X	HPLC-MS/MS	n.r.–122	[76]
WW	59	Ascorbic acid; Sodium thiosulfate	1000	SPE	Oasis HLB	HPLC-MS/MS	9–143	[77]
WW	20	Sodium thiosulfate	500	Online-SPE	Oasis HLB	HPLC-MS/MS	-	[78]
WW	20	Sodium thiosulfate	-	Direct injection	-	HPLC-MS/MS	-	[78]
SW	13	-	-	Passive sampling	PES membranes	LC-DAD	-	[79]
WW	21	-	1000	SPE	Oasis HLB	LC-HRMS	40 (average)	[80]
WW, SW	103 (of 300)	Formaldehyde	250	SPE innovative	Strata-X	UPLC-Q-TOF-MS/MS	-	[81]
WW	37	-	0.5	Online-SPE	PLRPs	HPLC-MS/MS	5–132	[82]
WW	20	-	2	Direct injection	-	HPLC-MS/MS	60–124	[83]
WW	12	-	20–100	SPE	Oasis HLB	HPLC-MS/MS	77–115	[84]
WW, SW	48	-	300–400	SPE	Oasis HLB Prime	GC-MS	>40	[85]
WW	38	-	100	SPE	Oasis MCX	HPLC-MS/MS	65–134	[86]
SW	33	-	200	SPE innovative	Oasis HLB, LC18 column	HPLC-MS/MS	50–106	[87]

Table 1. *Cont.*

Matrix *	No. of PPCPs	Preservation	Volume (mL)	Extraction Method	Sorbent or Cartridge	Detection	Recovery %	Reference
SP	48	Na$_4$EDTA	200	SPE	Oasis MCX	HPLC-MS/MS	71–122	[88]
WW	22	NaCl	100	Online SPE	DVB/CAR/PDMS	GC-MS	6–104	[89]
WW	19	-	250	SPE	Oasis HLB	LC-TOF/MS	5–111	[90]
WW	11	-	0.9	DLLME	-	HPLC-MS/MS	n.r.–124	[91]
WW, SW	40 (of 139)	-	1000	SPE	Oasis HLB	HPLC-MS/MS	n.r–99	[92]
WW, SW	41 (of 139)	-	1000	SPE	Bond-Elut ENV	HPLC-MS/MS	n.r–99	[92]
SW	10 (of 28)	-	500	Buckypaper Device	-	HPLC-MS/MS	n.r–102	[93]
SW	44	-	200	SPE	Strata-X	HPLC-MS/MS	85–100	[94]
SW	45	Na$_2$EDTA	1000	SPE	Strata-X	HPLC-MS/MS	38–112	[95]
WW	13	-	150–300	SPE	Oasis HLB	HPLC-MS/MS	40–115	[96]
SW	42	Na$_2$EDTA; ASA(DW)	50	SPE	Oasis HLB	HPLC-MS/MS	33–117	[97]
WW, SW	39 (of 80)	-	500–100	SPE	Oasis MCX; Oasis HLB	HPLC-MS/MS	31–131	[98]
SW	110 (of 1153)	Phosphate buffer	1000	Disk SPE	Glass microfiber, Empore™ SDB-XD, Empore™ AC	GC-TOF-MS/MS	-	[99]
WW	82	-	250	SPE	Oasis HLB	LC-Q-TOF-MS	66–149	[100]
SW	35	-	100–500	SPE	Oasis HLB	HPLC-MS/MS	2–132	[101]
WW, SW	10	-	50–100	SPE innovative	Oasis MCX, Oasis MAX	LC-HRMS	60–109	[102]
WW, SW	10	Sodium azide; ascorbic acid	200–1000	Disk SPE	Atlantic HLB	HPLC-MS/MS	48–122	[103]
WW, SW	10	Sodium azide ascorbic acid	200–1000	SPE	Oasis HLB	HPLC-MS/MS	1–110	[103]
WW	17	-	250	SPE	Oasis HLB	HPLC-MS/MS	<40%	[104]
SW	10	Citric acid	1000	SPE	C18	HPLC-MS/MS	97–101	[105]
WW	100	-	200	SPE	UCT XRDAH	LC-Q-TOF-MS/MS	-	[106]

* WW = Wastewater; SF = Surface water; SP = Swimming pool water. n.r. = not recovered.

2.1.1. Sorbents and Formats

There are different marketed sorbents that work principally with two distinct separation mechanisms: classical reversed phase chromatography (RP) or ion exchange chromatography (IC). Characteristics of these sorbents are summarized in Table 2. The sorbents used in RP are mostly of a polymeric nature and could be used for a large spectrum of PPCPs, including acidic, basic, and neutral compounds. RP sorbent was applied in 85% of SPE approaches (see Table 1). This is attributable to its ability to retain a wide range of different polarity compounds, a relevant characteristic for a multiresidue method that includes different chemical classes. Of these, two-thirds had a stationary phase characterized by polymeric sorbent that contained vinylpyrrolidone (Oasis® HLB, Strata®X and Cleanert® PEP). Another polymeric sorbent applied was marked by the presence of polystyrene-divinylbenzene (Isolute® ENV+, Chromabond® HR-X and Bond ElutTM ENV) that was applied to polar analytes, or the presence of octadecyl endcapped silica RP (Supelclean™ LC-18 SPE) that was used for nonpolar to moderately polar analytes from aqueous samples.

The IC sorbents were used in the so-called mixed-mode cartridges that combine the polymeric sorbent with an ionic exchanger that could be weak or strong. The IC can be of cations or anions, and this affects the target specificity. Cationic exchange sorbents (weak or strong) were designed to extract basic PPCPs, and anionic exchange sorbent (weak or strong) were to extract the acidic ones. However, weak polymeric cationic-exchangers (Oasis® MCX, Strata®X-CW or X-Drug B and UCT XTRACT® XRDAH) were the most prevalent.

Different studies applied attractive modifications to these traditional cartridges, to obtain the best recoveries for a large spectrum of compounds. Zhu et al. [52] developed, characterized and tested a hydrophilic resin based on poly(*N*-vinyl pyrrolidone-co-divinylbenzene) (NVP-co-DVB) that improved the average absolute recovery for 44 PPCPs, with respect to the use of HLB sorbents. Alternatively, Caban et al. [62] studied the modification of the columns through the application of additional sorbents on top of a polymeric HLB column to improve the SPE of 15 analytes (pharmaceuticals and estrogens) from water. PSAs (Primary and Secondary Amines) and alumina retained matrix components (e.g., humic and fulvic acids) without decreasing the analyte recovery. The solution was named triple-sorbents SPE. They were applied in order to reduce matrix effect. Similarly, Gago-Ferrero et al. [53] mixed four SPE materials simultaneously in an in-house cartridge. These materials included classical RP and ion mixed-mode sorbents (Oasis HLB, Isolute ENV+, Strata-X-AW and Strata-X-CV). Salas et al. [102] combined anionic and cationic exchange sorbents in the same cartridge to extract basic and acidic pharmaceuticals simultaneously. These minor improvements in the SPE procedure (without any relevant cost increment or enlargement of the procedure) can produce an important effect on the quantification of PPCPs, and increase the reliability and reproducibility of the results.

The format of the cartridge and the volume of samples that pass through the cartridge were other elements to take into consideration. The sorbent weight (mg), capacity (mL) and pore size (µm) can influence the efficiency of the columns. These parameters have a key role on the surface area, on which analytes interacted. The amount of sorbent ranged from 60 to 600 mg, and the most used was a quantity of 200 mg. The capacities most used were 3 and 6 mL. There is a global consensus on this point.

In SPE, the volume of water processed generally involves hundreds of milliliters. The capacity to detect lower amounts is one of the advantages of SPE. The matrix effect (ME) could be an element to take in consideration in the choice of volume because it is directly proportional to volume as well as the organic matter content of the water; in this case, clogging of the cartridge slows the process too much. For this reason, some studies chose different volumes of water (depending on its characteristics) with the same method. For example, influent wastewater samples were generally analyzed with smaller volumes compared to effluent samples [46,48].

Table 2. Mechanism, type of sorbent and target of the most used brand name of offline columns.

Brand Name	Mechanism	Sorbent	Target
Oasis HLB, HLB Prime	RP	divinylbenzene-co-N-vinylpyrrolidone	acidic, basic, and neutral compounds
STRATA-X	RP	styrene-divinylbenzene-co-N-vinylpyrrolidone	acidic, basic, and neutral compounds
Cleanert PEP	RP	divinylbenzene-co-N-vinylpyrrolidone-Urea	acidic, basic, and neutral compounds
Isolute ENV+	RP	polystyrene-divinylbenzene (PS-DVB)	polar compounds
Bond-Elut ENV	RP	polystyrene-divinylbenzene (PS-DVB)	polar compounds
Chromabond HR-X	RP	polystyrene-divinylbenzene (PS-DVB)	polar compounds
Oasis MCX	IC	mixed-mode CATION-exchange polymer-based	basic compounds, particularly strong bases
Oasis WAX	IC	mixed-mode ANION-exchange sorbent polymer-based	acidic compounds
Strata-X-CW	IC	mixed-mode CATION-exchange polymer-based	basic compounds, particularly strong bases
Strata-X-AW	IC	mixed-mode ANION-exchange sorbent Polymer-based	acidic compounds
Strata-X-Drug B	IC	mixed-mode strong CATION-exchange polymer-based	basic compounds, particularly strong bases
UCT XRDAH	IC	mixed-mode CATION-exchange polymer-based	basic compounds, particularly strong bases

Figure 2 compares the various volumes that were selected in the different studies. There are two ranges of volumes commonly chosen—between 200 and 250 mL (32% of the studies) and in interval ≥500 mL (42% of works). The volumes commonly used were 200, 250, 500 and 1000 mL. The latter is most commonly selected, but mainly for surface and drinking water.

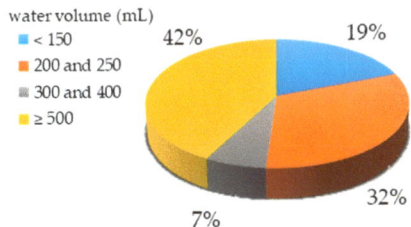

water volume (mL)
- < 150
- 200 and 250
- 300 and 400
- ≥ 500

42%   19%   32%   7%

Figure 2. Percentage of studies according to water volume.

2.1.2. SPE Activation, Washing and Elution

In SPE, the first step is the conditioning of the sorbent in order to favor the interaction of the sorbent with the analytes by a mechanism namely "solvation". RF and IC sorbents are usually activated by filling the column two or three times with a solvent miscible with water (e.g., methanol, acetonitrile) followed by the solvent in which the analyte is dissolved (pure matrix, e.g., water, buffer). Methanol and water are most frequent solvents to activate cartridges. Water used for activation can be pH-adjusted or spiked with salt, counter ions or metal sequestrators (sodium acetate buffer [106], monopotassium phosphate [95], sodium dodecyl sulphate [57], etc.) to promote several types of interactions with compounds.

Once the sample passed through the sorbent, another important step is to remove the impurities retained on the SPE packing. For this reason, a wash solution strong enough to remove these impurities, but weak enough to leave the analytes of interest, is passed through the cartridge. These solutions are water, pH-adjusted water, or in a few cases methanol-water (5:95, *v/v*). The wash was followed by cartridges air-drying by vacuum or pressure to remove the remaining water, although in few cases, the cartridges can be dried with nitrogen instead of air [37,48,55,61,63].

The RP SPE is the mechanism more commonly used to extract contaminants from water samples, as described previously. The hydrophobic or non-polar interactions between sorbent functional groups and analytes must be destroyed with an organic solvent or solvent combination of enough non-polar character. The most used elution solvents are methanol and acetonitrile. The pH modification during elution can improve recovery if the analyte is ionizable and the eluent favors its ionic form, and basic and acidic compounds become more polar [107].

Figure 3 shows the eluents most frequently applied with HLB sorbents. Only methanol was used in more than half of methods, with acid or basic pH adjustment in 13%, and with a mix of other solvents or followed by a second elution with different solvent, for example, acetonitrile, dichloromethane and acetone, in 27%. Only 7% of the methods presented an eluent mixes (mostly acetonitrile and acetone) without methanol. In the case of weak cationic-exchange sorbents, the eluent was a basic solution with NH_4OH in methanol or acetonitrile.

It is fundamental to consider not only the elution but also all steps that follow the elution. Once the PPCPs have been eluted, the extract is evaporated under a gentle stream of nitrogen in order to concentrate the analytes. The temperature used in this evaporation step is an important parameter that could affect the molecular stability of PPCPs. However, this temperature was not always reported in the studies, even though it can be responsible for the degradation of some PPCPs. The temperature range was between 25 and 50 °C, and the average was 39 °C.

Eluent

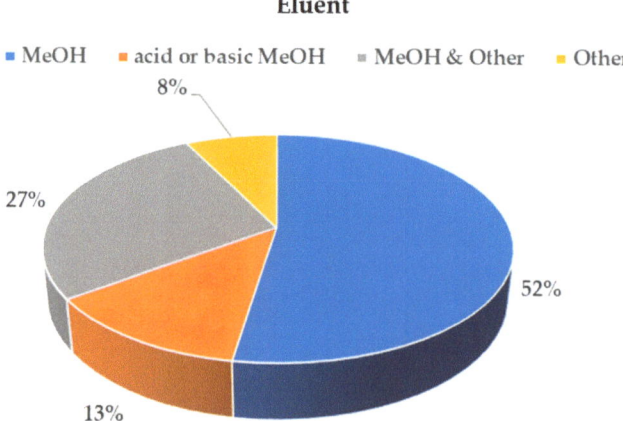

Figure 3. Percentage of studies according to the type of eluent used.

The dried extracts are reconstituted with one or more solvents. If the compounds are determined by HPLC-MS, the mixture of solvents is usually comparable in composition to the initial mobile phase. Again, the reconstituting solvent more commonly used was the methanol-water mix. The most common volumes used for reconstitution were 0.5 and 1 mL. In the case of GC analysis, the addition of a derivatization reagent was often required because PPCPs are mostly polar and/or thermolabile, and these compounds are not amenable for GC analysis without increasing the stability of analytes.

Lastly, the final extract is filtrated before the analysis. Different syringe filters were reported, such as nylon, PVDF (polyvinylidene fluoride), PTFE, RC, GHP, PP, etc. The size used were 0.20 and 0.22 μm. This procedure was an ulterior clean-up to remove the particles with a large size to optimize analysis, for example by reducing the presence of obstruction phenomena in column. For this step, it was fundamental to remember which combinations (solvent-filter) are safe and which are corrosive.

2.1.3. Water Sample Pretreatment before SPE

Regarding the characteristics of the water sample, the preservation of its integrity from the sampling point to the moment of the analysis was vital. To this end, in many cases, the water samples were spiked by preservative compounds or solutions. As described in the U.S. Library of Medicine (2017), a pharmaceutical preservative is referred to as "substances added to pharmaceutical preparations to protect them from chemical change or microbial action" [108]. In the same way, these preservatives can be added to the water sample to avoid the degradation of the PPCPs present in the sample. Preservatives can be natural or synthetic compounds and included buffers, bulking agents, chelating agents, antioxidants, antimicrobial agents, surfactants, etc. " [108,109]. The most used was EDTA—a chelating agent thanks to its four carboxyl groups and two nitrogen atoms that can form stable complexes with cations—which is able to improve the extraction efficiency of certain pharmaceuticals that also form complexes with metals, such as antibiotics, because they sequestrate the metals of the solution, liberating the PPCPs and increasing their recovery [110]. Generally, EDTA was added in the samples to a final concentration of 0.1% (1.000 g/L) or 0.05% (0.500 g/L), but it could reach a final concentration of 0.2, 2 or even 5%. Other preservatives used were antimicrobials, such as formaldehyde, NaCl, sodium azide and citric acid, and/or antioxidants, such as ascorbic acid and sodium thiosulfate. These antioxidants reduced any residual chlorine, chloramine and ozone that had been used as a disinfectant because they could react with some antibiotics [97]. Furthermore, more than half of the methods adjusted the pH of the sample to prevent degradation, ionization phenomena, or to achieve the optimum value for the extraction. For about 70% of methods, pH was adjusted between 2 and 3 units. In a few cases this adjustment was between 3.5 and 7. It was rarely adjusted by 9 and 10 units.

To ensure the proper quantification of the analytes is always very important. Therefore, samples were spiked by a solution of internal standard (IS) in more than 83% of the studies to obtain more reliable results, taking matrix effects into consideration. It was not always possible to use an isotopic reference for every compound, because the cost of the ISs are high, and are not available for some of the target analytes [57].

Another pretreatment widely used in water sample preparation was filtration, the role of which was to remove suspended substances, such as suspended particles, colloids and microorganisms from samples to prevent obstruction of the SPE cartridges or significant interferences in subsequent treatment processes [74]. In various studies, it was easy to find filters constituted of different materials, such as paper, nylon, PVDF (polyvinylidene fluoride) and cellulose membrane. The most frequently used to monitor the water pollution was a glass microfiber filter. The mesh filter range was from 0.20 to 1.60 μm; the size most commonly used was 0.45 μm. Sometimes filters with different sizes were coupled to remove particles at different levels. Centrifugation was an alternative to filtration, with the same goal but much less used [32]. In this case, the mass deposited on the bottom was removed and analysis was focused on supernatant.

2.2. Online SPE

SPE can be used offline (independently from the further chromatographic analysis) or online (directly connected to the chromatographic system). However, there are few difference in the components of the techniques between the two formats, with the exception of the valves system used to connect SPE online with the determination technique (commonly any type of HPLC-MS). In both, the main factors that affected the results of these technique were the formats and sorbents of stationary phase (cartridge) and the solvent(s) used for activation, washing and elution. Online SPE can be coupled to both, LC or GC. However, the preferred technique is online SPE liquid chromatography tandem mass spectrometry (SPE-LC-MS/MS), PPCPs and the SPE eluents are much more compatible with the mobile phase of LC than with that of GC.

To be functional, online systems require the use of 6 or 10 port valves to automate extraction and connection to the instrument. A schematic illustration of the analytical system is presented in Figure 4. The most common are home-made devices, but there are also commercial systems such as an automated sample processor [82]. These devices can be personalized and adapted to the particular analysis, for example, to obtain online cross, which allows the automatic cross-utilization of two SPE columns to speed up the analysis of pharmaceuticals in different water samples [40].

The columns for online-SPE mode were characterized by similar polymeric RP sorbents. The main differences, with offline-SPE, were the length of columns (generally 1–2 cm). OASIS HLB (2.1 × 30 mm, 10 μm) [40,50,78], MAYI-ODS (10 mm × 2.0 mm, 50 μm) [68], and PLRP columns [39,82] have been the most widely reported. However, sorbents based on alternative mechanisms, such as the so-called TurboFlow™, which mixes size exclusion chromatography with reversed sorbents, has also been reported to determine pharmaceuticals [46]. To this end, three TurboFlow™ columns (TFC) connected in series were used, i.e., Cyclone P–C18-P XL–Cyclone MAX in order to achieve a proper extraction and clean-up.

The online SPE systems present the advantage to provide automatic and efficient sample loading, clean-up, desorption, separation, and detection at the same time, to reduce the sample volume, save time and solvents, prevent sample contamination and PPCP loss, and improve the method performance. The reduction in sample volume that could decrease sensitivity is normally compensated for by the increase in sensitivity as all the analyte retained in the sorbent passes to the chromatographic column.

The disadvantage of these methods is that they are not very versatile to be adapted to different types and conditions of analysis, because sample pH, injection volume, and valve-switching time needs to be very carefully optimized to ensure appropriate method performance for target PPCPs [39]. Once established, they are more suitable for routine analysis.

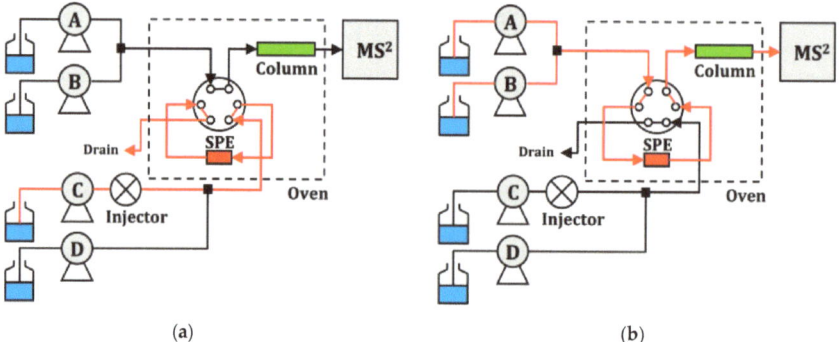

(a)   (b)

Figure 4. Configuration of the online SPE liquid chromatography tandem mass spectrometry (SPE-LC-MS/MS) system. I. separation column equilibration (pump A and B); II. sample loading and trapping (pump C); III. SPE desorption (pump A and B) and cleaning (pump D). Reproduced from [68] with permission from Elsevier. (**a**) Sample loading (Pump A, B, and C: On); (**b**) Sample analysis (Pump A, B, and D: On).

2.3. SPE Disk

The disks are a variant of cartridges that follow the same principle to retain an elute PPCPs. The disks have a higher diameter (commonly ca. 45 mm) and low height (a few millimeters). This format attempts to address several disadvantages of the cartridges, such as plugging due to the suspended particulate matter, high back-pressure that reduces flow rates, and improve retention kinetics of the analytes by using lower particle size. Generally, it was used with higher volumes of samples than SPE. The passage of the sample was much quicker (up to 100 mL/min). Moreover, the use of disks was advantageous for handling dirty samples. Only four studies used this approach (Table 1). There are disposable disks of many types of sorbents: C18, hydrophilic divinylbenzene (DVB), HLB or carbon.

Hydrophilic divinylbenzene (DVB) disks have been compared with Oasis HLB cartridges. Although the most apolar analytes (LogP ≥ 4) attained higher process efficiencies following Speedisk extraction, it could be noticed that in general, process efficiency was lower than for Oasis HLB extraction: 16 versus 59 analytes having a process efficiency >60% for Speedisk and Oasis HLB, respectively [71]. However, this study did not compare the same type of sorbent in both formats. Kafeenah et al. [103] did compare both formats using HLB sorbent. The method using disk SPE was better in terms of recovery, sensitivity, rapidness, and matrix effect, compared to the cartridge method.

The combination of different disks in order to improve recoveries has also been tested for GC-MS amenable analytes using, in sequence, a glass microfiber disk (GMF 150, 47 mm, Whatman), a styrene-divinylbenzene disk (Empore™ SDB-XD, 47 mm), and an active carbon disk (Empore™ AC, 47 mm) [99]. However, the same study recommended the use PS-2 and AC-2 Sep-Pak short cartridges for compounds analyzed by HPLC-MS.

The main disadvantages, as evidenced in the studies, are related with highest waste of samples and reagents used to active, wash and elute the sorbent.

2.4. Other Extraction Approaches

Other approaches have been reported to extract the PPCPs from water, and even though they are not as used as SPE, can be advantageous for some applications. These approaches are commonly focused on more environmentally friendly alternatives that reduce the use of materials, organic solvents and reagents; the so-called green chemistry. The simplest process is direct injection without any pre-concentration steps [78,83]. This was possible thanks to the excellent sensitivity of the HPLC-MS/MS. Botero-Coy et al. [83] included a simple dilution with water (×5) in order to reduce the matrix complexity. The most important problem in this method is its high matrix effects.

The microextractions, both solid and liquid, are an attractive alternative to SPE. The Dispersive Liquid-liquid Microextraction (DLLME) [37,47] has benefits related to a quick, easy cleaning and highly efficient pre-concentration procedure. Moreover, the sample volume required was very small, reducing the wastes. Two solvents were used, a dispersant and an extractor. The dispersant must be soluble in water and in the extractor, and the choice was methanol. As extractor (few microliters) both, the most traditional, chloroform, and the most recently introduced ionic liquids (1-Hexyl-3-methylimidazolium hexafluorophosphate) have been reported. The most important problem of this technique to determine PPCPs is that it is more efficient for non-polar compounds. Other approaches applied were the solid phase microextraction (SPME), which is attractive, because it enables extraction and clean-up in only one step, eliminating the problems associated with extensive use of solvents and equipment, since the analytes retained in the fiber can be thermally desorbed in the GC injector [111]. Another advantage of this technique is that the adsorbed analyte can be derivatized on-fiber, to transform it into a more volatile compound in order to make it more GC-MS amenable. Recently, this on-fiber derivatization and on-line thermal desorption has been applied for the extraction of 17 mL of water sample for the simultaneous determination of 22 pharmaceuticals and personal care products, including three transformation products, in sewage [89]. The fiber used was 2-cm long, 50/30-μm thick Divinylbenzene/Carboxen/Polydimethylsiloxane (DVB/CAR/PDMS). This method has the advantage of avoiding the use of organic solvent, since analytes are directly desorbed in the GC injector. SPME fiber can also be desorbed with a few mL of organic solvent to analyze the PPCPs by HPLC-MS. As in the method reported by Mijangos et al. [49], preconcentrated pharmaceuticals and personal care products in a disposable and low cost polyethersulfone (PES) sorbent are further desorbed in methanol.

Other approaches are based on testing new phases consisted of nanomaterials as more efficient sorbents. Tomai et al. [93] performed an SPE with oxidized buckypaper (BP) for Stir-Disc. The aim was to propose an SPE extraction device which combined the properties of carbon nanotubes and magnetic stirring with the main advantages of disk SPE. The concept was the same as classic SPE, but in this case the extraction device was immersed into the aqueous sample and left under magnetic stirring to permit analyte absorption on a BP membrane.

The last approach was the use of a PASSIL sampler [79]. The acronym PASSIL describes a device constituted of two PES membranes impregnated with an ionic liquid. After passive sampling, the receiving phase was eluted from the membranes and dissolved with acetonitrile.

3. Environmental Applications

Seventy-six studies have been selected (Table 1). Nine studies proposed the application of two different methods or approaches for sample preparation. In many cases, the method was applied to various aqueous matrices. The average number of PPCPs detected for each method was forty-one. The average number of PPCPs included in the studies reviewed increased over time, which may be justified by the growing interest in using methods that include as many substances as possible in the same analysis. In the studies published in 2018, the average number of PPCPs included was 34; in 2019 the average was 38 and up to May 2020 it was 60. In the Figure 5, the different studies are classified according to the number of compounds detected. Forty-six percent of studies covered between 10 and 25 PPCPs. The second range (between 26 and 50) included 30% of studies. Fifteen percent showed a 50 < PPCPs < 100 range. Lastly, a small portion (9%) included more than 100 PPCPs. A few methods reported a contaminants list characterized by a multiclass of compounds, not only PPCPS, but pesticides, drugs, flame retardants, etc., too. For these cases, only the total number of PPCPs was considered.

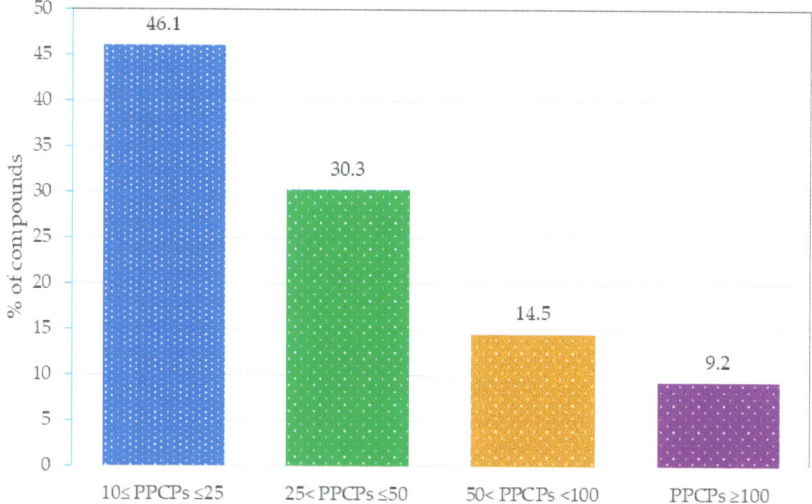

Figure 5. Percentage of studies according to the number of PPCPs analyzed.

Different types of water were collected. The principal parameters monitored were temperature, pH, EC (electrical conductivity, µS cm^{-1}), TDS (total dissolved solids, µg L^{-1}), DO (dissolved oxygen, mg L^{-1}), TSS (total suspended solids, mg L^{-1}) and BOD (biochemical oxygen demand, mg L^{-1}). Influent and effluent wastewater samples from hospitals or wastewater treatment plants (WWTPs) were the most analyzed (46% of studies). This matrix was marked by complexity due to the presence of numerous interferents. Some studies investigated the presence of PPCPs in raw wastewater at the treatment plant, some just the effluents, and some investigated both [35]. Most of these studies also studied the efficiency of the elimination of the PPCPs in the WWTPs [35]. All these studies identified the WWTP effluents as one source of PPCPs to the environment.

In the second block of studies, different matrices were regrouped into a single group: surface water (SW). The 40% of works analyzed and studied in this group investigated a large spectrum of water sources: streams, rivers, estuaries, lakes, seas, ground water, and urban and agricultural storm waters. Drinking and tap water (DW) constituted the third group, with an occurrence of 12% in the studies selected. DW was regulated by "The Drinking Water Directive 98/83/EC" that supervises the quality of water (for human consumption). It provided a general framework and a minimum value of 48 specific parameters that must be monitored regularly [24].

Lastly, two works included other two aqueous matrices: thermal and swimming pool water (SP). Chemicals in these matrices can come from different sources, such as bathers, who continuously release organic matter mainly through sweat and urine [25].

4. Conclusions and Future Trends

The review article focused on the extraction methods for PPCPs. Although these extraction methods are clearly dominated by offline solid-phase extraction using cartridges, there are significant knowledge gaps in accurately understanding the extraction mechanisms for PPCPs, including some metabolite and/or degradation products. Many of the most recent and innovative methods are based on the combination of sorbents with different chemical-physical properties either in the same cartridge, in parallel, or even in series. These modifications are considered to be small steps, but nevertheless, they represent a great advance by improving the extraction of a group of compounds with very different polarities. Multiresidue methods able to cover more than 100 compounds are already a reality. Rapid, lower cost, and eco-friendly sample preparation techniques are urgently needed. Automated, online preconcentration, and clean-up steps prior to the instrument analysis would be the future of the technique. Solvent-free microextraction methods are the trend of extractions such as DLLME. However, these techniques are rarely used, as they work especially well with non-polar PPCPs, but most are polar.

Most of the environmental studies carried out so far have two aspects, (i) analytical including validation of the methods that mainly improves the accuracy in the quantification and the elimination of the matrix effects, and (ii) the environmental aspect in which the whole cycle of the water is covered, including the identification of the sources of these compounds to the environment, the efficiency in the elimination, and the influence of environmental factors such as seasonality. All these studies have contributed to an important advance of knowledge about the distribution and hazards of PPCPs. A gap detected in these studies is the lack of knowledge about the mixtures of PPCPs found in the environment and on the different metabolites and/or degradation products that can be present. It is expected that, in the near future, there will be an increase in knowledge in these fields.

Author Contributions: Conceptualization, D.S. and Y.P.; resources, D.S.; writing—original draft preparation, D.S.; writing—review and editing, Y.P.; funding acquisition, Y.P. All authors have read and agreed to the published version of the manuscript.

Funding: This research was funded by the Spanish Ministry of Science, Innovation and Universities and the ERDF (European Regional Development Fund) through the project CICLIC—subproject WETANPACK (RTI2018-097158-B-C31) and by the Generalitat Valenciana through the project ANTROPOCEN@ (PROMETEO/2018/155) and D. Sadutto was funded by Generalitat Valenciana through a Santiago Grisolia grant: "GRISOLIAP/2018/102, Ref CPI-18-118."

Conflicts of Interest: The authors declare no conflict of interest.

References

1. Pereira, A.M.P.T.; Silva, L.J.G.; Lino, C.M.; Meisel, L.M.; Pena, A. A critical evaluation of different parameters for estimating pharmaceutical exposure seeking an improved environmental risk assessment. *Sci. Total Environ.* **2017**, *603–604*, 226–236. [CrossRef] [PubMed]
2. Liu, J.; Dan, X.; Lu, G.; Shen, J.; Wu, D.; Yan, Z. Investigation of pharmaceutically active compounds in an urban receiving water: Occurrence, fate and environmental risk assessment. *Ecotoxicol. Environ. Saf.* **2018**, *154*, 214–220. [CrossRef]
3. Parezanović, G.Š.; Lalic-Popovic, M.; Golocorbin-Kon, S.; Vasovic, V.; Milijašević, B.; Al-Salami, H.; Mikov, M. Environmental Transformation of Pharmaceutical Formulations: A Scientific Review. *Arch. Environ. Contam. Toxicol.* **2019**, *77*, 155–161. [CrossRef] [PubMed]
4. Papageorgiou, M.; Kosma, C.; Lambropoulou, D. Seasonal occurrence, removal, mass loading and environmental risk assessment of 55 pharmaceuticals and personal care products in a municipal wastewater treatment plant in Central Greece. *Sci. Total Environ.* **2016**, *543*, 547–569. [CrossRef] [PubMed]
5. Carmona, E.; Andreu, V.; Picó, Y. Multi-residue determination of 47 organic compounds in water, soil, sediment and fish—Turia River as case study. *J. Pharm. Biomed. Anal.* **2017**, *146*, 117–125. [CrossRef]
6. aus der Beek, T.; Weber, F.-A.; Bergmann, A.; Hickmann, S.; Ebert, I.; Hein, A.; Küster, A. Pharmaceuticals in the environment—Global occurrences and perspectives. *Environ. Toxicol. Chem.* **2016**, *35*, 823–835. [CrossRef]
7. Palma, P.; Fialho, S.; Lima, A.; Novais, M.H.; Costa, M.J.; Montemurro, N.; Pérez, S.; de Alda, M.L. Pharmaceuticals in a Mediterranean Basin: The influence of temporal and hydrological patterns in environmental risk assessment. *Sci. Total Environ.* **2020**, *709*, 136205. [CrossRef] [PubMed]
8. Barra Caracciolo, A.; Patrolecco, L.; Grenni, P.; Di Lenola, M.; Ademollo, N.; Rauseo, J.; Rolando, L.; Spataro, F.; Plutzer, J.; Monostory, K.; et al. Chemical mixtures and autochthonous microbial community in an urbanized stretch of the River Danube. *Microchem. J.* **2019**, *147*, 985–994. [CrossRef]
9. Zhang, L.; Carvalho, P.N.; Bollmann, U.E.; Ei-taliawy, H.; Brix, H.; Bester, K. Enhanced removal of pharmaceuticals in a biofilter: Effects of manipulating co-degradation by carbon feeding. *Chemosphere* **2019**, *236*, 124303. [CrossRef]
10. Jeeva, M.P.; Usha, K.A.; Charuvila, T.A. Use of Antibiotics in Animals and Its Possible Impacts in the Environment. In *Handbook of Research on Social Marketing and Its Influence on Animal Origin Food Product Consumption*; Diana, B., Dora, M., Talia, R., Eds.; IGI Global: Hershey, PA, USA, 2018; pp. 77–91.

11. Xiang, L.; Wu, X.-L.; Jiang, Y.-N.; Yan, Q.-Y.; Li, Y.-W.; Huang, X.-P.; Cai, Q.-Y.; Mo, C.-H. Occurrence and risk assessment of tetracycline antibiotics in soil from organic vegetable farms in a subtropical city, south China. *Environ. Sci. Pollut. Res.* **2016**, *23*, 13984–13995. [CrossRef]

12. Du, X.; Bayliss, S.C.; Feil, E.J.; Liu, Y.; Wang, C.; Zhang, G.; Zhou, D.; Wei, D.; Tang, N.; Leclercq, S.O.; et al. Real time monitoring of Aeromonas salmonicida evolution in response to successive antibiotic therapies in a commercial fish farm. *Environ. Microbiol.* **2019**, *21*, 1113–1123. [CrossRef]

13. Yang, Y.; Ok, Y.S.; Kim, K.-H.; Kwon, E.E.; Tsang, Y.F. Occurrences and removal of pharmaceuticals and personal care products (PPCPs) in drinking water and water/sewage treatment plants: A review. *Sci. Total Environ.* **2017**, *596–597*, 303–320. [CrossRef]

14. Kumar, R.; Sarmah, A.K.; Padhye, L.P. Fate of pharmaceuticals and personal care products in a wastewater treatment plant with parallel secondary wastewater treatment train. *J. Environ. Manag.* **2019**, *233*, 649–659. [CrossRef] [PubMed]

15. Li, W.C. Occurrence, sources, and fate of pharmaceuticals in aquatic environment and soil. *Environ. Pollut.* **2014**, *187*, 193–201. [CrossRef]

16. Maryam, B.; Büyükgüngör, H. Wastewater reclamation and reuse trends in Turkey: Opportunities and challenges. *J. Water Process. Eng.* **2019**, *30*, 100501. [CrossRef]

17. Salgueiro-González, N.; Muniategui-Lorenzo, S.; López-Mahía, P.; Prada-Rodríguez, D. Trends in analytical methodologies for the determination of alkylphenols and bisphenol A in water samples. *Anal. Chim. Acta* **2017**, *962*, 1–14. [CrossRef]

18. Lin, T.; Yu, S.; Chen, W. Occurrence, removal and risk assessment of pharmaceutical and personal care products (PPCPs) in an advanced drinking water treatment plant (ADWTP) around Taihu Lake in China. *Chemosphere* **2016**, *152*, 1–9. [CrossRef] [PubMed]

19. Praveena, S.M.; Mohd Rashid, M.Z.; Mohd Nasir, F.A.; Sze Yee, W.; Aris, A.Z. Occurrence and potential human health risk of pharmaceutical residues in drinking water from Putrajaya (Malaysia). *Ecotoxicol. Environ. Saf.* **2019**, *180*, 549–556. [CrossRef] [PubMed]

20. Krakkó, D.; Licul-Kucera, V.; Záray, G.; Mihucz, V.G. Single-run ultra-high performance liquid chromatography for quantitative determination of ultra-traces of ten popular active pharmaceutical ingredients by quadrupole time-of-flight mass spectrometry after offline preconcentration by solid phase extraction from drinking and river waters as well as treated wastewater. *Microchem. J.* **2019**, *148*, 108–119.

21. Fatta-Kassinos, D.; Vasquez, M.I.; Kümmerer, K. Transformation products of pharmaceuticals in surface waters and wastewater formed during photolysis and advanced oxidation processes—Degradation, elucidation of byproducts and assessment of their biological potency. *Chemosphere* **2011**, *85*, 693–709. [CrossRef]

22. Boras, J.A.; Vaqué, D.; Maynou, F.; Sà, E.L.; Weinbauer, M.G.; Sala, M.M. Factors shaping bacterial phylogenetic and functional diversity in coastal waters of the NW Mediterranean Sea. *Estuar. Coast. Shelf Sci.* **2015**, *154*, 102–110. [CrossRef]

23. Matilainen, A.; Sillanpää, M. Removal of natural organic matter from drinking water by advanced oxidation processes. *Chemosphere* **2010**, *80*, 351–365. [CrossRef]

24. Rushing, B.; Wooten, A.; Shawky, M.; Selim, M.I. Comparison of LC-MS and GC-MS for the Analysis of Pharmaceuticals and Personal Care Products in Surface Water and Treated Wastewaters. *Spectroscopy* **2016**, *14*, 8–14.

25. Kachhawaha, A.S.; Nagarnaik, P.M.; Labhasetwar, P.; Banerjee, K. A Review of Recently Developed LC-MS/MS Methods for the Analysis of Pharmaceuticals and Personal Care Products in Water. *J. Aoac Int.* **2020**, *103*, 9–22. [CrossRef]

26. Knoll, S.; Rosch, T.; Huhn, C. Trends in sample preparation and separation methods for the analysis of very polar and ionic compounds in environmental water and biota samples. *Anal. Bioanal. Chem.* **2020**, *412*, 6149–6165. [CrossRef]

27. Matich, E.K.; Soria, N.G.C.; Aga, D.S.; Atilla-Gokcumen, G.E. Applications of metabolomics in assessing ecological effects of emerging contaminants and pollutants on plants. *J. Hazard. Mater.* **2019**, *373*, 527–535. [CrossRef]

28. Sereshti, H.; Duman, O.; Tunc, S.; Nouri, N.; Khorram, P. Nanosorbent-based solid phase microextraction techniques for the monitoring of emerging organic contaminants in water and wastewater samples. *Microchim. Acta* **2020**, *187*, 1–35. [CrossRef] [PubMed]

29. Büyüktiryaki, S.; Keçili, R.; Hussain, C.M. Functionalized nanomaterials in dispersive solid phase extraction: Advances & prospects. *TrAC Trends Anal. Chem.* **2020**, *127*, 115893.

30. Wei, X.; Wang, Y.; Chen, J.; Xu, F.; Liu, Z.; He, X.; Li, H.; Zhou, Y. Adsorption of pharmaceuticals and personal care products by deep eutectic solvents-regulated magnetic metal-organic framework adsorbents: Performance and mechanism. *Chem. Eng. J.* **2020**, *392*, 124808. [CrossRef]

31. Font, G.; Mañes, J.; Moltó, J.C.; Picó, Y. Solid-phase extraction in multi-residue pesticide analysis of water. *J. Chromatogr. A* **1993**, *642*, 135–161. [CrossRef]

32. Zhang, Y.; Duan, L.; Wang, B.; Liu, C.S.; Jia, Y.; Zhai, N.; Blaney, L.; Yu, G. Efficient multiresidue determination method for 168 pharmaceuticals and metabolites: Optimization and application to raw wastewater, wastewater effluent, and surface water in Beijing, China. *Environ. Pollut.* **2020**, *261*, 114113. [CrossRef]

33. Hong, B.; Yu, S.; Niu, Y.; Ding, J.; Lin, Q.; Lin, X.; Hu, W. Spectrum and environmental risks of residual pharmaceuticals in stream water with emphasis on its relation to epidemic infectious disease and anthropogenic activity in watershed. *J. Hazard. Mater.* **2020**, *385*, 121594. [CrossRef]

34. Fan, X.; Gao, J.; Li, W.; Huang, J.; Yu, G. Determination of 27 pharmaceuticals and personal care products (PPCPs) in water: The benefit of isotope dilution. *Front. Environ. Sci. Eng.* **2020**, *14*, 8. [CrossRef]

35. Papageorgiou, M.; Zioris, I.; Danis, T.; Bikiaris, D.; Lambropoulou, D. Comprehensive investigation of a wide range of pharmaceuticals and personal care products in urban and hospital wastewaters in Greece. *Sci. Total Environ.* **2019**, *694*, 133565. [CrossRef]

36. Guzel, E.Y.; Cevik, F.; Daglioglu, N. Determination of pharmaceutical active compounds in Ceyhan River, Turkey: Seasonal, spatial variations and environmental risk assessment. *Hum. Ecol. Risk Assess. Int. J.* **2019**, *25*, 1980–1995. [CrossRef]

37. Koçoğlu, E.S.; Sözüdoğru, O.; Komesli, O.T.; Yılmaz, A.E.; Bakırdere, S. Simultaneous determination of drug active compound, hormones, pesticides, and endocrine disruptor compounds in wastewater samples by GC-MS with direct calibration and matrix matching strategies after preconcentration with dispersive liquid-liquid microextraction. *Environ. Monit. Assess.* **2019**, *191*, 653.

38. Gumbi, B.P.; Moodley, B.; Birungi, G.; Ndungu, P.G. Target, Suspect and Non-Target Screening of Silylated Derivatives of Polar Compounds Based on Single Ion Monitoring GC-MS. *Int. J. Environ. Res. Public Health* **2019**, *16*, 4022. [CrossRef]

39. Zhong, M.; Wang, T.; Qi, C.; Peng, G.; Lu, M.; Huang, J.; Blaney, L.; Yu, G. Automated online solid-phase extraction liquid chromatography tandem mass spectrometry investigation for simultaneous quantification of per-and polyfluoroalkyl substances, pharmaceuticals and personal care products, and organophosphorus flame retardants in environmental waters. *J. Chromatogr. A* **2019**, *1602*, 350–358.

40. Liang, Y.; Liu, J.; Zhong, Q.; Yu, D.; Yao, J.; Huang, T.; Zhu, M.; Zhou, T. A fully automatic cross used solid-phase extraction online coupled with ultra-high performance liquid chromatography–tandem mass spectrometry system for the trace analysis of multi-class pharmaceuticals in water samples. *J. Pharm. Biomed. Anal.* **2019**, *174*, 330–339. [CrossRef]

41. Miossec, C.; Lanceleur, L.; Monperrus, M. Multi-residue analysis of 44 pharmaceutical compounds in environmental water samples by solid-phase extraction coupled to liquid chromatography-tandem mass spectrometry. *J. Sep. Sci.* **2019**, *42*, 1853–1866. [CrossRef]

42. Vreys, N.; Amé, M.; Filippi, I.; Cazenave, J.; Valdés, M.; Bistoni, M. Effect of Landscape Changes on Water Quality and Health Status of Heptapterus mustelinus (Siluriformes, Heptapteridae). *Arch. Environ. Contam. Toxicol.* **2019**, *76*, 453–468. [CrossRef] [PubMed]

43. Xie, H.; Hao, H.; Xu, N.; Liang, X.; Gao, D.; Xu, Y.; Gao, Y.; Tao, H.; Wong, M. Pharmaceuticals and personal care products in water, sediments, aquatic organisms, and fish feeds in the Pearl River Delta: Occurrence, distribution, potential sources, and health risk assessment. *Sci. Total Environ.* **2019**, *659*, 230–239. [CrossRef]

44. Abdallah, M.A.-E.; Nguyen, K.-H.; Ebele, A.J.; Atia, N.N.; Ali, H.R.H.; Harrad, S. A single run, rapid polarity switching method for determination of 30 pharmaceuticals and personal care products in waste water using Q-Exactive Orbitrap high resolution accurate mass spectrometry. *J. Chromatogr. A* **2019**, *1588*, 68–76. [CrossRef]

45. Fatoki, O.S.; Opeolu, B.O.; Genthe, B.; Olatunji, O.S. Multi-residue method for the determination of selected veterinary pharmaceutical residues in surface water around Livestock Agricultural farms. *Heliyon* **2018**, *4*, e01066. [CrossRef]

46. Pérez-Alvarez, I.; Islas-Flores, H.; Gómez-Oliván, L.M.; Barceló, D.; De Alda, M.L.; Solsona, S.P.; Sánchez-Aceves, L.; SanJuan-Reyes, N.; Galar-Martínez, M. Determination of metals and pharmaceutical compounds released in hospital wastewater from Toluca, Mexico, and evaluation of their toxic impact. *Environ. Pollut.* **2018**, *240*, 330–341. [CrossRef]

47. Marube, L.C.; Caldas, S.S.; Santos, E.O.D.; Michaelsen, A.; Primel, E.G. Multi-residue method for determination of thirty-five pesticides, pharmaceuticals and personal care products in water using ionic liquid-dispersive liquid-liquid microextraction combined with liquid chromatography-tandem mass spectrometry. *J. Braz. Chem. Soc.* **2018**, *29*, 1349–1359. [CrossRef]

48. Krizman-Matasic, I.; Kostanjevecki, P.; Ahel, M.; Terzic, S. Simultaneous analysis of opioid analgesics and their metabolites in municipal wastewaters and river water by liquid chromatography-tandem mass spectrometry. *J. Chromatogr. A* **2018**, *1533*, 102–111. [CrossRef]

49. Mijangos, L.; Ziarrusta, H.; Olivares, M.; Zuloaga, O.; Möder, M.; Etxebarria, N.; Prieto, A. Simultaneous determination of 41 multiclass organic pollutants in environmental waters by means of polyethersulfone microextraction followed by liquid chromatography–tandem mass spectrometry. *Anal. Bioanal. Chem.* **2018**, *410*, 615–632. [CrossRef]

50. Pivetta, R.C.; Rodrigues-Silva, C.; Ribeiro, A.R.; Rath, S. Tracking the occurrence of psychotropic pharmaceuticals in Brazilian wastewater treatment plants and surface water, with assessment of environmental risks. *Sci. Total Environ.* **2020**, *727*, 138661. [CrossRef] [PubMed]

51. Gopal, C.M.; Bhat, K.; Praveenkumarreddy, Y.; Shailesh; Kumar, V.; Basu, H.; Joshua, D.I.; Singhal, R.K.; Balakrishna, K. Evaluation of selected pharmaceuticals and personal care products in water matrix using ion trap mass spectrometry: A simple weighted calibration curve approach. *J. Pharm. Biomed. Anal.* **2020**, *185*, 113214. [CrossRef]

52. Zhu, F.; Yao, Z.; Ji, W.; Liu, D.; Zhang, H.; Li, A.; Huo, Z.; Zhou, Q. An efficient resin for solid-phase extraction and determination by UPLCMS/MS of 44 pharmaceutical personal care products in environmental waters. *Front. Environ. Sci. Eng.* **2020**, *14*, 51. [CrossRef]

53. Gago-Ferrero, P.; Bletsou, A.A.; Damalas, D.E.; Aalizadeh, R.; Alygizakis, N.A.; Singer, H.P.; Hollender, J.; Thomaidis, N.S. Wide-scope target screening of >2000 emerging contaminants in wastewater samples with UPLC-Q-ToF-HRMS/MS and smart evaluation of its performance through the validation of 195 selected representative analytes. *J. Hazard. Mater.* **2020**, *387*, 121712. [CrossRef]

54. Castillo Meza, L.; Piotrowski, P.; Farnan, J.; Tasker, T.L.; Xiong, B.; Weggler, B.; Murrell, K.; Dorman, F.L.; Vanden Heuvel, J.P.; Burgos, W.D. Detection and removal of biologically active organic micropollutants from hospital wastewater. *Sci. Total Environ.* **2020**, *700*, 134469. [CrossRef] [PubMed]

55. Nantaba, F.; Wasswa, J.; Kylin, H.; Palm, W.-U.; Bouwman, H.; Kümmerer, K. Occurrence, distribution, and ecotoxicological risk assessment of selected pharmaceutical compounds in water from Lake Victoria, Uganda. *Chemosphere* **2020**, *239*, 124642. [CrossRef]

56. Fernandes, M.J.; Paíga, P.; Silva, A.; Llaguno, C.P.; Carvalho, M.; Vázquez, F.M.; Delerue-Matos, C. Antibiotics and antidepressants occurrence in surface waters and sediments collected in the north of Portugal. *Chemosphere* **2020**, *239*, 124729. [CrossRef]

57. Sadutto, D.; Álvarez-Ruiz, R.; Picó, Y. Systematic assessment of extraction of pharmaceuticals and personal care products in water and sediment followed by liquid chromatography–tandem mass spectrometry. *Anal. Bioanal. Chem.* **2020**, *412*, 113–127. [CrossRef] [PubMed]

58. Jakab, G.; Szalai, Z.; Michalkó, G.; Ringer, M.; Filep, T.; Szabó, L.; Maász, G.; Pirger, Z.; Ferincz, Á.; Staszny, Á.; et al. Thermal baths as sources of pharmaceutical and illicit drug contamination. *Environ. Sci. Pollut. Res.* **2020**, *27*, 399–410. [CrossRef]

59. Afsa, S.; Hamden, K.; Lara Martin, P.A.; Mansour, H.B. Occurrence of 40 pharmaceutically active compounds in hospital and urban wastewaters and their contribution to Mahdia coastal seawater contamination. *Environ. Sci. Pollut. Res.* **2020**, *27*, 1941–1955. [CrossRef] [PubMed]

60. Guedes-Alonso, R.; Montesdeoca-Esponda, S.; Pacheco-Juárez, J.; Sosa-Ferrera, Z.; Santana-Rodríguez, J. A Survey of the Presence of Pharmaceutical Residues in Wastewaters. Evaluation of Their Removal using Conventional and Natural Treatment Procedures. *Molecules* **2020**, *25*, 1639. [CrossRef]

61. Hou, F.; Tian, Z.; Peter, K.T.; Wu, C.; Gipe, A.D.; Zhao, H.; Alegria, E.A.; Liu, F.; Kolodziej, E.P. Quantification of organic contaminants in urban stormwater by isotope dilution and liquid chromatography-tandem mass spectrometry. *Anal. Bioanal. Chem.* **2019**, *411*, 7791–7806. [CrossRef]

62. Caban, M.; Lis, H.; Kobylis, P.; Stepnowski, P. The triple-sorbents solid-phase extraction for pharmaceuticals and estrogens determination in wastewater samples. *Microchem. J.* **2019**, *149*, 103965. [CrossRef]

63. Maasz, G.; Mayer, M.; Zrinyi, Z.; Molnar, E.; Kuzma, M.; Fodor, I.; Pirger, Z.; Takács, P. Spatiotemporal variations of pharmacologically active compounds in surface waters of a summer holiday destination. *Sci. Total Environ.* **2019**, *677*, 545–555. [CrossRef]

64. Tran, N.H.; Reinhard, M.; Khan, E.; Chen, H.; Nguyen, V.T.; Li, Y.; Goh, S.G.; Nguyen, Q.B.; Saeidi, N.; Gin, K.Y.-H. Emerging contaminants in wastewter, stormwater runoff, and surface water: Application as chemical markers for diffuse sources. *Sci. Total Environ.* **2019**, *676*, 252–267. [CrossRef]

65. Xu, M.; Huang, H.; Li, N.; Li, F.; Wang, D.; Luo, Q. Occurrence and ecological risk of pharmaceuticals and personal care products (PPCPs) and pesticides in typical surface watersheds, China. *Ecotoxicol. Environ. Saf.* **2019**, *175*, 289–298. [CrossRef]

66. Reis, E.O.; Foureaux, A.F.S.; Rodrigues, J.S.; Moreira, V.R.; Lebron, Y.A.R.; Santos, L.V.S.; Amaral, M.C.S.; Lange, L.C. Occurrence, removal and seasonal variation of pharmaceuticals in Brasilian drinking water treatment plants. *Environ. Pollut.* **2019**, *250*, 773–781. [CrossRef]

67. Coelho, M.M.; Lado Ribeiro, A.R.; Sousa, J.C.G.; Ribeiro, C.; Fernandes, C.; Silva, A.M.T.; Tiritan, M.E. Dual enantioselective LC–MS/MS method to analyse chiral drugs in surface water: Monitoring in Douro River estuary. *J. Pharm. Biomed. Anal.* **2019**, *170*, 89–101. [CrossRef]

68. Hong, Y.; Lee, I.; Lee, W.; Kim, H. Mass-balance-model-based evaluation of sewage treatment plant contribution to residual pharmaceuticals in environmental waters. *Chemosphere* **2019**, *225*, 378–387. [CrossRef] [PubMed]

69. de Oliveira, J.A.; Izeppi, L.J.P.; Loose, R.F.; Muenchen, D.K.; Prestes, O.D.; Zanella, R. A multiclass method for the determination of pharmaceuticals in drinking water by solid phase extraction and ultra-high performance liquid chromatography-tandem mass spectrometry. *Anal. Methods* **2019**, *11*, 2333–2340. [CrossRef]

70. Peng, Y.; Gautam, L.; Hall, S.W. The detection of drugs of abuse and pharmaceuticals in drinking water using solid-phase extraction and liquid chromatography-mass spectrometry. *Chemosphere* **2019**, *223*, 438–447. [CrossRef]

71. Vanryckeghem, F.; Huysman, S.; Van Langenhove, H.; Vanhaecke, L.; Demeestere, K. Multi-residue quantification and screening of emerging organic micropollutants in the Belgian Part of the North Sea by use of Speedisk extraction and Q-Orbitrap HRMS. *Mar. Pollut. Bull.* **2019**, *142*, 350–360. [CrossRef]

72. Singh, R.R.; Angeles, L.F.; Butryn, D.M.; Metch, J.W.; Garner, E.; Vikesland, P.J.; Aga, D.S. Towards a harmonized method for the global reconnaissance of multi-class antimicrobials and other pharmaceuticals in wastewater and receiving surface waters. *Environ. Int.* **2019**, *124*, 361–369. [CrossRef]

73. Semreen, M.H.; Shanableh, A.; Semerjian, L.; Alniss, H.; Mousa, M.; Bai, X.; Acharya, K. Simultaneous determination of pharmaceuticals by solid-phase extraction and liquid chromatography-tandem mass spectrometry: A case study from sharjah sewage treatment plant. *Molecules* **2019**, *24*, 633. [CrossRef]

74. Chen, L.; Lin, H.; Li, H.; Wang, M.; Qiu, B.; Yang, Z. Influence of filtration during sample pretreatment on the detection of antibiotics and non-steroidal anti-inflammatory drugs in natural surface waters. *Sci. Total Environ.* **2019**, *650*, 769–778. [CrossRef] [PubMed]

75. Česen, M.; Ahel, M.; Terzić, S.; Heath, D.J.; Heath, E. The occurrence of contaminants of emerging concern in Slovenian and Croatian wastewaters and receiving Sava river. *Sci. Total Environ.* **2019**, *650*, 2446–2453. [CrossRef]

76. Paíga, P.; Correia, M.; Fernandes, M.J.; Silva, A.; Carvalho, M.; Vieira, J.; Jorge, S.; Silva, J.G.; Freire, C.; Delerue-Matos, C. Assessment of 83 pharmaceuticals in WWTP influent and effluent samples by UHPLC-MS/MS: Hourly variation. *Sci. Total Environ.* **2019**, *648*, 582–600. [CrossRef]

77. Lv, J.; Zhang, L.; Chen, Y.; Ye, B.; Han, J.; Jin, N. Occurrence and distribution of pharmaceuticals in raw, finished, and drinking water from seven large river basins in China. *J. Water Health* **2019**, *17*, 477–489. [CrossRef]

78. Chauveheid, E.; Scholdis, S. Removal of pharmaceuticals by a surface water treatment plant. *Water Supply* **2019**, *19*, 1793–1801. [CrossRef]

79. Męczykowska, H.; Stepnowski, P.; Caban, M. Impact of humic acids, temperature and stirring on passive extraction of pharmaceuticals from water by trihexyl(tetradecyl)phosphonium dicyanamide. *Microchem. J.* **2019**, *144*, 500–505. [CrossRef]

80. Pemberton, J.A.; Lloyd, C.E.M.; Arthur, C.J.; Johnes, P.J.; Dickinson, M.; Charlton, A.J.; Evershed, R.P. Untargeted characterisation of dissolved organic matter contributions to rivers from anthropogenic point sources using direct-infusion and high-performance liquid chromatography/Orbitrap mass spectrometry. *Rapid Commun. Mass Spectrom.* **2019**, *34*, e8618. [CrossRef]

81. Arsand, J.B.; Hoff, R.B.; Jank, L.; Dallegrave, A.; Galeazzi, C.; Barreto, F.; Pizzolato, T.M. Wide-Scope Determination of Pharmaceuticals and Pesticides in Water Samples: Qualitative and Confirmatory Screening Method Using LC-qTOF-MS. *Water Air Soil Pollut.* **2018**, *229*, 399. [CrossRef]

82. López-García, E.; Mastroianni, N.; Postigo, C.; Barceló, D.; López de Alda, M. A fully automated approach for the analysis of 37 psychoactive substances in raw wastewater based on on-line solid phase extraction-liquid chromatography-tandem mass spectrometry. *J. Chromatogr. A* **2018**, *1576*, 80–89. [CrossRef]

83. Botero-Coy, A.M.; Martínez-Pachón, D.; Boix, C.; Rincón, R.J.; Castillo, N.; Arias-Marín, L.P.; Manrique-Losada, L.; Torres-Palma, R.; Moncayo-Lasso, A.; Hernández, F. An investigation into the occurrence and removal of pharmaceuticals in Colombian wastewater. *Sci. Total Environ.* **2018**, *642*, 842–853. [CrossRef]

84. Li, W.-L.; Zhang, Z.-F.; Ma, W.-L.; Liu, L.-Y.; Song, W.-W.; Li, Y.-F. An evaluation on the intra-day dynamics, seasonal variations and removal of selected pharmaceuticals and personal care products from urban wastewater treatment plants. *Sci. Total Environ.* **2018**, *640–641*, 1139–1147. [CrossRef]

85. Česen, M.; Heath, D.; Krivec, M.; Košmrlj, J.; Kosjek, T.; Heath, E. Seasonal and spatial variations in the occurrence, mass loadings and removal of compounds of emerging concern in the Slovene aqueous environment and environmental risk assessment. *Environ. Pollut.* **2018**, *242*, 143–154. [CrossRef] [PubMed]

86. González-Mariño, I.; Castro, V.; Montes, R.; Rodil, R.; Lores, A.; Cela, R.; Quintana, J.B. Multi-residue determination of psychoactive pharmaceuticals, illicit drugs and related metabolites in wastewater by ultra-high performance liquid chromatography-tandem mass spectrometry. *J. Chromatogr. A* **2018**, *1569*, 91–100. [CrossRef]

87. Asghar, M.A.; Zhu, Q.; Sun, S.; Peng, Y.E.; Shuai, Q. Suspect screening and target quantification of human pharmaceutical residues in the surface water of Wuhan, China, using UHPLC-Q-Orbitrap HRMS. *Sci. Total Environ.* **2018**, *635*, 828–837. [CrossRef]

88. Fantuzzi, G.; Aggazzotti, G.; Righi, E.; Predieri, G.; Castiglioni, S.; Riva, F.; Zuccato, E. Illicit drugs and pharmaceuticals in swimming pool waters. *Sci. Total Environ.* **2018**, *635*, 956–963. [CrossRef] [PubMed]

89. López-Serna, R.; Marín-de-Jesús, D.; Irusta-Mata, R.; García-Encina, P.A.; Lebrero, R.; Fdez-Polanco, M.; Muñoz, R. Multiresidue analytical method for pharmaceuticals and personal care products in sewage and sewage sludge by online direct immersion SPME on-fiber derivatization—GCMS. *Talanta* **2018**, *186*, 506–512. [CrossRef] [PubMed]

90. Al-Qaim, F.F.; Mussa, Z.H.; Yuzir, A. Development and validation of a comprehensive solid-phase extraction method followed by LC-TOF/MS for the analysis of eighteen pharmaceuticals in influent and effluent of sewage treatment plants. *Anal. Bioanal. Chem.* **2018**, *410*, 4829–4846. [CrossRef]

91. Diuzheva, A.; Balogh, J.; Jekő, J.; Cziáky, Z. Application of liquid-liquid microextraction for the effective separation and simultaneous determination of 11 pharmaceuticals in wastewater samples using high-performance liquid chromatography with tandem mass spectrometry. *J. Sep. Sci.* **2018**, *41*, 2870–2877. [CrossRef]

92. Tröger, R.; Klöckner, P.; Ahrens, L.; Wiberg, K. Micropollutants in drinking water from source to tap—Method development and application of a multiresidue screening method. *Sci. Total Environ.* **2018**, *627*, 1404–1432. [CrossRef]

93. Tomai, P.; Martinelli, A.; Morosetti, S.; Curini, R.; Fanali, S.; Gentili, A. Oxidized Buckypaper for Stir-Disc Solid Phase Extraction: Evaluation of Several Classes of Environmental Pollutants Recovered from Surface Water Samples. *Anal. Chem.* **2018**, *90*, 6827–6834. [CrossRef]

94. Klančar, A.; Trontelj, J.; Roškar, R. Development of a Multi-Residue Method for Monitoring 44 Pharmaceuticals in Slovene Surface Water by SPE-LC-MS/MS. *Water Air Soil Pollut.* **2018**, *229*, 192. [CrossRef]

95. Yao, B.; Yan, S.; Lian, L.; Yang, X.; Wan, C.; Dong, H.; Song, W. Occurrence and indicators of pharmaceuticals in Chinese streams: A nationwide study. *Environ. Pollut.* **2018**, *236*, 889–898. [CrossRef] [PubMed]

96. Wiest, L.; Chonova, T.; Bergé, A.; Baudot, R.; Bessueille-Barbier, F.; Ayouni-Derouiche, L.; Vulliet, E. Two-year survey of specific hospital wastewater treatment and its impact on pharmaceutical discharges. *Environ. Sci. Pollut. Res.* **2018**, *25*, 9207–9218. [CrossRef]

97. Monteiro, M.A.; Spisso, B.F.; Ferreira, R.G.; Pereira, M.U.; Grutes, J.V.; de Andradec, B.R.; d'Avila, L.A. Development and validation of liquid chromatography-tandem mass spectrometry methods for determination of beta-lactams, macrolides, fluoroquinolones, sulfonamides and tetracyclines in surface and drinking water from Rio de Janeiro, Brazil. *J. Braz. Chem. Soc.* **2017**, *29*, 801–813. [CrossRef]

98. Castiglioni, S.; Davoli, E.; Riva, F.; Palmiotto, M.; Camporini, P.; Manenti, A.; Zuccato, E. Mass balance of emerging contaminants in the water cycle of a highly urbanized and industrialized area of Italy. *Water Res.* **2018**, *131*, 287–298. [CrossRef]

99. Chau, H.T.C.; Kadokami, K.; Duong, H.T.; Kong, L.; Nguyen, T.T.; Nguyen, T.Q.; Ito, Y. Occurrence of 1153 organic micropollutants in the aquatic environment of Vietnam. *Environ. Sci. Pollut. Res.* **2018**, *25*, 7147–7156. [CrossRef]

100. Souza, F.S.; Da Silva, V.V.; Rosin, C.K.; Hainzenreder, L.; Arenzon, A.; Pizzolato, T.; Jank, L.; Féris, L.A. Determination of pharmaceutical compounds in hospital wastewater and their elimination by advanced oxidation processes. *J. Environ. Sci. Health Part A* **2018**, *53*, 213–221. [CrossRef]

101. Rivera-Jaimes, J.A.; Postigo, C.; Melgoza-Alemán, R.M.; Aceña, J.; Barceló, D.; López de Alda, M. Study of pharmaceuticals in surface and wastewater from Cuernavaca, Morelos, Mexico: Occurrence and environmental risk assessment. *Sci. Total Environ.* **2018**, *613–614*, 1263–1274. [CrossRef]

102. Salas, D.; Borrull, F.; Fontanals, N.; Marcé, R.M. Combining cationic and anionic mixed-mode sorbents in a single cartridge to extract basic and acidic pharmaceuticals simultaneously from environmental waters. *Anal. Bioanal. Chem.* **2018**, *410*, 459–469. [CrossRef]

103. Kafeenah, H.I.S.; Osman, R.; Bakar, N.K.A. Disk solid-phase extraction of multi-class pharmaceutical residues in tap water and hospital wastewater, prior to ultra-performance liquid chromatographic-tandem mass spectrometry (UPLC-MS/MS) analyses. *Rsc Adv.* **2018**, *8*, 40358–40368. [CrossRef]

104. Kanama, K.M.; Daso, A.P.; Mpenyana-Monyatsi, L.; Coetzee, M.A.A. Assessment of Pharmaceuticals, Personal Care Products, and Hormones in Wastewater Treatment Plants Receiving Inflows from Health Facilities in North West Province, South Africa. *J. Toxicol.* **2018**, *2018*, 3751930. [CrossRef] [PubMed]

105. da Silva, D.C.; Oliveira, C.C. Development of Micellar HPLC-UV Method for Determination of Pharmaceuticals in Water Samples. *J. Anal. Methods Chem.* **2018**, *2018*, 9143730. [CrossRef]

106. Bade, R.; White, J.M.; Gerber, C. Qualitative and quantitative temporal analysis of licit and illicit drugs in wastewater in Australia using liquid chromatography coupled to mass spectrometry. *Anal. Bioanal. Chem.* **2018**, *410*, 529–542. [CrossRef]

107. Trinh, A.; Marlatt, L.; Bell, D.S. Controlling SPE Selectivity Through pH and Organic Modifier Manipulation. Reporter EU 2020, 21. Available online: https://www.sigmaaldrich.com/technical-documents/articles/reporter-eu/controlling-spe-selectivity.html (accessed on 6 November 2020).

108. Moldenhauer, J. *Disinfection and Decontamination: A Practical Handbook*; CRC Press: Boca Raton, FL, USA, 2018.

109. Kumari, P.K.; Akhila, S.; Rao, Y.S.; Devi, B.R. Alternative to Artificial Preservatives. *Syst. Rev. Pharm.* **2019**, *10*, 99–102.

110. Paíga, P.; Santos, L.H.M.L.M.; Delerue-Matos, C. Development of a multi-residue method for the determination of human and veterinary pharmaceuticals and some of their metabolites in aqueous environmental matrices by SPE-UHPLC-MS/MS. *J. Pharm. Biomed. Anal.* **2017**, *135*, 75–86. [CrossRef]

111. Basaglia, G.; Pietrogrande, M.C. Optimization of a SPME/GC/MS Method for the Simultaneous Determination of Pharmaceuticals and Personal Care Products in Waters. *Chromatographia* **2012**, *75*, 361–370. [CrossRef]

Publisher's Note: MDPI stays neutral with regard to jurisdictional claims in published maps and institutional affiliations.

Review

Sponges and Sponge-Like Materials in Sample Preparation: A Journey from Past to Present and into the Future

Theodoros G. Chatzimitakos * and Constantine D. Stalikas

Laboratory of Analytical Chemistry, Department of Chemistry, University of Ioannina, 45110 Ioannina, Greece; cstalika@uoi.gr
* Correspondence: Chatzimitakos@outlook.com; Tel.: +30-26510-08725

Academic Editor: Joselito P. Quirino
Received: 19 July 2020; Accepted: 10 August 2020; Published: 12 August 2020

Abstract: Even though instrumental advancements are constantly being made in analytical chemistry, sample preparation is still considered the bottleneck of analytical methods. To this end, researchers are developing new sorbent materials to improve and replace existing ones, with the ultimate goal to improve current methods and make them more efficient and effective. A few years ago, an alternative trend was started toward sample preparation: the use of sponge or sponge-like materials. These materials possess favorable characteristics, such as negligible weight, open-hole structure, high surface area, and variable surface chemistry. Although their use seemed promising, this trend soon reversed, due to either the increasing use of nanomaterials in sample preparation or the limited scope of the first materials. Currently, with the development of new materials, such as melamine sponges, along with the advancement in nanotechnology, this topic was revived, and various functionalizations were carried out on such materials. The new materials are used as sorbents in sample preparation in analytical chemistry. This review explores the development of such materials, from the past to the present and into the future, as well as their use in analytical chemistry.

Keywords: polyurethane foams; melamine sponges; carbon foams; sample preparation

1. Introduction

Many scientists consider sample preparation as an integral part of the analytical process, which enhances the quality of the obtained results. Others consider sample preparation as the bottleneck of analytical chemistry since labor-intensive steps are required, thus limiting the productivity [1,2]. Nevertheless, more and more emphasis is placed on sample preparation, trying to face the challenges of various matrices in analysis. Since the basic principles of sample preparation remain the same, researchers are developing new sorbent materials to improve and replace existing ones, with the ultimate goal to improve current methods and make them more efficient and effective [3].

To this end, back in the 1990s began the use of sponge or sponge-like materials (e.g., foams) in sample preparation [4–6]. These materials possess favorable characteristics, such as negligible weight, open-hole structure, high surface area, and variable surface chemistry [7]. As a result, a new trend was started and many reports were published, with the peak reached in the years 1997–2003 [7]. The key merit of these materials that increased their popularity and started the trend was their ability to be compressed into mini-columns so that they could be used under the solid-phase extraction (SPE) principle. Although their use seemed promising, this trend was soon reversed and the number of reports declined significantly. The reasons behind the reversal of the trend are not clear. One possible explanation is the fact that the use of nanomaterials in sample preparation began around the 2000s and, since then, it skyrocketed, as can be seen in Figure 1.

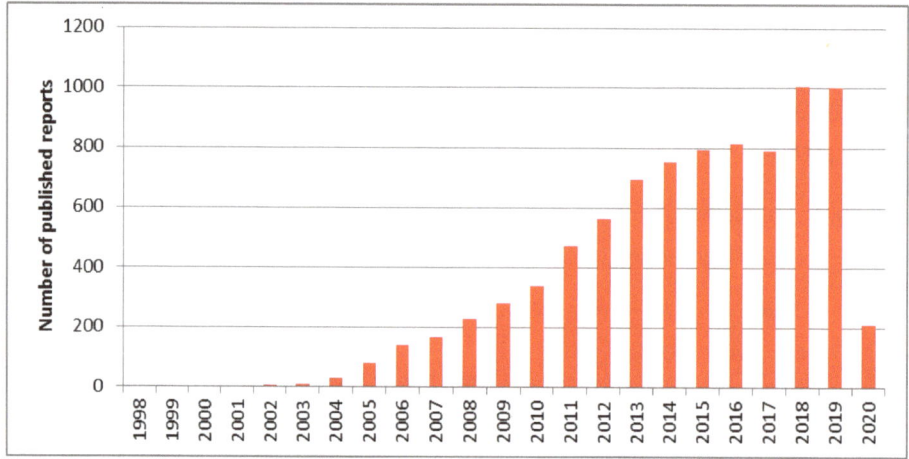

Figure 1. Number of published reports on the use of nanomaterials in sample preparation; source: PubMed.

The trend, as detailed below, started with the use of polyurethane foams (PUFs). By definition, PUFs are plastic materials in which gas (in the form of numerous small bubbles (cells)) replace a proportion of the solid phase [8]. Depending on the percentage of the volume the bubbles occupy, the geometrical shape of the bubbles varies from spherical to quasi-spherical polyhedral, thus changing the properties (such as elasticity) of the PUFs. Another important parameter that affects their properties is the synthesis method. PUFs can be fabricated via the reaction either of isocyanates with hydroxyl compounds (resulting in polyester or polyether PUFs) or of isocyanates with water [7]. Either way, the chemical composition of PUFs differs in terms of polar and non-polar groups, thus making them suitable for the sorption of compounds with different properties [9]. While PUFs are the most widely produced foam materials, currently, advancements are being made and materials with similar structure and physical properties are being fabricated. Melamine sponges (MeS) and carbon foams are two typical examples. Melamine sponges are three-dimensional copolymers (composed of formaldehyde, melamine, and sodium bisulfite) with low density, high porosity, and an open-hole structure [10,11]. Their negligible cost and the presence of functional groups, which make them amenable to functionalization, increased their popularity [12]. Likewise, carbon foams are materials with open-cell structures; they have high surface area and tend to be highly hydrophobic [13].

Currently, with the development of new sponge or foam materials, along with the advancement of nanotechnology, the abovementioned topic was revived, and various uses of these materials were proposed. Furthermore, functionalizations were carried out to alter their applicability, so that they could be used as sorbents in sample preparation in analytical chemistry. The recent reports in this field are scanty and sparse, but the use of sponge and foam materials in analytical chemistry is once again an up-and-coming trend. This review explores the development and use of such materials, from past to present, and it highlight futures perspectives on their use in analytical chemistry.

2. The Beginning of the Trend

Currently, PUFs are mostly known in analytical chemistry, owing to their widespread use as passive samplers for the collection of volatile compounds [14–19]. Their low cost and ease of handling, as well as the fact that they can accumulate particulate matter, make their use favorable. However, PUFs were not always used in such a way. The first reported method that employed PUFs for absorption was published 50 years ago by Bowen, but it was not until the 1990s that the use of PUFs in sample preparation procedures started to become more popular [20]. From that time onward and especially in the next two decades from 1990 to 2010, PUFs were used for the preconcentration of common metal species, such as iron [21,22], copper [23], zinc [6], and nickel [24], and rare metals, such as germanium [4], thorium [25], thallium [26], and uranium [27] from aqueous matrices. In most cases, the PUFs were used bare, without any functionalization, necessitating the addition of organic or inorganic ligands in the sample solution to form metal complexes that could be sorbed onto PUFs. In the case of aqueous samples, no sample pretreatment was carried out, while, for other matrices (such as metal granules, alloys, dried shrimp, fruits etc.), the samples were digested with hot, concentrated nitric acid prior to analysis. A typical example of this concept was the addition of sodium molybdate in a germanium-containing solution so that molybdogermanate could be formed, which was then extracted onto PUFs [4]. The PUFs were left in the sample solution for 1 h (to reach equilibrium). Similar to the above case is the use of polyether-type PUFs for the extraction of gallium (in the form of gallium chloride) from alumina, aluminum alloys, and residues from the aluminum industry [5]. Another example is the use of salicylate for complexation with U^{5+} (in an acidic environment) so that the produced complex could be extracted into PUFs [27]. In all three cases, the PUFs could be analyzed directly, without conducting an elution step, by analyzing the PUFs directly using X-ray fluorescence, which is an asset for the overall time spent for analysis of a sample.

In another study, a method for the determination of molybdenum in iron-based matrices was developed [28]. The method was based on the use of ascorbic acid to reduce Mo^{5+} to Mo^{4+} and Fe^{3+} to Fe^{2+} and then on the employment of thiocyanates to form metal complexes which were extracted into PUFs. The reducing step was important to avoid interferences from iron since Fe^{2+} does not form thiocyanate complexes. The molybdenum thiocyanate complexes were efficiently extracted, even in the presence of ten times as high as the concentration of other metal species (such as copper, cobalt, and zinc), as evidenced by the high recoveries achieved by analyzing pure iron and steel samples. Later on, another study was published, in which the experimental parameters were studied, so that the above-described principle was used for the determination of molybdenum in water and plant leaf samples (digested with the addition of boiling concentrated nitric acid and hydrogen peroxide) [29]. It was found that sorption kinetics were fast since a high flow rate was used (up to 10 mL per min), which was an advantage for the analysis of high sample volumes in a short time. Moreover, it was suggested that the elution should not be carried out with nitric acid solution more concentrated than 3 mol·L^{-1}, since the PUF structure is altered and, thus, the PUFs cannot be reused. The recoveries were satisfactory and good accuracy was reported by analyzing certified reference materials. In another study, the use of thiocyanates for the formation of complexes with zinc was proposed, so that they could be extracted onto PUFs [6]. In this method, many metal species, such as calcium, aluminum, and nickel, as well as anions such as chloride, sulfate, nitrate, etc., do not affect the extraction. However, Fe^{3+}, Cu^{2+}, Co^{2+}, Hg^{2+}, Ga^{3+}, and Pb^{2+} are co-extracted with this method. To avoid their presence, the authors proposed the reduction of Fe^{3+} to Fe^{2+} with ascorbic acid and the use of citrate to mask the copper and cobalt species. The method was developed to extract zinc from cadmium-rich matrices. However, cadmium can also form complexes with thiocyanates. To avoid this, the pH of the solution was adjusted to 3. For the three other metal species, the authors proposed an elution clean-up step with water, which does not elute the mercury, gallium, and lead complexes. The use of water as an elution solvent was also an added advantage since organic solvents were avoided.

In a similar context, Abbas proposed the use of molybdate for the formation of the respective complexes with phosphates and arsenates, which were then reduced to molybdenum blue (using antimony as a catalyst and ascorbic acid as a reducing agent) and adsorbed into PUFs [30]. The sorbed complexes were then eluted, and the absorbance of the eluent was measured photometrically. However, since both species form molybdenum blue, it was difficult to determine their concentration in the same sample. This is a common problem for the detection of arsenate, which exists at lower concentrations in water samples, compared with phosphates. The author claimed that, by conducting extractions at two pH values (i.e., 0.9 and 1.2, adjusted with sulfuric acid), the formation of molybdenum blue by arsenates was totally inhibited at the solution with pH 0.9 and the recorded absorbance value was only due to phosphomolybdenum blue. Following simple calculations using the two recorded absorbances, the concentrations of both phosphates and arsenates were obtained. Based on the definite formation of the Fe^{3+}–thiocyanate complexes, it was proposed that the complexes could be extracted in PUFs [22]. The importance of adding hydrochloric acid (until pH was close to 1.3) to the sample solution so that the formation of Fe^{3+}–OH^- complexes was avoided was strongly emphasized. The authors proposed that the adsorption of the Fe^{3+}–thiocyanate complexes was completed in three steps. Firstly, the solute reaches near the boundary layer film of the adsorbent surface. The second step is film diffusion, which is the diffusion of the complex through the boundary film. The third step is intraparticle diffusion, which is the diffusion of the complex into the porous PUFs. To complete these steps and achieve reproducible results, a 90-min extraction time was proposed. Capitalizing on the same complex formation principle, Casella developed an on-line solid-phase extraction system using PUFs for Fe^{3+} determination in acidic water samples [21]. Following spectrophotometric measurements, highly satisfactory relative standard deviations (between 1.2% and 1.5%) and low detection limits (0.45 or 0.75 $\mu g \cdot L^{-1}$, depending on the preconcentration time) were achieved. PUFs combined with thiocyanates were also been proposed for the on-line detection of nickel [31]. The use of thiocyanates was suggested, to form complexes with other interfering ions, so that, ultimately, these complexes could be removed by adsorbing into the PUFs, while nickel, which does not form a complex with thiocyanates, could pass through the PUF mini-column. After reacting with 4-(2-pyridylazo)-resorcinol, nickel was determined spectrophotometrically. The two above studies highlighted the potential of PUFs mini-columns for the development of low-cost, on-line systems.

All of the abovementioned studies made use of bare, non-functionalized PUFs. However, there were certain cases where PUFs were loaded with selective reagents for the determination of various ions. This was done to counterbalance two main disadvantages of PUFs: lack of selectivity and low sorption capacity [32]. To make feasible the solid-phase extraction and determination of Ru^{3+}, the use of PUFs functionalized with 3-hydroxy-2-methyl-1,4-naphthoquinone-4-oxime was proposed [33]. The developed PUFs were highly selective and made feasible the extraction of 1 μg of Ru^{3+}, even in the presence of a high excess of other ions, such as Ba^{2+}, Zn^{2+}, Cr^{3+}, etc. The only metals that could interfere with Ru^{3+} extraction were easily masked with common reagents (e.g., Ni^{2+} was masked by 1% KCN solution; Fe^{3+} and V^{5+} were masked by the addition of one crystal of potassium fluoride and sodium fluoride, respectively). The recoveries were above 98%, highlighting the prospect of functionalized PUFs being exploited. Similarly, the functionalization of PUFs with 9,10-phenanthaquinone monoethylthiosemicarbazone was proposed, so that the functionalization reagent could form a highly stable and colored complex with Ti^{3+} [34]. By using these functionalized PUFs and spectrophotometric detection, recoveries between 99.2% and 100.2% were achieved for zinc granulates and lead foil samples, highlighting the great potential of the method. Similarly, 2-ethylhexylphosphonic acid was proposed as a reagent to functionalize PUFs, to produce selective PUFs for thorium [25]. Thorium ions were extracted based on a cation exchange mechanism. Therefore, highly acidic ambiance was avoided, since hydrogen ions were competitively extracted. Finally, a report was published for the functionalization of PUFs with ammonium hexamethylenedithiocarbamate [35]. The functionalized PUFs were used as sorbents for arsenic, bismuth, mercury, antimony, selenium, and tin. Since, As^{5+}, Sb^{5+}, and Se^{6+} do not form complexes with ammonium hexamethylene dithiocarbamate, a reduction step was necessary prior to extraction.

The cases reported so far concerned the extraction of metal ions in (functionalized) PUFs. However, there are two published reports which pertained to the extraction of polycyclic aromatic hydrocarbons (PAHs) in PUFs. In the first study, the authors developed a single-pass flow-through extraction method and examined the suitability of PUFs for the extraction of 18 PAHs from diesel exhaust samples [36]. In the second study, PUFs were suggested for the extraction of four PAHs from water samples [37]. The authors examined the mechanism via which the PAHs and PUFs interact. Firstly, they found that PAH sorption was not dependent on the chemical structure of the PUFs. Whether the polyester, the polyether, or their co-polymer was used, sorption remained unchanged. However, it was evident that the hydrophobicity of the PUFs greatly influenced their sorption potential. A more hydrophobic PUF presented a greater sorption potential. Both of these studies highlight the effectiveness of bare PUFs for PAH extraction, without laborious functionalization steps. A summary of the analytical methods developed, between 1990 and 2010, based on PUFs is given in Table 1.

Table 1. Summary of the analytical methods developed, between 1990 and 2005, based on polyurethane foams (PUFs), SP: spectrophotometry; ETAAS: electrothermal atomic absorption spectrometry; XRF: X-ray fluorescence spectrometry; AAS: atomic absorption spectrometry; ICP-AES: inductively coupled plasma atomic emission spectrometry; PAH: polycyclic aromatic hydrocarbon.

Analyte	Ligand	Sample	Extraction Time (min)	Analytical Technique	LOD (Limit of Detection) $\mu g \cdot L^{-1}$	Reference
Ge	Molybdate	Water	60	XRF	70	[4]
Zn^{2+}	Thiocyanate	Cadmium-rich matrices	10	SP	20	[6]
I, Hg, Au, Fe, Sb, Th, Mo, Re, U, benzene, chloroform, phenol	-	Water	30	SP	-	[20]
Fe^{3+}	Thiocyanate	Water and rice flour	50	SP	450	[21]
Fe^{3+}	Thiocyanate	Water	5	SP	-	[22]
Cu^{2+}	Diethyl dithiocarbamate	Water	40	ETAAS	-	[23]
Zn^{2+}	Thiocyanate	Aluminum matrices	10	ICP-AES	0.02	[24]
Th	2-Ethylhexylphosphonic acid	Water	30	XRF	4.0	[25]
U	Salicylate	Water	50	XRF	5.5	[27]
Mo	Thiocyanate	Steel and pure iron	10	ICP-AES	0.9	[28]
Mo	Thiocyanate	Water, peach, apple, and citrus leaves	-	ETAAS	0.08	[29]
AsO_4^{3-} and PO_4^{3-}	Molybdate	Water	-	SP	5.4 and 1.66	[30]
Ni	Thiocyanate	Silicates and alloys	-	SP	77	[31]
Ru^{3+}	3-Hydroxy-2-methyl-1, 4-naphthoquinonexime	Water	3–5	SP	20	[33]
Tl^{1+} and Tl^{3+}	9, 10 Phenathaquinone monomethyl thlio semicarbazone	Water and lead solutions	30	SP	2.76	[34]
As, Bi, Hg, Sb, Se, Sn,	Dithiocarbamate	Water	-	ETAAS	0.06–0.3	[35]
PAHs	-	Diesel exhaust	-	GC/MS	-	[36]
PAHs	-	Water	30	Solid-matrix spectrofluorimetry	0.02	[37]
Cu^{2+}	-	Dried shrimp		SP	1.2	[38]
Co, Fe, Zn, Cd, Ni, Hg	Thiocyanates (for Co, Fe, Zn), 1-(2-pyridylazo)-2-naphthol (for Ni, Hg, Cd)	Water	180	SP	-	[39]
Co^{2+}	thiocyanate	Water	5	SP	-	[40]
PAHs	-	ambient air	960	GC/MS	-	[41]
As, Bi, Pb, Sb, Sn, Se, Hg	Dithiocarbamate	Waste Water and seawater	60	ICP-AES	0.03–30	[42]
chlorobenzenes		ambient air	1320	GC/MS	-	[43]
Two-ring aromatic hydrocarbons, chlorinated phenols, guaiacols, and benzenes	-	Ambient air	720	GC/electron capture detection	200–500	[44]
Cd, Co, Cu, Hg, Ni, and Pb	Hexamethylene ammonium hexamethylene dithiocarbamate	Oxalic acid	-	AAS or ICP-AES	0.1–0.3	[45]
Zn	Thiocyanate	Water	5	SP	0.9	[46]
Ni^{2+}	Dimethylglyoxime	Water	15	SP	0.5	[47]
acaricides	-	Water	10	SP	-	[48]
Co^{2+}	Thiocyanate	Water	2	Gamma-spectrometer	-	[49]
Al	Thiocyanate	Rhyolite, syenite, andesite, basalt, iron ore	30	SP	30	[50]

Table 1. *Cont.*

Analyte	Ligand	Sample	Extraction Time (min)	Analytical Technique	LOD (Limit of Detection) µg·L^{-1}	Reference
Dimethote, azodrine, lannate	-	Water	10	SP	-	[51]
As	-	Water	60	XRF	36	[52]
Polychlorinated biphenyls, PAHs, and n-alkanes	-	Diesel exhaust, cigarette smoke, and roofing tar volatiles	<20	GC/MS	-	[53]
Ag	-	Water	20	SP	-	[54]
Ascorbic acid	Molybdosilicic heteropolyacid	Fruit juices and pharmaceutical preparations	6.5	SP	0.6–40	[55]
Zn, Hg, In	Thiocyanate	Water	30			[56]

3. The Revival of the Trend

3.1. Recent Uses of Polyurethane Foams

After a marked decline in the number of publications with the use of sponge and sponge-like materials in sample preparation, after 2010, this trend was revived, and more and more studies were published. As with the reports of the previous years, some PUF-based sorbents were developed for metal species, albeit to a lower extent [9,32,57,58]. For instance, the use of Eriochrome Black T as a complexing agent for Cu^{2+} was recently proposed [9]. Eriochrome Black T was selected since, in acidic conditions, the metal–Eriochrome Black T complex formation constants for copper and iron are higher than those for other metals (such as cobalt, zinc, etc.). The formed complexes were sorbed on PUFs and, based on this, an SPE procedure was developed for Cu^{2+} extraction from water samples. At acidic ambiance, the copper–Eriochrome Black T complex was in its neutral form and was extracted more efficiently. This was justified by proposing a solvent-like mechanism, where PUFs acted as a polymeric solvent, which was able to retain neutral substances or substances with very low charge density. Although Eriochrome Black T forms more stable complexes with iron than with copper, the iron complexes could not be sorbed efficiently on PUFs, because, at acidic ambiance, nitrogen atoms of PUFs are protonated and repulse the cationic iron–Eriochrome Black T complex. Capitalizing on the affinity of thializodin-4-ones for heavy metal ions, a spirothializodine analogue (3-sulfonamoyl-phenyl-spiro[4-oxo-thiazolidin-2,2′steroid]) was synthesized to functionalize PUFs with it so that PUFs were rendered selective and able to preconcentrate Cd^{2+} from water samples containing iodide ions [57]. Iodide ions were added to the sample solution so that the anionic complex [CdI$_4$]$^{2-}$ could be formed, which formed a ternary ion that could be sorbed onto PUFs. In another study, an SPE procedure for the separation of Au^{2+} traces from geological samples was developed [32]. To make this feasible, acid hydrolysis of the PUFs was carried out, and then, using glutaraldehyde as a linking arm, cytosine was added onto the PUFs. Cytosine was selected due to its low cost and its nitrogen and oxygen atoms that render it a good ligand for metal ions. To carry out the extraction, the pH of the sample solution was adjusted to 1 so that protonation of the binding sites of the chelators could take place, while avoiding metal precipitation from hydroxides. The use of hydrochloric acid ensured the formation of the chloro-anionic species (AuCl$_4^-$) which are readily adsorbed onto the imine groups of the cytosine-modified PUFs, via electrostatic interactions (formation of ion-pair complexes). The functionalized PUFs exhibited high selectivity, sensitivity, and high adsorption capacity for Au^{2+}. PUFs were also functionalized with β-naphthol and used as sorbents for Fe^{3+}, Cu^{2+}, Cr^{3+}, Co^{2+}, and Mg^{2+} [58]. The developed PUFs efficiently adsorbed the metal species, following a simple SPE procedure. In the case that methylene blue was used to functionalize PUFs, penicillins could be extracted from pharmaceuticals and milk samples (previously deproteinated with the addition of acetic acid), following a simple SPE procedure [59]. The developed procedure was sensitive to pH changes since they affected the formation of ion pairs between the antibiotics and the methylene blue-functionalized PUFs. At pH values lower than 8, antibiotics are primarily in their neutral form; hence, they cannot form ion pairs. At pH above 9.5, a competition between the antibiotics and the hydroxyl ions for the positively charged centers of the sorbent was recorded. The method developed exhibited adequate accuracy and good precision. More importantly, the reusability of the prepared PUFs was examined, and it was found that they can last six months, after performing 15 sorption–desorption cycles each day, which is a great asset of the developed material.

Finally, PUFs were functionalized with graphene oxide (GO) to combine the extractive properties of both [60]. The preparation process is very easy (stirring PUFs in GO solution and then drying at room temperature), but it takes more than 24 h to be completed. The epoxy groups of the GO link with the carboxyl and amino groups of the PUFs, while the formation of hydrogen bonds further stabilizes the coating. The prepared sponges were used to extract sulfonamides from milk samples (after protein precipitation with acetonitrile). From analyzing the Fourier-transform infrared (FT-IR) spectra of the sorbent prior and after the extraction, the authors found that amides where formed, as a result of the sulfonamides amine groups and the carboxylate groups of the sorbent. Moreover, the formation of hydrogen bonds was validated and contributed to the overall sorption of sulfonamides. A summary of the analytical methods, developed between 2005 and today, based on PUFs is given in Table 2.

Table 2. Summary of the analytical methods developed, between 2005 and the present, based on PUFs. HPLC-UV: high-performance liquid chromatography with ultraviolet detection.

Analyte	Functionalization Agent	Sample	Extraction Time (min)	Analytical Technique	LOD (Limit of Detection) $\mu g \cdot L^{-1}$	Reference
Cu^{2+}	Eriochrome Black T	Water	30	Flame atomic absorption spectrometry	20–100	[9]
Cd^{2+}	3-Sulfonamoyl-phenyl-spiro[4-oxo-thiazolidin-2,2'steroid]	Industrial wastewater	20	Flame atomic absorption spectrometry	30–100	[57]
Fe^{3+}, Cu^{2+}, Cr^{3+}, Co^{2+}, and Mn^{2+}	β-Naphthol	Water		Flame atomic absorption spectroscopy		[58]
Penicillin G, amoxicillin, and ampicillin	Methylene blue	Pharmaceuticals and milk	-	Flow injection analysis/solid-phase extraction	12, 15, and 19	[59]
Sulfathiazole, sulfamethizole, sulfadiazine, and sulfanilamide	Graphene oxide	Cow milk	15	HPLC-UV	50	[60]
Au^{3+}	Cytosine	Geological samples	-	Inductively coupled plasma optical emission spectrometry	0.006	[32]

3.2. Development of Carbon-Based Foams

All the above studies highlight the potential of PUFs in sample preparation. Currently, the trend seems to have shifted toward other sponge-like materials. One such material is carbon-based foam. In one study, melamine–formaldehyde polymer foams were annealed at 800 °C, under a nitrogen atmosphere, to produce carbon foams [61]. The obtained carbon foams retained the initial three-dimensional (3D) interconnected network of the initial foams, and they were composed of nitrogen and carbon atoms, while exhibiting moderate hydrophobicity. If the temperature during synthesis was increased to more than 1000 °C, more hydrophobic carbon foams would be obtained. The carbon foams were used for the extraction of phenolic endocrine-disrupting compounds (i.e., bisphenol A, 4-*tert*-octylphenol, and 4-*n*-nonylphenol) from water samples, since they have a good affinity for moderately polar phenols, owing to their hydrophobicity. The recoveries were found to be above 90% for experiments conducted using well water, leachates, and wastewater. The use of the synthesized carbon foams resulted in enhanced preconcentration factors (i.e., 24–38), compared to PUFs (preconcentration factors: 11–15) and MeS (preconcentration factors: 7–12). In two other studies, GO/polypyrrole foams were developed and used in pipette-tip SPE procedures [62,63]. In the first case, the polypyrrole (by polymerization of pyrrole in the presence of FeCl$_3$) was firstly prepared and then mixed with a GO solution for 24 h [62]. Although the final material was obtained in the form of a powder, the scanning electron images revealed that the morphology was loose three-dimensional foam. The GO/polypyrrole foams were used to extract three auxins from papaya juice. In the second case, the polymerization of the pyrrole was conducted in the presence of GO and, thus, the amount of time needed for the preparation of the GO/polypyrrole foams was significantly reduced (from 48 h needed in the previous case to 12 h) [63]. The above GO/polypyrrole foams were used to extract seven sulfonamides from milk and honey samples. Acetonitrile and hexane were successively added to the samples to remove proteins and fat, respectively. The proposed method consumes a very small amount of adsorbent (3 mg) and the extraction step is completed in 3 min, which are significant assets of the developed method. The relative standard deviations (RSDs) achieved with the proposed method were low (<1.1% for intra-day analyses and <1.9% for inter-day analyses) when analyzing water samples. However, the recoveries were less satisfactory for real samples (62.3–109.0% and 66.6–106.9% for honey and milk samples, respectively) and the RSDs were significantly higher (<11.2% and <10.8% for honey and milk samples, respectively). Despite the significant advantages of the method, further improvements are needed to enhance its performance. Another case was the freeze-drying of a GO dispersion to obtain a GO sponge which was reduced to form a graphene sponge, by following a reduction step using hydrazine [64]. The SEM images revealed that the GO sponge had a more compact structure, compared to graphene sponge, which was attributed to the interactions of the oxygen-containing groups. In both cases, a smooth structure and a porous three-dimensional open-hole structure were observed. Using the graphene sponges under the principle of solid-phase extraction, the authors were able to extract six organic ultraviolet (UV) filters (i.e., 2-(2′-hydroxy-5′-methylphenyl) benzotriazole, 2-(2*H*-benzotriazol-2-yl)-4,6-bis(1-methyl-1-phenylethyl)phenol, 2-*tert*-butyl-6-(5-chloro-2*H*-benzotriazol-2-yl)-4-methylphenol, 2-(2′-hydroxy-3′, 5′-di-*tert*-butylphenyl)- 5-chlorobenzotriazole, 2-(2′-hydroxy-3′,5′-dipentylphenyl) benzotriazolel, and 2-(2*H*-benzotriazol-2-yl)-4-(1,1,3,3-tetramethylbutyl) phenol) from water and personal care products (e.g., skin cream, sunscreen etc.). A wide linear range (20.0 to 1000 μg·L^{-1}), was recorded for the developed method. The synthesized graphene sponges could be reused more than 60 times, which counterbalances their lengthy preparation (five days are needed to prepare 4.5 g of graphene sponge, starting from pristine graphite oxide to synthesize GO and then the sponges). In another study, the authors heated a mixture of zinc nitrate and sucrose in a crucible at 120 °C for 2 min, then at 180 °C for 5 min, and then at 1100 °C for 3 h, combining synthesis and calcination in a single step [65]. During the heating step, various gases are released (such as CO$_2$, N$_2$, and H$_2$O vapors) that "blow" the heated mixture and form the foam-like structure. The SEM images revealed that the foam material contains many "bubbles", some intact and others broken, which facilitate the interaction of the foam material with analytes, not only on the surface but also in the internal holes, cavities, and channels (Figure 2). The carbon foam developed was used in a stir-bar-supported micro-solid-phase extraction procedure for the extraction of five polyaromatic hydrocarbons from wastewater samples. The performance of the carbon foam was similar to that of multiwalled carbon nanotubes and graphene; however, since its synthesis is faster and cheaper, it is a good alternative to the other two nanomaterials. A summary of sample preparation procedures developed based on various carbon-based foams is given in Table 3.

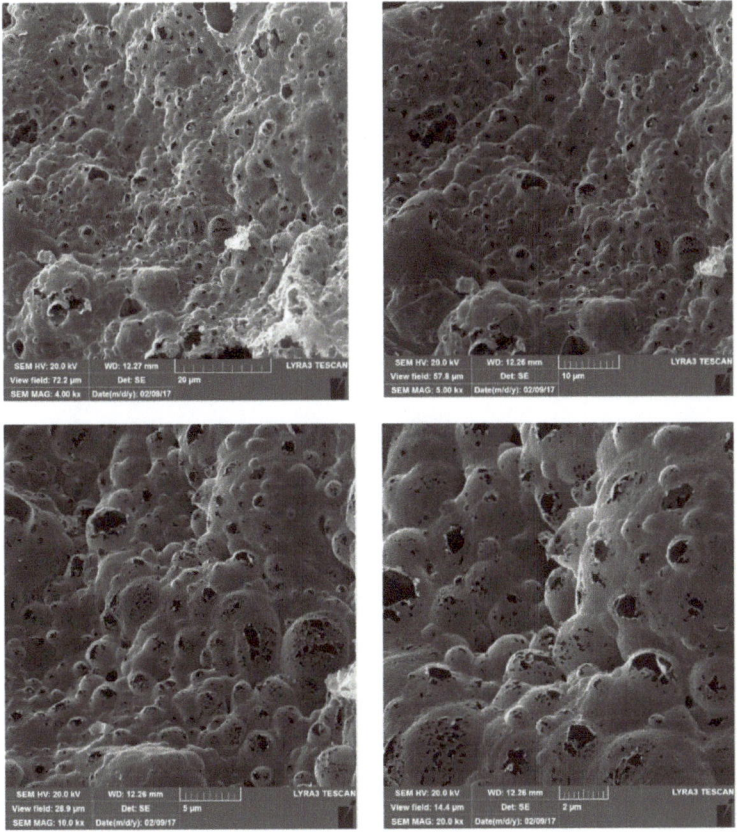

Figure 2. Field-emission (FE)-SEM images of carbon foam. Source: Reproduced from [65], with permission from Elsevier.

Table 3. Analytical methods developed based on carbon-based foams. RSD: relative standard deviation.

Precursors for Carbon Foam	Analyte	Sample	Analytical Technique	LOD ($\mu g \cdot L^{-1}$)	Recoveries (%)	RSD (%)	Reference
Carbonization of melamine sponges	bisphenol A, 4-*tert*-octylphenol, and 4-*n*-nonylphenol	Well water, rainwater, and wastewater	Sequential injection analysis	0.02–0.04	>89.0	2.8–6.3	[61]
Graphene oxide/polypyrrole	Sulfathiazole, sulfapyridine, sulfamethizole, sulfadoxine, sulfisoxazole, sulfamethoxazole, and sulfadimethoxine	Honey and milk	HPLC-UV	0.00104–0.00150	62.3–109.0	>11.2	[63]
Graphene oxide/polypyrrole	Indole-3-butyric acid, indole-3-propionic acid, and 1-naphthaleneacetic acid	Papaya juice	HPLC-UV	0.0012–0.0017	89.4–105.6	<3	[62]
Graphene oxide	Organic UV filters (UV-P, UV-234, UV-326, UV-327, UV-328, and UV-329)	Water and cosmetic products	HPLC-UV	0.02–0.08	89–105	<8.1	[64]
Zinc nitrate and sucrose	Naphthalene, biphenyl, acenaphthene, fluorene, and phenanthrene	Wastewater	GC/MS	0.29–8.4	91.8–102	3.8–10.9	[65]

3.3. Combinations of Carbon-Based Foams with Metals

Instead of using solely carbon-based foams, in some cases, carbon-based foams functionalized with metals were developed. One good example is the study published by Sajid et al., who prepared carbon foam with zinc oxide nanoparticles incorporated in the network [13]. To do so, they heated a mixture of sucrose and zinc nitrate at 110 °C. The concept is similar to that discussed above. However, in this case, an annealing step was not employed. The resulting product had a foamy structure, as evidenced by scanning electron microscopy images, while zinc oxide nanoparticles were visible all over the surface (Figure 3). The presence of zinc oxide nanoparticles was further confirmed by X-ray diffraction (XRD) spectra. The authors also calcinated the produced zinc oxide nanoparticle-incorporated carbon foam by heating at 900 °C. Although the calcined product exhibited a better crystalline structure, its sorption performance was worse, compared with the as-synthesized foam. The developed product was used as a sorbent for the extraction of 15 organochlorine pesticides from milk samples (without sample pretreatment). In another study, the authors combined chitosan and metal–organic frameworks to prepare foams [66]. The foams were prepared by an ice-templating procedure where proper amounts of the metal–organic framework, chitosan, and glutaraldehyde were mixed, and then the mixture was placed into a mold. After freezing at −20 °C, the material was freeze-dried to form the porous foams. The authors prepared six such foams, using different metal–organic frameworks (i.e., MIL-53(Al)/chitosan, MIL-53(Fe)/chitosan, MIL-101(Cr)/chitosan, MIL-101(Fe)/chitosan, UiO-66(Zr)/chitosan, and MIL-100(Fe)/chitosan), and they examined their sorptive performance for a mixture of five parabens in water. The results were conclusive that the best sorbent was MIL-53(Al)/chitosan foam.

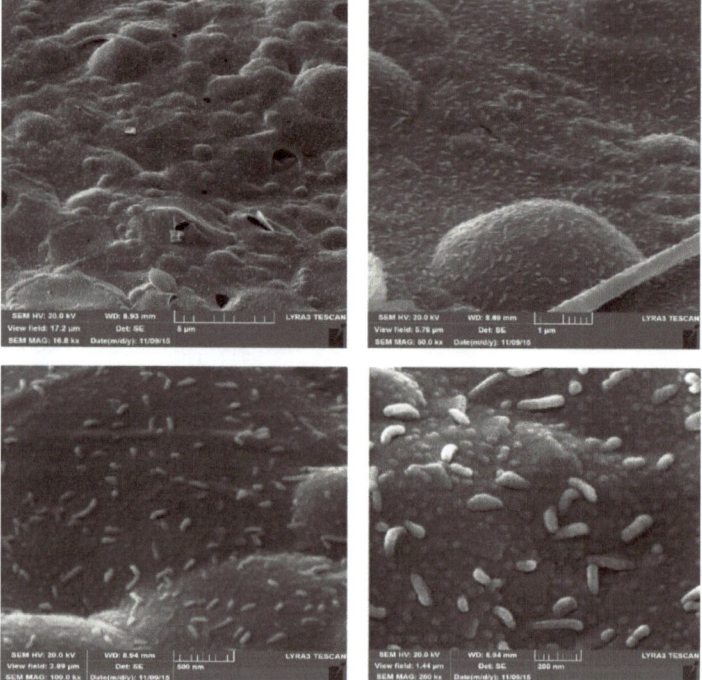

Figure 3. SEM images at different magnifications showing the distribution of ZnO nanoparticles over the surface of carbon foam. Source: Reproduced from [12], with permission from Elsevier.

Metal–organic frameworks exhibit drawbacks in aqueous-phase adsorption due to low stability in water (coordination bonds are likely to collapse). For this reason, the use of a zeolitic imidazolate framework-8 (ZIF-8) in combination with a GO sponge was proposed [67]. An ice-templating procedure was employed to obtain a product, similar to the previous case; however, to obtain a functional material, a calcination step at 800 °C was necessary. Without the calcination step, the material had a mono-dispersion rhombic dodecahedral structure, similar to that of ZIF-8. When the material was calcined at 800 °C, a rich open-hole structure could be observed, which could not be achieved at lower temperatures. The ZIF-8/GO sponge was used for the extraction of five sex hormones in defatted and deproteinated milk and milk products. The developed method exhibited wide linear ranges (10.0–3000 mg·L^{-1}) with remarkable linear correlation coefficients ($R^2 > 0.9998$). Excellent repeatability (intra-day RSDs < 0.39% and inter-day RSDs < 3.86%) and good recoveries (83.8–108.4%) were the two most significant advantages of the developed procedure. Finally, in another study, the authors employed a somewhat "reversed" concept, where, instead of functionalizing a carbon-based foam with some metal, they functionalized nickel foams with polydopamine [68]. This was achieved by placing nickel foam in a dopamine solution prior to its self-polymerization. Owing to the presence of catechol and quinine groups on the surface of the prepared foam, good affinity with Sudan dyes was expected. As a proof of concept, the developed foam was used in a solid-phase microextraction procedure for four Sudan dyes from diluted tomato sauce and hotpot seasoning samples. A summary of the analytical methods developed based on carbon-based foams functionalized with metals is given in Table 4.

Table 4. Analytical methods developed based on carbon foams with metals.

Sorbent	Analyte	Sample	Analytical Technique	LOD (µg·L^{-1})	Recoveries (%)	RSD (%)	Reference
Zinc oxide-incorporated carbon foam	Organochlorine pesticides	Milk	GC/MS	0.19–1.64	85.1–100.7	2.3–10.2	[13]
Metal organic framework/chitosan foams	Parabens	Water	UPLC–MS/MS	0.09–0.45	78.75–102.1	<7.4	[66]
Zeolitic imidazolate framework-8@graphene oxide sponge	Sex hormones	Milk and milk products	HPLC	520–2110	83.8–108.4	<0.39	[67]
Nickel foam functionalized with polydopamine	Sudan dyes	Tomato sauce and hotpot sample	Ion mobility spectrometry	0.005–0.25	81%–91.3	<15.5	[68]

3.4. Development of Functionalized Melamine Sponges

Until recently, MeS was an unexplored material in sample preparation. In our laboratory, we modified MeS with graphene (GMeS) in a straightforward way and used it, for the first time, in sample preparation [10]. Previous studies attempted to modify MeS with graphene, using multiple steps and sophisticated equipment, resulting in time-consuming methods. We achieved the modification by dipping MeS cubes into a GO solution, containing hydrazine, before irradiating the cubes with microwaves for 2 min and then drying. The as-prepared GMeS contained G sheets through their structure (Figure 4) and were rendered hydrophobic (Figure 5). The prepared GMeS were used for the extraction and preconcentration of sulfonamides from deproteinated milk and eggs, as well as lake water samples (based on π–π and hydrophobic interactions), and a method validated according to the SANCO/12571/2013 guideline was developed. Low limits of quantification (between 0.31 µg·kg^{-1} and 1.3 µg·kg^{-1} for the food samples and between 0.10 µg·L^{-1} and 0.29 µg·L^{-1} for lake water samples), and high enrichment factors for milk and lake water samples (94–100) were some of the figures merit of the developed procedure. Following this study, next, we proposed the decoration of MeS with copper sheets, so as to prepare a sorbent, selective and suitable for sulfonamide extraction, based on the affinity of copper for sulfonamides [11]. The synthesis was based on the addition of hydrazine in a heated solution of copper acetate, in which MeS was placed, and stirring the mixture for 30 min (Figure 6).

After washing, the copper-decorated MeS could be used, directly, without the need for a drying step. Owing to the size of the copper-decorated MeS, sulfonamides could be extracted following a radically different mechanism. Their sorption was based on the fact that sulfonamides acted as bridges between two copper ions, via their aromatic amine nitrogen and the nitrogen atom of the heterocycle. This mechanism renders the sorbent selective and efficient for sulfonamides. Using the prepared sponges, we developed a method for sulfonamide determination in deproteinated milk and water samples, validated according to the Commission Decision 657/2002/EC. The method exhibited a wide linear range (0.05–150 $\mu g \cdot L^{-1}$) and high enrichment factors (234–463 for water samples), which render it suitable for the routine analysis of sulfonamides. In another study, we functionalized MeS with urea–formaldehyde co-oligomers [12]. Instead of adopting an acid-catalyzed polymerization step for the preparation of urea–formaldehyde oligomers, we employed a base-catalyzed step. This resulted in the formation of a more hydrophobic product, which does not exhibit the typical resin structure of the acid-catalyzed polymer and consists mainly of oligomers. The prepared sponges were found to be suitable for the extraction of six different classes of compounds (i.e., non-steroidal anti-inflammatory drugs, benzophenones, parabens, phenols, pesticides, and musks). The developed method had low limits of quantification (0.03 and 1.0 $\mu g \cdot L^{-1}$), wide linear ranges, and excellent recoveries.

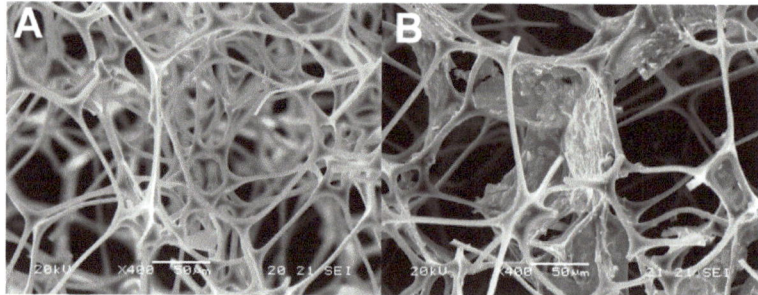

Figure 4. SEM images of MeS before (**A**) and after (**B**) functionalization with graphene (GMeS). Source: Reproduced from [10], with permission from Elsevier.

Figure 5. Images of (**A**) water droplet on the surface of a GMeS; (**B**) contact angle of a water droplet; (**C**) MeS and GMes in a glass beaker with water; (**D**) GMeS immersed in water; (**E**) GMeS on top of a dandelion flower. Source: Reproduced from [10], with permission from Elsevier.

Figure 6. Copper mirror formed on the surface of the reaction glass beaker during the decoration of MeS (**A**). The produced CuMeS (**B**) and its behavior in water (**C,D**). Source: Reproduced from [11], with permission from Elsevier.

In another study, graphene-modified MeS were functionalized with β-cyclodextrin [69]. The synthesis procedure consisted of multiple steps: Firstly, MeS was dipped into a GO solution and dried. Then, the sponges were placed into a solution of β-cyclodextrin (previously modified with aminopropyl tetraethoxysilane), removed after 2 h, and left to dry overnight. The modification procedure was repeated once more, resulting in the final material. The sponges were used for the extraction of flavonoids. The developed material is new and has some advantages, but the synthesis is time-consuming. MeS functionalized with β-cyclodextrin and graphene was also proposed [70]. An MeS cube was added into a β-cyclodextrin and graphene dispersion, and then ammonia solution was added to adjust pH to 10. After adding hydrazine and heating, the modified MeS was dried. The developed sponges could be used for the extraction of malachite green. The presence of β-cyclodextrin was found to significantly affect the adsorption, since sponges prepared with lower amounts of β-cyclodextrin had lower sorption efficiency. The sponges were employed to extract malachite green from fresh crayfish and squid extracts (samples were homogenized and acetonitrile was added to extract malachite green). Another type of functionalization for MeS reported was the use of carboxylated multi-walled carbon nanotubes and the metal–organic framework MIL-101(Cr) [71]. The synthesis was based on mixing carboxylated multi-walled carbon nanotubes, MIL-101(Cr), and polyvinylidene difluoride, and then immersing MeS into the final solution. The modified sponges were used in an SPE procedure for the extraction of six triazines from corn extracts (corn was crushed to fine powder, and hexane was used to extract the compounds).

To coat MeS with polyaniline a new procedure was developed [72]. Since polyaniline is a polymer with four different states, it can serve well as a sorbent in sample preparation. The synthesis of the sponges was very simple. After dipping the MeS into an aniline solution and freezing them at 4 °C for 30 min, a chilled solution of ammonium persulfate was added, and the mixture was stirred for 30 s. Then, the mixture was left at 4 °C for 4.5 h and, after rinsing and drying, the sponges were ready to be used. Stirring was avoided during synthesis so that the formation of polyaniline agglomerates in the MeS could not be present. The modified MeS was found to be suitable for the extraction of perfluorooctanoic acid and perfluorooctane sulfonate from deproteinated (with acetonitrile) human urine and serum. A similar simple procedure was followed in another study, were silanization of MeS with trichloromethylsilane was completed in 10 min [73]. The sponges were rendered hydrophobic after the silanization, which made it suitable for the extraction of benzene, toluene, ethylbenzene, *m*-xylene, and *o*-xylene. The adsorption was based mainly on hydrophobic interactions, since benzene was adsorbed less efficiently than *m*-xylene ($K_{o/w}$ for *m*-xylene is 10 times higher than that of benzene). Following a needle-trap extraction method, an analytical procedure was developed which exhibited low limits of detection (0.005–0.0010 µg·L^{-1}). In a more complex study, layered double hydroxides were developed on the surface of MeS [74]. To do so, the Co(II)/2-methylimidazole porous coordination polymers were firstly immobilized on MeS and served as a source of Co, so that Ni–Co layered double hydroxides could be synthesized on the MeS. Three phenolic acids (gallic, *p*-hydroxybenzoic, and caffeic acid) were used to examine the suitability of the developed sponges in sample preparation. It is noteworthy that the developed layered double hydroxides were dissolved during the elution step to obtain the analytes. Owing to the good analytical figures of merit, a simple and effective analytical method was developed. Compared to the extraction with bare layered double hydroxides, the composite material had superior performance, due to the increased surface area. This is probably one of the reasons that the use of sponges and sponge-like materials is again a trend, since the surface area of existing compounds can be increased, by "depositing" them onto sponges. Finally, Liu et al. formed silica monoliths on the surface of MeS [75]. The sponges were dipped into a hydrolyzed mixture of tetramethoxysilane and vinyltrimethoxysilane, containing polyethylene glycol, urea, and acetic acid, and then heated. To render the sponges suitable for the extraction of dipeptides, sponges were modified with 3-mercapto-1-propanesulfonic acid so that sulfonate groups could interact with the free amine groups of the peptides. The silica-monolith-functionalized MeS could also be easily used for other applications, by altering the synthesis mixture. For instance, the addition of β-cyclodextrin during synthesis makes the sponges effective for the sorption of 4,4'-sulfonyldiphenol. Analytical methods developed based on modified melamine sponges are summarized in Table 5.

Table 5. Analytical methods developed based on melamine sponges. MWCNT: multi-walled carbon nanotube.

Functionalization Moieties	Analyte	Sample	Analytical Technique	LOD (μg·L^{-1})	Recoveries (%)	RSD (%)	Reference
β-Cyclodextrin/ graphene oxide	Flavonoids	*Lycium barbarum*	HPLC	0.5–2	77.9–102.6	3.5–6.8	[69]
Polyvinylidene difluoride-MIL-101(Cr)/MWCNTs-	Triazines	Corn	HPLC–MS/MS	0.01–0.04	90.3–116.5	1.08–12.32	[71]
Graphene	Sulfonamides	Milk, eggs, and lake water	HPLC-UV	0.03–0.44	90–108	<10.1	[10]
Copper sheets	Sulfonamides	Milk and lake water	HPLC-UV	0.075–0.35 and 0.009–0.019	88–97 and 89–102	6.8–9.9	[11]
Urea–formaldehyde co-oligomers	Fenbufen, flurbiprofen, benzophenone-8, butylparaben, cumylphenol, 4-octylphenol, chlorpyrifos, trifluralin, deltamethrin, tonalide	Lake water	HPLC-UV	0.01–0.33	92–100	5.6–8.4	[12]
β-Cyclodextrin and graphene oxide	Malachite green	Crayfish and squid	HPLC-Vis	0.21	88.6–100.8	-	[70]
Polyaniline	Perfluorooctanoic acid and perfluorooctane sulfonate	Human serum and urine	HPLC/MS	-	79–91	5.5–8.2	[72]
Trichloromethylsilane	Benzene, toluene, ethylbenzene, *m*-xylene, and *o*-xylene	Hookah, gulf water, and petrochemical wastewater	GC/MS	0.005–0.0010	91–105	<13	[73]
Ni–Co layered double hydroxides	Gallic acid, *p*-hydroxybenzoic acid, and caffeic acid	Fruit juices	HPLC-UV	0.15–0.35	89.7–95.3	<10	[74]
Silica monolith	Dipeptides (Tyr–Gly, Phe–Gly, Tyr–Val, Tyr–Ala, 3-I-Tyr–Ala, 3,5-dI-Tyr–Ala)	Water	HPLC–MS/MS	0.00002–0.0013	100	2–3	[75]

3.5. Use of Natural Sponges

The use of natural sponges has many benefits, including renewability, low cost, environmental friendliness, etc. Therefore, the use of natural sponges is highly promising. It is known that sea sponges are used as biomarkers to monitor the contamination of water with heavy metals [76]. Therefore, they are suitable for the preconcentration of metals. Capitalizing on this principle, a sea sponge was used to fill a column, which was used to extract copper, iron, lead, and nickel [76]. Due to the complex composition of the sponge and the plenty functional groups, no complexation agents were needed to sorb the metals, as in the case of PUFs. Moreover, the developed method is environmentally friendly, as it uses natural sea sponge. Later, the same groups published another study, where they proposed the use of sea sponges for the adsorption of Ponceau 4R and Sudan Orange G dyes [77]. The sea sponges served as an excellent sorbent since very good analytical figures of merit were recorded and, more importantly, very good recoveries were obtained from analyzing real samples without any pretreatment (i.e., peach-flavored drink powder, fruit-flavored mint candy, flavored rock candy, rosehip-flavored drink powder, fruit-flavored soft drinks, tomato paste, and pepper). No interference was recorded from ions commonly present in food products (such as iron, copper, potassium, calcium, etc.) or other food dyes (chocolate brown, tartrazine, sunset yellow, brilliant blue, patent blue V), even though, in their previous study, they found that metals could readily be adsorbed from the sea sponges. These two studies highlight the potential of sea sponges in sample preparation.

Luffa sponge is another natural sponge, obtained from the ripened fruits of *Luffa cylindrica* [78]. It is composed of lignin, cellulose, hemicellulose, and smaller quantities of pectin and proteins. Owing to its many functional groups, it can also serve as an excellent sorbent material. This is evidenced by three recent reports. In the first report, *Luffa* sponges were used as a sorbent for phosphopeptides from protein digests [78]. The sponges exhibited exceptional selectivity for phosphopeptides, over other non-phosphopeptides, making feasible their detection, even when their concentration was 100 times lower than other non-phosphopeptides. In the second report, the authors used *Luffa* sponges for the selective extraction of chromium (III) [79]. Selectivity over chromium (VI) was ensured by adjusting the pH of the solution to 4.0. The analytical method developed exhibited high accuracy as evidenced by analyzing certified reference materials (certified value: 300.00 ± 0.5, found value: 299.57 ± 0.006). In the final report, the ionic liquid 1-hexadecyl-3-methylimidazolium bromide was deposited on *Luffa* sponges by physisorption [80]. The modification was carried out to render *Luffa* sponges suitable for the determination of four benzoylurea insecticides in water and tea beverage samples. Before the addition of the ionic liquid, an alkali treatment of the *Luffa* sponges was carried out to remove hemicellulose and lignin and to make the sponges more hydrophilic.

4. Conclusions

Herein, we discussed the use and selected applications of sponges and sponge-like materials in analytical chemistry. This trend started with the use of PUFs for metal ions sorption. Although quite a few articles were published on this topic due to the limitations of PUF use, the narrow applicability (only metal ions), and the advancement of nanotechnology, this trend soon declined. However, in the last decade, this trend was not only reversed, but more and more researchers also aimed to develop new sorbent materials, based on sponges. Currently, many different sponge materials exist, such as PUFs, MeS, carbon foams, sea sponges, *Luffa* sponges, and others. This makes it easier to develop more advanced or selective sorbents than in the past. The next step in this field was the combination of nanomaterials with sponges. This resulted in the development of sorbents with more advanced characteristics and wider sorption capabilities, rendering sponges suitable for the sorption of small organic molecules, such as antibiotics and pesticides. In the future, large-scale synthesis of the materials should be examined, so as to result in commercially available and reliable sorbent materials. Moreover, the use of other nanomaterials should also be examined in order to make even wider the gamut of potential sorbents, as well as to render them more suitable for specific applications. Whether this trend will come to its own in sample preparation or not remains to be seen in the near future. Until then, the development of sponge-based sorbents will continue to improve, resulting in exceptional materials that could significantly alter existing methods.

Author Contributions: Conceptualization, T.G.C. and C.D.S.; writing—review and editing, T.G.C. and C.D.S. All authors have read and agreed to the published version of the manuscript.

Funding: This research received no external funding.

Conflicts of Interest: The authors declare no conflict of interest.

References

1. Wen, Y.; Chen, L.; Li, J.; Liu, D.; Chen, L. Recent advances in solid-phase sorbents for sample preparation prior to chromatographic analysis. *TrAC-Trends Anal. Chem.* **2014**, *59*, 26–41. [CrossRef]

2. Kuhtinskaja, M.; Koel, M. Smart Materials and Green Analytical Chemistry. In *Handbook of Smart Materials in Analytical Chemistry*; Guardia, M., Esteve-Turrillas, F.A., Eds.; John Wiley & Sons Ltd.: Hoboken, NJ, USA, 2019; pp. 503–530.

3. Chatzimitakos, T.; Stalikas, C. Carbon-based nanomaterials functionalized with ionic liquids for microextraction in sample preparation. *Separations* **2017**, *4*, 14. [CrossRef]

4. Khan, A.S.; Chow, A. X-ray fluorescence spectrometric determination of germanium after extraction with polyurethane foam. *Anal. Chim. Acta* **1990**, *238*, 423–426. [CrossRef]

5. Carvalho, M.S.; Medeiros, J.A.; Nóbrega, A.W.; Mantovano, J.L.; Rocha, V.P.A. Direct determination of gallium on polyurethane foam by X-ray fluorescence. *Talanta* **1995**, *42*, 45–47. [CrossRef]

6. Santiago De Jesus, D.; Souza De Carvalho, M.; Spínola Costa, A.C.; Costa Ferreira, S.L. Quantitative separation of zinc traces from cadmium matrices by solid-phase extraction with polyurethane foam. *Talanta* **1998**, *46*, 1525–1530. [CrossRef]

7. Lemos, V.A.; Santos, M.S.; Santos, E.S.; Santos, M.J.S.; dos Santos, W.N.L.; Souza, A.S.; de Jesus, D.S.; das Virgens, C.F.; Carvalho, M.S.; Oleszczuk, N.; et al. Application of polyurethane foam as a sorbent for trace metal pre-concentration-A review. *Spectrochim. Acta-Part B At. Spectrosc.* **2007**, *62*, 4–12. [CrossRef]

8. Braun, T.; Navratil, J.D.; Farag, A.B. *Polyurethane Foam Sorbents in Separation Science*; CRC Press: Boca Raton, FL, USA, 2018.

9. Soriano, S.; Cassella, R.J. Solid-phase extraction of Cu(II) using polyurethane foam and eriochrome black T as ligand for its determination in waters by flame atomic absorption spectrometry. *J. Braz. Chem. Soc.* **2013**, *24*, 1172–1179. [CrossRef]

10. Chatzimitakos, T.; Samanidou, V.; Stalikas, C.D. Graphene-functionalized melamine sponges for microextraction of sulfonamides from food and environmental samples. *J. Chromatogr. A* **2017**, *1522*, 1–8. [CrossRef]

11. Chatzimitakos, T.G.; Stalikas, C.D. Melamine sponge decorated with copper sheets as a material with outstanding properties for microextraction of sulfonamides prior to their determination by high-performance liquid chromatography. *J. Chromatogr. A* **2018**, *1554*, 28–36. [CrossRef]

12. García-Valverde, M.T.; Chatzimitakos, T.; Lucena, R.; Cárdenas, S.; Stalikas, C.D. Melamine sponge functionalized with urea-formaldehyde co-oligomers as a sorbent for the solid-phase extraction of hydrophobic analytes. *Molecules* **2018**, *23*, 2595. [CrossRef]

13. Sajid, M.; Basheer, C.; Mansha, M. Membrane protected micro-solid-phase extraction of organochlorine pesticides in milk samples using zinc oxide incorporated carbon foam as sorbent. *J. Chromatogr. A* **2016**, *1475*, 110–115. [CrossRef] [PubMed]

14. Shoeib, M.; Harner, T. Characterization and comparison of three passive air samplers for persistent organic pollutants. *Environ. Sci. Technol.* **2002**, *36*, 4142–4151. [CrossRef]

15. Strandberg, B.; Julander, A.; Sjöström, M.; Lewné, M.; Koca Akdeva, H.; Bigert, C. Evaluation of polyurethane foam passive air sampler (PUF) as a tool for occupational PAH measurements. *Chemosphere* **2018**, *190*, 35–42. [CrossRef] [PubMed]

16. Estellano, V.H.; Pozo, K.; Harner, T.; Corsolini, S.; Focardi, S. Using PUF disk passive samplers to simultaneously measure air concentrations of persistent organic pollutants (POPs) across the Tuscany region, Italy. *Atmos. Pollut. Res.* **2012**, *3*, 88–94. [CrossRef]

17. Jaward, F.M.; Farrar, N.J.; Harner, T.; Sweetman, A.J.; Jones, K.C. Passive air sampling of polycyclic aromatic hydrocarbons and polychlorinated naphthalenes across Europe. *Environ. Toxicol. Chem.* **2004**, *23*, 1355–1364. [CrossRef] [PubMed]

18. Motelay-Massei, A.; Harner, T.; Shoeib, M.; Diamond, M.; Stern, G.; Rosenberg, B. Using passive air samplers to assess urban-rural trends for persistent organic pollutants and polycyclic aromatic hydrocarbons. 2. Seasonal trends for PAHs, PCBs, and organochlorine pesticides. *Environ. Sci. Technol.* **2005**, *39*, 5763–5773. [CrossRef]

19. Santiago, E.C.; Cayetano, M.G. Polycyclic aromatic hydrocarbons in ambient air in the Philippines derived from passive sampler with polyurethane foam disk. *Atmos. Environ.* **2007**, *41*, 4138–4147. [CrossRef]

20. Bowen, H.J.M. Absorption by polyurethane foams; new method of separation. *J. Chem. Soc. A Inorg. Phys. Theor. Chem.* **1970**, 1082–1085. [CrossRef]
21. Cassella, R.J. On-line solid phase extraction with polyurethane foam: Trace level spectrophotometric determination of iron in natural waters and biological materials. *J. Environ. Monit.* **2002**, *4*, 522–527. [CrossRef]
22. de Almeida, G.N.; de Sousa, L.M.; Pereira Netto, A.D.; Cassella, R.J. Characterization of solid-phase extraction of Fe(III) by unloaded polyurethane foam as thiocyanate complex. *J. Colloid Interface Sci.* **2007**, *315*, 63–69. [CrossRef]
23. Sant'Ana, O.D.; Jesuino, L.S.; Cassella, R.J.; Carvalho, M.S.; Santelli, R.E. Solid Phase Extraction of Cu(II) as Diethyldithiocarbamate (DDTC) Complex by Polyurethane Foam. *J. Braz. Chem. Soc.* **2003**, *14*, 728–733. [CrossRef]
24. De Jesus, D.S.; Das Graças Korn, M.; Ferreira, S.L.C.; Carvalho, M.S. Separation method to overcome the interference of aluminum on zinc determination by inductively coupled plasma atomic emission spectroscopy. *Spectrochim. Acta Part B At. Spectrosc.* **2000**, *55*, 389–394. [CrossRef]
25. De Carvalho, M.S.; Domingues, M.D.L.F.; Mantovano, J.L.; Da Cunha, J.W.S.D. Preconcentration method for the determination of thorium in natural water by wavelength dispersive X-ray fluorescence spectrometry. *J. Radioanal. Nucl. Chem.* **2002**, *253*, 253–256. [CrossRef]
26. Trokhimenko, O.M.; Sukhan, V.V.; Nabivanets, B.I.; Ishchenko, V.B. Sorption preconcentration of thallium(I) on polyurethane foam modified with molybdophosphate. *J. Anal. Chem.* **2000**, *55*, 626–629. [CrossRef]
27. Carvalho, M.S.; Maria de Lourdes, F.D.; Mantovano, J.L.; Filho, E.Q.S. Uranium determination at ppb levels by X-ray fluorescence after its preconcentration on polyurethane foam. *Spectrochim. Acta Part B At. Spectrosc.* **1998**, *53*, 1945–1949. [CrossRef]
28. Costa Ferreira, S.L.; Costa Dos Santos, H.; Santiago De Jesus, D. Molybdenum determination in iron matrices by ICP-AES after separation and preconcentration using polyurethane foam. *Fresenius. J. Anal. Chem.* **2001**, *369*, 187–190. [CrossRef]
29. Ferreira, S.L.C.; Dos Santos, H.C.; Campos, R.C. The determination of molybdenum in water and biological samples by graphite furnace atomic spectrometry after polyurethane foam column separation and preconcentration. *Talanta* **2003**, *61*, 789–795. [CrossRef]
30. Abbas, M.N.D. Solid phase spectrophotometric determination of traces of arsenate and phosphate in water using polyurethane foam sorbent. *Anal. Lett.* **2003**, *36*, 1231–1244. [CrossRef]
31. Ferreira, S.L.C.; De Jesus, D.S.; Cassella, R.J.; Costa, A.C.S.; De Carvalho, M.S.; Santelli, R.E. An on-line solid phase extraction system using polyurethane foam for the spectrophotometric determination of nickel in silicates and alloys. *Anal. Chim. Acta* **1999**, *378*, 287–292. [CrossRef]
32. Xue, D.; Wang, H.; Liu, Y.; Shen, P.; Sun, J. Cytosine-functionalized polyurethane foam and its use as a sorbent for the determination of gold in geological samples. *Anal. Methods* **2016**, *8*, 29–39. [CrossRef]
33. El-Shahawi, M.S.; Almehdi, M. Qualitative, semi-quantitative and spectrophotometric determination of ruthenium(III) by solid-phase extraction with 3-hydroxy-2-methyl-1,4-naphthoquinone-4-oxime-loaded polyurethane foam columns. *J. Chromatogr. A* **1995**, *697*, 185–190. [CrossRef]
34. Abou-El-Sherbini, K.S.; Mostafa, G.A.E.; Hassanien, M.M. A new selective chromogenic reagent for the spectrophotometric determination of thallium(I) and (III) and its separation using flotation and the solid-phase extraction on polyurethane foam. *Anal. Sci.* **2003**, *19*, 1269–1275. [CrossRef] [PubMed]
35. Arpadjan, S.; Vuchkova, L.; Kostadinova, E. Sorption of arsenic, bismuth, mercury, antimony, selenium and tin on dithiocarbamate loaded polyurethane foam as a preconcentration method for their determination in water samples by simultaneous inductively coupled plasma atomic emission spectrometry and electrothermal atomic absorption spectrometry. *Analyst* **1997**, *122*, 243–246. [CrossRef]
36. Maddalena, R.L.; McKone, T.E.; Kado, N.Y. Simple and rapid extraction of polycyclic aromatic hydrocarbons collected on polyurethane foam adsorbent. *Atmos. Environ.* **1998**, *32*, 2497–2503. [CrossRef]
37. Dmitrienko, S.G.; Gurariy, E.Y.; Nosov, R.E.; Zolotov, Y.A. Solid-phase extraction of polycyclic aromatic hydrocarbons from aqueous samples using polyurethane foams in connection with solid-matrix spectrofluorimetry. *Anal. Lett.* **2001**, *34*, 425–438. [CrossRef]
38. Azevedo Lemos, V.; Alves De Jesus, A.; Moreira Gama, E.; Teixeira David, G.; Terumi Yamaki, R. On-line solid phase extraction and determination of copper in food samples using polyurethane foam loaded with Me-BTANC. *Anal. Lett.* **2005**, *38*, 683–696. [CrossRef]

39. Maloney, M.P.; Moody, G.J.; Thomas, J.D.R. Extraction of metals from aqueous solution with polyurethane foam. *Analyst* **1980**, *105*, 1087–1097. [CrossRef]

40. Hasany, S.M.; Saeed, M.M.; Ahmed, M. Adsorption isotherms and thermodynamic profile of Co(II)-SCN complex uptake on polyurethane foam. *Sep. Sci. Technol.* **2000**, *35*, 379–394. [CrossRef]

41. de Raat, W.K.; Schulting, F.L.; Burghardt, E.; de Meijere, F.A. Application of polyurethane foam for sampling volatile mutagens from ambient air. *Sci. Total Environ.* **1987**, *63*, 175–189. [CrossRef]

42. Vuchkova, L.; Arpadjan, S. Behaviour of the dithiocarbamate complexes of arsenic, antimony, bismuth, mercury, lead, tin and selenium in methanol with a hydride generator. *Talanta* **1996**, *43*, 479–486. [CrossRef]

43. Zaranski, M.T.; Patton, G.W.; McConnell, L.L.; Bidleman, T.F.; Mulik, J.D. Collection of Nonpolar Organic Compounds from Ambient Air Using Polyurethane Foam-Granular Adsorbent Sandwich Cartridges. *Anal. Chem.* **1991**, *63*, 1228–1232. [CrossRef] [PubMed]

44. McConnell, L.L.; Bidleman, T.F. Collection of two-ring aromatic hydrocarbons, chlorinated phenols, guaiacols, and benzenes from ambient air using polyurethane foam/Tenax-GC cartridges. *Chemosphere* **1998**, *37*, 885–898. [CrossRef]

45. Alexandrova, A.; Arpadjan, S. Column solid phase extraction as preconcentration method for trace element determination in oxalic acid by atomic absorption spectrometry and inductively coupled plasma atomic emission spectrometry. *Anal. Chim. Acta* **1995**, *307*, 71–77. [CrossRef]

46. De Jesus, D.S.; Cassella, R.J.; Ferreira, S.L.C.; Costa, A.C.S.; De Carvalho, M.S.; Santelli, R.E. Polyurethane foam as a sorbent for continuous flow analysis: Preconcentration and spectrophotometric determination of zinc in biological materials. *Anal. Chim. Acta* **1998**, *366*, 263–269. [CrossRef]

47. Lee, D.W.; Halmann, M. Selective Separation of Nickel(ll) by Dimethylglyoxime-Treated Polyurethane Foam. *Anal. Chem.* **1976**, *48*, 2214–2217. [CrossRef]

48. El-Shahawi, M.S.; Aldhaheri, S.M. Preconcentration and separation of acaricides by polyether based polyurethane foam. *Anal. Chim. Acta* **1996**, *320*, 277–287. [CrossRef]

49. Hamon, R.F.; Chow, A. Extraction of cobalt from thiocyanate solutions with polyurethane foam. *Talanta* **1984**, *31*, 963–973. [CrossRef]

50. Cassella, R.J.; Santelli, R.E.; Branco, A.G.; Lemos, V.A.; Ferreira, S.L.C.; De Carvalho, M.S. Selectivity enhancement in spectrophotometry: On-line interference suppression using polyurethane foam minicolumn for aluminum determination with Methyl Thymol Blue. *Analyst* **1999**, *124*, 805–808. [CrossRef]

51. Farag, A.B.; El-Shahawi, M.S. Removal of organic pollutants from aqueous solution. V. Comparative study of the extraction, recovery and chromatographic separation of some organic insectisides using unloaded polyurethane foam columns. *J. Chromatogr. A* **1991**, *552*, 371–379. [CrossRef]

52. Khan, A.S.; Chow, A. Determination of arsenic by polyurethane foam extraction and X-ray fluorescence. *Talanta* **1984**, *31*, 304–306. [CrossRef]

53. Hawthorne, S.B.; Krieger, M.S.; Miller, D.J. Supercritical Carbon Dioxide Extraction of Polychlorinated Biphenyls, Polycyclic Aromatic Hydrocarbons, Heteroatom-Containing Polycyclic Aromatic Hydrocarbons, and n-Alkanes from Polyurethane Foam Sorbents. *Anal. Chem.* **1989**, *61*, 736–740. [CrossRef]

54. Hasany, S.M.; Saeed, M.M.; Ahmed, M. Sorption of traces of silver ions onto polyurethane foam from acidic solution. *Talanta* **2001**, *54*, 89–98. [CrossRef]

55. Dmitrienko, S.G.; Goncharova, L.V.; Zhigulev, A.V.; Nosov, R.E.; Kuzmin, N.M.; Zolotov, Y.A. Sorption-photometric determination of ascorbic acid using molybdosilicic heteropolyacid and polyurethane foam after microwave irradiation. *Anal. Chim. Acta* **1998**, *373*, 131–138. [CrossRef]

56. Braun, T.; Abbas, M.N. Unloaded polyurethane foams as solid extractants for some metal thiocyanate complexes from aqueous solution. *Anal. Chim. Acta* **1982**, *134*, 321–326. [CrossRef]

57. Makki, M.S.T.; Abdel-Rahman, R.M.; Alfooty, K.O.; El-Shahawi, M.S. Thiazolidinone steroids impregnated polyurethane foams as a solid phase extractant for the extraction and preconcentration of cadmium(II) from industrial wastewater. *E-J. Chem.* **2011**, *8*, 887–895. [CrossRef]

58. El-sharief, F.M.; Asweisi, A.A.; Bader, N. Separation of Some Metal Ions Using β-Naphthol Modified Polyurethane Foam. *Asian J. Nanosci. Mater.* **2019**. [CrossRef]

59. El-Shahat, M.F.; Burham, N.; Azeem, S.M.A. Flow injection analysis-solid phase extraction (FIA-SPE) method for preconcentration and determination of trace amounts of penicillins using methylene blue grafted polyurethane foam. *J. Hazard. Mater.* **2010**, *177*, 1054–1060. [CrossRef]

60. Maggira, M.; Deliyanni, E.A.; Samanidou, V.F. Synthesis of graphene oxide based sponges and their study as sorbents for sample preparation of cow milk prior to HPLC determination of sulfonamides. *Molecules* **2019**, *24*, 2086. [CrossRef]

61. Ghani, M.; Maya, F.; Cerdà, V. Automated solid-phase extraction of organic pollutants using melamine-formaldehyde polymer-derived carbon foams. *RSC Adv.* **2016**, *6*, 48558–48565. [CrossRef]

62. Wang, L.; Wang, M.; Yan, H.; Yuan, Y.; Tian, J. A new graphene oxide/polypyrrole foam material with pipette-tip solid-phase extraction for determination of three auxins in papaya juice. *J. Chromatogr. A* **2014**, *1368*, 37–43. [CrossRef]

63. Qi, M.; Tu, C.; Li, Z.; Wang, W.; Chen, J.; Wang, A.J. Determination of Sulfonamide Residues in Honey and Milk by HPLC Coupled with Novel Graphene Oxide/Polypyrrole Foam Material-Pipette Tip Solid Phase Extraction. *Food Anal. Methods* **2018**, *11*, 2885–2896. [CrossRef]

64. Wang, X.; Wang, J.; Du, T.; Kou, H.; Du, X.; Lu, X. Determination of six benzotriazole ultraviolet filters in water and cosmetic samples by graphene sponge-based solid-phase extraction followed by high-performance liquid chromatography. *Anal. Bioanal. Chem.* **2018**, *410*, 6955–6962. [CrossRef] [PubMed]

65. Jillani, S.M.S.; Sajid, M.; Alhooshani, K. Evaluation of carbon foam as an adsorbent in stir-bar supported micro-solid-phase extraction coupled with gas chromatography–mass spectrometry for the determination of polyaromatic hydrocarbons in wastewater samples. *Microchem. J.* **2019**, *144*, 361–368. [CrossRef]

66. Li, S.; Jia, M.; Guo, H.; Hou, X. Development and application of metal organic framework/chitosan foams based on ultrasound-assisted solid-phase extraction coupling to UPLC-MS/MS for the determination of five parabens in water. *Anal. Bioanal. Chem.* **2018**, *410*, 6619–6632. [CrossRef] [PubMed]

67. Wang, X.; Wang, J.; Du, T.; Kou, H.; Du, X.; Lu, X. Application of ZIF-8-graphene oxide sponge to a solid phase extraction method for the analysis of sex hormones in milk and milk products by high-performance liquid chromatography. *New J. Chem.* **2019**, *43*, 2783–2789. [CrossRef]

68. Sun, T.; Wang, M.; Wang, D.; Du, Z. Solid-phase microextraction based on nickel-foam@polydopamine followed by ion mobility spectrometry for on-site detection of Sudan dyes in tomato sauce and hot-pot sample. *Talanta* **2020**, *207*, 120244. [CrossRef]

69. Hou, X.; Lu, X.; Niu, P.; Tang, S.; Wang, L.; Guo, Y. β-Cyclodextrin-modified three-dimensional graphene oxide-wrapped melamine foam for the solid-phase extraction of flavonoids. *J. Sep. Sci.* **2018**, *41*, 2207–2213. [CrossRef]

70. Gao, Z.; Li, Y.; Ma, Y.; Ji, W.; Chen, T.; Ma, X.; Xu, H. Functionalized melamine sponge based on β-cyclodextrin-graphene oxide as solid-phase extraction material for rapidly pre-enrichment of malachite green in seafood. *Microchem. J.* **2019**, *150*, 104167. [CrossRef]

71. Qin, Z.; Jiang, Y.; Piao, H.; Li, J.; Tao, S.; Ma, P.; Wang, X.; Song, D.; Sun, Y. MIL-101(Cr)/MWCNTs-functionalized melamine sponges for solid-phase extraction of triazines from corn samples, and their subsequent determination by HPLC-MS/MS. *Talanta* **2020**, *211*, 120676. [CrossRef]

72. Qi, L.; Gong, J. Facile in-situ polymerization of polyaniline-functionalized melamine sponge preparation for mass spectrometric monitoring of perfluorooctanoic acid and perfluorooctane sulfonate from biological samples. *J. Chromatogr. A* **2020**, *1616*, 460777. [CrossRef]

73. Bagheri, H.; Zeinali, S.; Baktash, M.Y. A single–step synthesized supehydrophobic melamine formaldehyde foam for trace determination of volatile organic pollutants. *J. Chromatogr. A* **2017**, *1525*, 10–16. [CrossRef] [PubMed]

74. Ghani, M.; Frizzarin, R.M.; Maya, F.; Cerdà, V. In-syringe extraction using dissolvable layered double hydroxide-polymer sponges templated from hierarchically porous coordination polymers. *J. Chromatogr. A* **2016**, *1453*, 1–9. [CrossRef] [PubMed]

75. Liu, Z.; Jiang, P.; Huang, G.; Yan, X.; Li, X.F. Silica Monolith Nested in Sponge (SiMNS): A Composite Monolith as a New Solid Phase Extraction Material for Environmental Analysis. *Anal. Chem.* **2019**, *91*, 3659–3666. [CrossRef] [PubMed]

76. Karatepe, A.; Soylak, M. Sea sponge as a low cost biosorbent for solid phase extraction of some heavy metal ions and determination by flame atomic absorption spectrometry. *J. AOAC Int.* **2014**, *97*, 1689–1695. [CrossRef]

77. Karatepe, A.; Akalin, C.; Soylak, M. Solid-phase extraction of some food dyes on sea sponge column and determination by UV–vis spectrophotometer. *Desalin. Water Treat.* **2016**, *57*, 25822–25829. [CrossRef]

78. Dai, L.; Sun, Z.; Zhou, P. Modification of luffa sponge for enrichment of phosphopeptides. *Int. J. Mol. Sci.* **2020**, *21*, 101. [CrossRef]
79. Neto, J.A.D.S.; Oliveira, J.D.A.N.; Siqueira, L.M.C.; Alves, V.N. Selective extraction and determination of chromium concentration using luffa cylindrica fibers as sorbent and detection by FAAS. *J. Chem.* **2019**, *2019*, 1679419. [CrossRef]
80. Wang, H.; Liu, C.; Huang, X.; Jia, C.; Cao, Y.; Hu, L.; Lu, R.; Zhang, S.; Gao, H.; Zhou, W.; et al. Ionic liquid-modified luffa sponge fibers for dispersive solid-phase extraction of benzoylurea insecticides from water and tea beverage samples. *New J. Chem.* **2018**, *42*, 8791–8799. [CrossRef]

© 2020 by the authors. Licensee MDPI, Basel, Switzerland. This article is an open access article distributed under the terms and conditions of the Creative Commons Attribution (CC BY) license (http://creativecommons.org/licenses/by/4.0/).

Article

Application of Deep Eutectic Solvents and Ionic Liquids in the Extraction of Catechins from Tea

Sylwia Bajkacz [1,2,*], Jakub Adamek [2,3] and **Anna Sobska [1]**

[1] Department of Inorganic, Analytical Chemistry and Electrochemistry, Faculty of Chemistry, Silesian University of Technology, Krzywoustego 6, 44-100 Gliwice, Poland; anna.sobska@gmail.com
[2] Biotechnology Center of Silesian University of Technology, Krzywoustego 8, 44-100 Gliwice, Poland
[3] Department of Organic and Bioorganic Chemistry and Biotechnology, Faculty of Chemistry, Silesian University of Technology, Krzywoustego 4, 44-100 Gliwice, Poland; jakub.adamek@polsl.pl
* Correspondence: sylwia.bajkacz@polsl.pl; Tel.: +48-32-237-18-18

Academic Editor: Victoria Samanidou
Received: 18 June 2020; Accepted: 10 July 2020; Published: 14 July 2020

Abstract: This work aimed to comprehensively evaluate the potential and effectiveness of deep eutectic solvents (DESs) in the extraction of seven catechins from various tea samples. Different combinations of DES were used, consisting of Girard's reagent T (GrT) in various mixing ratios with organic acids and choline chloride. The yields of the DES extractions were compared with those from ionic liquids and conventional solvent. DES contained malic acid, as the hydrogen bond donors showed a good solubility of catechins with different polarities. In the second part of the study, a solid-phase extraction (SPE) method was applied to the extraction of catechins from tea infusions. The method was applied to the determination of selected catechins in tea leaves and tea infusions. Furthermore, we demonstrated that the proposed procedure works well in the simultaneous monitoring of these polyphenols, which makes it a useful tool in the quality control of tea.

Keywords: green extraction; deep eutectic solvents; ionic liquids; catechins; tea leaves

1. Introduction

Tea is the most popular beverage consumed in the world, and contains significant levels of polyphenols, especially catechins. Based on species, season, horticultural conditions, and degree of oxidation during the manufacturing process in tea samples, different catechins can be present [1,2].

Recent findings indicate that flavonoids, in particular catechins, possess rather potent antioxidant properties, which may result in numerous health benefits. Based on the present results, it can be said that regular consumption of green tea can reduce the incidence of cancer, including colon, pancreatic, and stomach cancers, as well as other diseases [3,4]. Thus, it is recommended to eat products that contain large amounts of catechins, to which undoubtedly include tea [5]. The extraction and isolation of catechins from tea leaves have been achieved using numerous methods [6,7] such as solid–liquid extraction (SLE) with different solvents (e.g., methanol, acetone, ethanol, acetonitrile, water, acetate, *n*-butanol and *n*-hexane) [8–11], dispersive liquid–liquid microextraction (DLLME) [12] microwave assisted extraction (MAE) [13], supercritical fluid extraction (SFE) [14,15], and ultrahigh-pressure extraction (UPE) [16].

Owing to catechins' potential for improving human health and extending the shelf-life of food products, an efficient and safe extraction system, preferably using green solvents, is necessary. This will promote accurate quantification of catechins in tea and tea products, as well as creating an efficient step for the isolation of individual catechins [17].

In recent years, ionic liquids (ILs) and deep eutectic mixtures have shown great potential in extraction processes relevant to several scientific and technological activities. ILs possesses many advantageous properties, such as chemical and thermal stability, nonflammability, high ionic conductivity, and a wide electrochemical potential window. These unique properties have triggered extensive studies into ILs as solvents or co-catalysts in various reactions including organic catalysis, inorganic synthesis, biocatalysis, and polymerization [18]. Furthermore, they have also been successfully applied in various areas of analytical chemistry, especially in a separation of analytes. To the best of our knowledge, the application of IL or catechin extraction from tea samples has yet to be examined [19,20].

Deep eutectic solvents (DESs) are a group of emerging solvents with excellent properties including negligible volatility at room temperature, non-inflammability, and high viscosity, as well as being environmentally benign [21]. The advantage of DESs is also the possibility of the extraction of a wide range of non-polar and polar compounds [22]. The increasing range of available DESs is due to the majority being easy to prepare, inexpensive, and biodegradable, which has provoked their application in wide and in diverse fields of science [23]. Recently, some procedures of catechin extraction from tea samples have been described using DESs [24–29].

In this study, we aim to (1) evaluate the ability of new DESs and ILs to extract catechins from tea, (2) optimize a supported solid–liquid extraction (SLE) parameters using DES as the solvent, (3) develop a fast and sensitive UHPLC-UV method for comprehensive analysis of seven catechins (catechin (C), epicatechin (EC), epigallocatechin (EGC), epicatechin gallate (ECG), epigallocatechin gallate (EGCG), gallocatechin (GC), gallocatechin gallate (GCG)) in tea samples, and (4) determine the catechins in tea leaves using the DES–SLE–UHPLC–UV method and infusion tea using the solid-phase extraction (SPE)–UHPLC–UV method.

2. Results and Discussion

2.1. Chromatographic Separation

In the first part of the study, a rapid UHPLC–UV method for the determination of catechins in tea matrices was developed. Fast separation is crucial for analyzing vast numbers of samples in order to save both solvent and time. For this purpose, a mixture of seven catechin standards was analyzed in the reversed-phase system, using three different columns (Poroshell 120 SB-C18 (100 × 4.6 mm, 2.7 µm), Poroshell XDB-C18 (50 × 2.1 mm, 1.8 µm), Poroshell 120 EC-C18 (50 × 3.0 mm, 1.8 µm)). For each column, different gradient profiles were tested, with the aim of obtaining the standards retention times and peak widths. Among all columns used, the best average resolution in the shortest time of analysis was obtained using Poroshell 120 SB-C18 (100 × 4.6 mm, 2.7 µm) column. The Poroshell 120 SB-C18 column, due to the unique, superficially porous particles and 2.7 µm particle size providing a shorter diffusion path for solutes, minimized peak broadening at high flow rates and had a high permeability. Moreover, this column enabled robust symmetrical peaks and a low column backpressure to be obtained, as well as a more sensitive method to be used, compared to other columns.

The mobile phase and gradient were optimized using acetic acid, formic acid trifluoroacetic acid in water (mobile phase A), and methanol or acetonitrile (mobile phase B). Preliminary experiments determined that 0.05% TFA in water and acetonitrile were optimal conditions [30]. The optimized elution, with a total run time of 8 min, consisted of one consecutive acetonitrile gradient step, each with increasing slopes (from 10 to 60% in 5.5 min), and a final 2.4-min re-equilibration time. The column temperature was kept at 25 °C. The detection of catechins was performed at $\lambda = 270$ nm.

Figure S1 displays UHPLC–UV chromatogram of catechin standards. The order of the catechins elution is as following, non-epi-forms without gallate GC and C, epi-forms without gallate EGC and EC, epi-forms with gallate EGCG, non-epi-forms with gallate GCG, and finally ECG [31].

2.2. SLE Extraction Parameters

2.2.1. Screening of Ionic Liquids, Deep Eutectic Solvents and Conventional Solvents

The ideal solvent type and extraction method for target compounds extracted from the sample was vital for the optimization of the process. High extraction yields can be obtained by decreasing solution viscosity, since at lower viscosity the solvent can easily penetrate the sample matrix. Thus far, for the extraction of catechins from tea samples, several organic solvents such as ethyl acetate, *n*-hexane, and petroleum ether have been used [7].

In this study, the extraction efficiency of five ionic liquids ([C$_4$MIM]NO$_3$, [C$_4$MIM]Cl, [C$_4$MIM]HSO$_4$, [C$_4$MIM]BF$_4$, [C$_4$MIM]Br), eight DESs (containing malic acid, citric acid, and L-lactic acid as hydrogen bond donors) and three conventional solvents (methanol, water, mixture of methanol:water (1:1; *v/v*)) were tested for the isolation of selected catechins from tea. The results are shown in Figure 1.

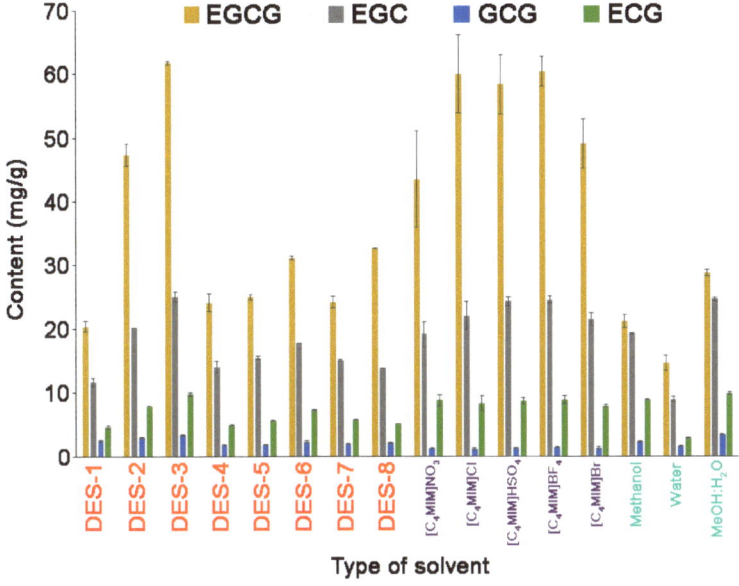

Figure 1. A comparison of the deep eutectic solvents (DESs), ionic liquids (ILs) and conventional solvent extraction efficiency of catechins.

In the extraction of DES-3, comparable yields were observed for the mixture of methanol:water and [C$_4$MIM]BF$_4$, which were obviously higher than those of methanol, water, [C$_4$MIM]NO$_3$ and [C$_4$MIM]Cl. Solvatochrome test ILs based on tetrafluoroboric acid and DES-3 are characterized as highly polar, which influences the effective extraction of catechins. In addition, [C$_4$MIM]BF$_4$ and DES-3 have a higher acidity than conventional solvents used for comparison. In the case of malic acid as HBD, in DES, high extraction yields were observed, due to stronger multi-interactions, including π–π, ionic/charge–charge and hydrogen bonding with targeted compounds. Moreover, the efficiency of the extraction can also be dependent on the role of the HBD:HBA ratio. The evidence emerging from examinations shows that by changing the molar ratio of malic acid:GrT from 1:2 to 2:1 the extraction efficiency of catechins from green tea shows a significant increase. In more acidic solvents (for example DESs based on carboxylic acid), catechins are more stable; therefore, the extraction efficiency is greater compared with water or methanol. Conventional organic solvents are usually volatile and toxic. In consideration of sustainability, biodegradability and pharmaceutically acceptable toxicity, the DES-based extraction method proposed in this study is efficient, non-toxic and eco-friendly, and can be used as a green substitute in the extraction of catechins from tea samples. According to the above results, DES-3 with malic acid:GrT (2:1) was selected as the optimal solvent.

2.2.2. Effect of Water Content in DES

After selecting the optimal solvent for catechin extraction, the procedures were performed with DES-3 and different water content (from 10% to 75% (v/v)). It was found that this factor has a marked impact on extraction yields (Figure S2). A higher water content reduced the viscosity and increased the polarity of the solvent mixture. Moreover, the addition of water affects the structure of eutectic solvents, which can be observed in FT-IR analyses of free-solvent components, DESs, and DESs with different water contents. For example, obtaining DES-3 from free components is related to the formation of a specific supramolecular structure based on hydrogen bonds. This is evidenced by the broadening and shifts of characteristic vibration bands (especially stretching vibrations: ν_{OH}, $\nu_{C=O}$, ν_{C-O}) in the IR spectra (Figure 2).

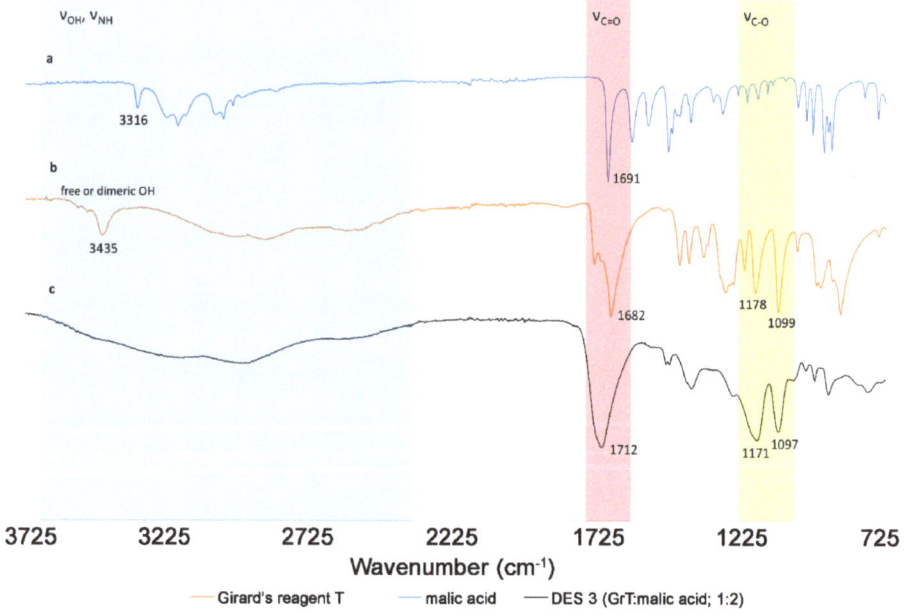

Figure 2. Changes in FT-IR spectra of free components (**a**) Girard's reagent, (**b**) malic acid and (**c**) the formed DES-3 (GrT:malic acid, 1:2).

Modifications to the DES-3 structure after water addition are reflected in the FT-IR spectra (Figure 3). According to FT-IR results, upon increasing water content from 30% to 75% (v/v), significant modifications to the supramolecular structure occur, which decrease the hydrogen bond interconnections between solvent and target bioactive catechins, resulting in the lower extraction yields of analytes.

Therefore, 30% (v/v) water in DES-3 gave the highest extraction yields and was utilized in further optimization tests.

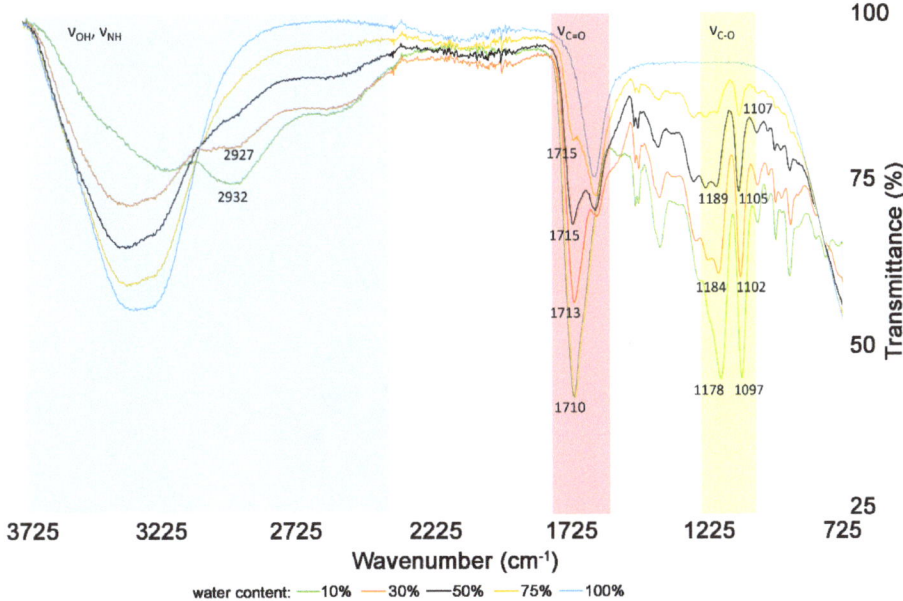

Figure 3. The effect of water content on the supramolecular structure of DES-3 based on FT-IR spectra.

2.2.3. Effect of Time, Temperature and Solid/Liquid Ratio

In order to optimize certain extraction conditions, response surface methodology (RSM) was adopted using Statistica 12 software package. Central composite design (CCD) was used to conduct the experiments. The effect and interaction of three parameters, specifically extraction time (6.5–73.5 min), temperature (6.5–48.4 °C), and solid/liquid ratio (1:2–1:12), were investigated. Peak area was adopted as the response function.

For visualization of the obtained results, three-dimensional (3D) RSM was applied. Figure 4 shows 3D plots of the response surface for the extraction efficiency of selected catechins, as related to extraction time (X_1), temperature (X_2), and solid/liquid ratio (X_3), respectively.

Based on the obtained results, an increase in extraction time from 40 to 50 min and in temperature to 50 °C enhanced the extraction yields of target compounds. When the time was constant at 50 min, and with an increase in temperature and solid/liquid ratio, the extraction yield of catechins increased within a certain range. The mass transfer of analytes from the tea samples to the DES solvent can be easier at higher temperatures, because of the decreased physical adsorption and chemical interactions between analytes and matrices. However, when the solid/liquid ratio and temperature exceed a certain value, the extraction yield declines.

As illustrated in Figure 4, the extraction yields of target compounds significantly increase with an increase in solid/liquid ratio, especially with a short time period. The extraction efficiency of catechins improve the solid/liquid ratio increase to 1:10 and extraction time to 50 min; however, the extraction yields decrease when the liquid/solid ratio exceeds 1:12. The extraction yield of target compounds increases with an increasing extraction temperature over a short extraction time. When the extraction temperature exceeds 50 °C with a longer extraction time, the extraction yield plateaus. Finally, the following DES extraction conditions were selected: extraction time: 50 min; extraction temperature: 50 °C; solid/liquid ratio: 1:10.

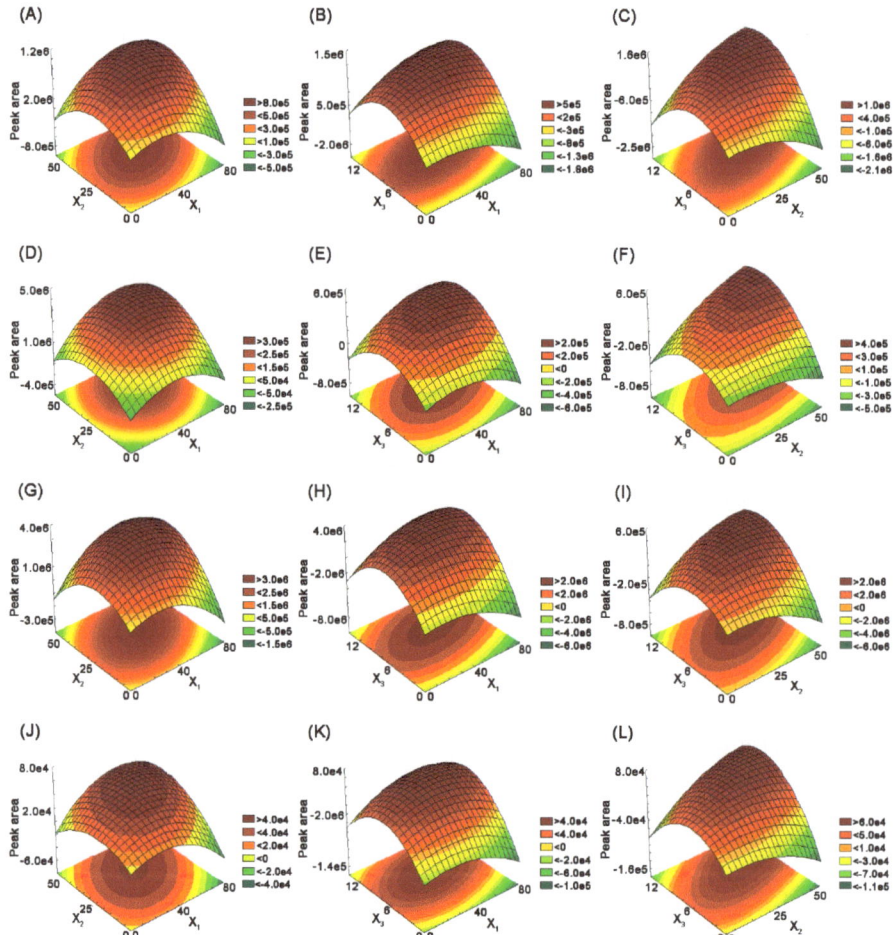

Figure 4. Response surface plots of (**A, D, G, J**) extraction time (X_1) and temperature (X_2) at a constant solid/liquid ratio (X_3) of 1:7, (**B**) extraction time (X_1) and solid/liquid ratio (X_3) at a constant extraction temperature (X_2) of 27.5 °C, (**C**) extraction temperature (X_2) and solid/liquid ratio (X_3) at a constant time (X_1) of 40 min (for (**A, B, C**)—Epigallocatechin (EGC), (**D, E, F**)—Epigallocatechin gallate (EGCG), (**G, H, I**)—Gallocatechin gallate (GCG), (**J, K, L**)—Epicatechin gallate (ECG)).

ANOVA was performed to evaluate the optimal conditions of DES and the relationship between the response and variables. A second-order polynomial equation for the extraction yield (Y) and variables was obtained by a multiple regression analysis of the experimental data. An ANOVA analysis is shown in Table S5. An F-test was examined in order to investigate the significance of each coefficient, allowing for the determination of the p-value. The F-value of lack of fit ($p = 0.1$) was not significant, which supported this model, giving an accurate representation of the experimental data. The coefficient of determination (R^2) was higher than 0.8721, showing that the experimental data were in accordance with predicted values. Furthermore, Table S5 indicates that linear coefficients (X_1, X_2 and X_3) and cross product coefficients (X_1X_2, X_1X_3, X_2X_3) are considered to be significant ($p < 0.05$), depending on compounds. Based on these results (Table S5), it can be stated that, depending on the compounds, the following effects have a significant impact on the electrochemical conversion efficiency: for EGC

X_3, X_2, $X_3{}^2$, $X_1{}^2$, X_2X_3, X_1X_3, X_1, for EGCG: X_3, $X_3{}^2$, X_2X_3, $X_2{}^2$, $X_1{}^2$, X_1X_3, for GCG: X_3, $X_3{}^2$, X_2X_3, $X_1{}^2$, for ECG: X_3, $X_3{}^2$, X_2X_3, $X_2{}^2$, X_1X_3, X_2, X_1X_2.

$$Y_{EGC} = 239295 + 1394X_1 - 90X_1{}^2 + 5223X_2 - 280X_2{}^2 + 64268X_3 - 7985X_3{}^2 + 96.4X_1X_2 + 783.7X_1X_3 + 1508X_2X_3; R^2 = 0.9383 \tag{1}$$

$$Y_{EGCG} = 602911 - 24536X_1 - 683X_1{}^2 - 10429X_2 - 3075X_2{}^2 + 807328X_3 - 95278X_3{}^2 + 1338X_1X_2 + 7245X_1X_3 + 18202X_2X_3; R^2 = 0.8940 \tag{2}$$

$$Y_{GCG} = 19620 + 109X_1 - 16X_1{}^2 - 212X_2 - 38X_2{}^2 + 14103X_3 - 1515X_3{}^2 + 21X_1X_2 + 94X_1X_3 + 271X_2X_3; R^2 = 0.8907 \tag{3}$$

$$Y_{ECG} = 88166 - 5663X_1 - 200X_1{}^2 - 6567X_2 - 646X_2{}^2 + 216979X_3 - 26344X_3{}^2 + 346X_1X_2 + 2080X_1X_3 + 5259X_2X_3; R^2 = 0.8721 \tag{4}$$

As shown in Figure S3, the plot of experimental values for extraction efficiency vs. those calculated from Equations (1)–(4) showed a good fit (i.e., EGC, EGCG, GCG, ECG). As a result of the full factorial design, a Pareto chart was drawn for selected catechins to visualize the estimated effects of the main variables and their interactions. Figure S4 shows the Pareto graphic analysis for selected catechins. The Pareto chart gives a graphical presentation of these effects and it allows for an assessment of both the magnitude and importance of an effect. In the Pareto charts, the bars (variables) that graphically exceed the significance line exert a statistically significant influence on the obtained results.

2.3. Validation of Method

The results of the developed UHPLC–UV method were validated in terms of their selectivity, linearity, limits of detection (LOD), limits of quantification (LOQ), precision, accuracy and recovery.

Selectivity was assessed using a matrix comparison method. The results of chromatographic testing showed that the chromatographic peaks of catechins were clearly separated in tea samples. Hence, this method is considered selective and appropriate for the identification and quantitative analysis of catechins in various tested tea samples.

Linear calibration curves were obtained over the range 1–40 µg/g (Table S6) (based on the peak area of analytes, corrected with the IS). All analytes showed good linearity with a coefficient of determination (R^2) ranging from 0.9983 to 0.9990 for the seven catechin standards.

The developed method's sensitivity was assessed by determining the limits of detection (LOD) and quantification (LOQ). LOD and LOQ were calculated at signal-to-noise ratios (S/N) of 3:1 and 10:1, respectively. LOD values were 0.33 µg/g for each catechin, while LOQ values were 1.0 µg/g for all analytes.

High (HQC = 35 µg/g), medium (MQC = 20 µg/g) and low (LQC = 2.5 µg/g) concentrations of the quality control (QC) samples were analyzed in triplicate within one day (intra-day precision and accuracy) and on three different days (inter-day precision and accuracy). Precision was expressed as the percentage of relative standard deviations (%RSD) and accuracy as the relative error (%RE). As shown in Table S7, RSD and RE obtained by the proposed approach was <11% and between −10–9.8%, respectively.

The recoveries with relative standard deviations (RSDs) for each catechin were measured by spiking blank yellow tea samples in six replicates at three different spiked levels (2.5; 20 and 35 µg/g). The recoveries ranged from 67.6 to 109%, with RSDs of 2.5–8.9% (Table S7).

The validation data showed that the proposed method provides good linearity, sensitivity, selectivity, accuracy, precision and recovery for the simultaneous analysis of seven catechins.

2.4. Application of the Developed Method to Real Samples

The new method was applied to analyze nine green, nine black, and two fruit tea samples, which were obtained from Polish markets. In Table 1, we present the obtained results. Due to the high content of catechins in the tested samples, some extracts were diluted (20-, 100- or 200-fold).

The content of catechins in tea varies due to the method of cultivation and treatment of the leaves. The highest level of catechins was determined in green teas, which is consistent with the literature data [8,32]. Their chemical composition is closest to the composition of the fresh tea plant leaves, since in the processing stage, the leaves are not fermented. The lowest concentration of catechins was determined in fruit teas, because they are products generated from dried fruits. Based on the literature data, a high concentration of EGCG, ECG, EGC and EC in tea samples was reported. Interestingly, our research showed that, in each type of tea, a different catechin predominates. In green tea, the highest content of the strongest antioxidant, was EGCG (27.7–63.1 mg/g). In good-quality leaves, this compound makes up 50% of all catechins. In black tea, it was found that the highest content was gallocatechin (6.3–22.8 mg/g), whereas in fruit it was epigallocatechin (8.07–8.4 mg/g). The results indicate that the quantitative differentiation of catechins occurs not only between different species of teas but also for different producers and forms of the same tea. Figure 5 shows UHPLC–UV chromatograms obtained for extracts of leaves of (A) black tea, (B) green tea, and (C) fruit tea. Figure S5 shows the representative multiple reaction monitoring (MRM) chromatograms of the analyzed extracts of green tea.

Table 1. Content of catechins determined in leaves of tea samples.

Sample		GC	EGC	C	EC	EGCG	GCG	ECG
		\multicolumn{7}{	c	}{Concentration (mg/g of Dry Weight [a])}
Black Tea	1	10.2 ± 0.77 [b]	4.18 ± 0.56	1.68 ± 0.13	1.70 ± 0.20	4.51 ± 0.12	1.06 ± 0.83	6.37 ± 0.55
	2	17.1 ± 1.50	3.39 ± 0.33	0.93 ± 0.01	2.15 ± 0.20	3.68 ± 0.19	0.22 ± 0.02	6.10 ± 0.11
	3	13.4 ± 0.03	3.47 ± 0.27	1.15 ± 0.06	1.25 ± 0.11	4.57 ± 0.34	0.53 ± 0.07	4.56 ± 0.48
	4	17.5 ± 0.15	7.35 ± 0.97	0.77 ± 0.04	1.90 ± 0.18	5.40 ± 0.24	0.52 ± 0.02	7.03 ± 0.50
	5	22.8 ± 1.08	5.41 ± 0.08	0.98 ± 0.01	1.50 ± 0.15	5.18 ± 0.23	0.37 ± 0.01	6.74 ± 0.17
	6	6.34 ± 0.13	3.76 ± 0.07	1.35 ± 0.02	2.05 ± 0.05	8.91 ± 0.23	0.24 ± 0.03	4.68 ± 0.10
	7	5.80 ± 0.08	2.25 ± 0.19	1.41 ± 0.01	4.29 ± 0.19	1.94 ± 0.04	0.89 ± 0.02	4.03 ± 0.02
	8	9.67 ± 0.05	0.65 ± 0.05	0.70 ± 0.03	0.34 ± 0.03	0.37 ± 0.06	0.40 ± 0.02	0.96 ± 0.03
	9	6.64 ± 0.08	1.17 ± 0.02	0.74 ± 0.06	0.44 ± 0.04	0.34 ± 0.04	0.39 ± 0.02	0.87 ± 0.01
Green Tea	10	5.99 ± 0.46	18.7 ± 0.39	0.46 ± 0.07	3.87 ± 0.31	41.0 ± 0.06	5.86 ± 0.25	9.35 ± 0.42
	11	9.58 ± 0.29	15.9 ± 0.43	1.07 ± 0.03	3.17 ± 0.23	41.2 ± 1.64	5.68 ± 0.09	11.0 ± 0.45
	12	17.6 ± 0.45	21.5 ± 0.73	3.43 ± 0.13	2.84 ± 0.12	42.1 ± 2.80	3.72 ± 0.10	14.4 ± 0.73
	13	9.90 ± 0.65	17.1 ± 0.05	2.37 ± 0.10	3.60 ± 0.08	27.2 ± 0.45	1.86 ± 0.02	17.0 ± 0.48
	14	7.09 ± 0.11	12.4 ± 0.34	1.75 ± 0.02	8.21 ± 0.45	46.8 ± 2.29	8.13 ± 0.07	9.91 ± 0.20
	15	7.30 ± 0.32	17.1 ± 0.18	2.59 ± 0.05	1.85 ± 0.16	53.8 ± 0.95	4.42 ± 0.08	15.5 ± 1.03
	16	11.5 ± 0.93	10.9 ± 0.48	3.64 ± 0.11	1.39 ± 0.17	53.5 ± 2.25	3.75 ± 0.01	11.2 ± 0.15
	17	5.80 ± 0.11	18.5 ± 0.31	1.10 ± 0.03	0.84 ± 0.06	63.1 ± 1.04	5.36 ± 0.02	11.1 ± 0.34
	18	19.8 ± 0.06	23.4 ± 1.07	0.86 ± 0.01	1.28 ± 0.05	53.2 ± 0.68	6.44 ± 0.13	8.01 ± 0.21
Fruit Tea	19	ND [c]	8.40 ± 0.01	3.55 ± 0.04	0.26 ± 0.02	0.16 ± 0.04	0.16 ± 0.04	0.23 ± 0.01
	20	ND	8.07 ± 0.08	2.67 ± 0.07	0.25 ± 0.02	0.21 ± 0.06	ND	0.20 ± 0.01

[a] Each value is the mean (µg/g of dry weight) of three replications; [b] SD relative standard deviation; [c] not detectable (ND).

As part of the study, catechins in tea infusions were also determined. SPE extraction was used to isolate the analytes from the sample, and then analyzed using UHPLC–UV. Catechins were extracted from tea leaves into a water solution to varying degrees. It was estimated that during the traditional method of brewing, approximately 60–70% of compounds found in a dry product passed into the brew. Tested infusions of green, black and fruit tea contained, respectively, 42.5–66.7 mg, 6.9–29.4 mg,

7.5–10.4 mg catechins, in 1 g of product. Based on the results obtained, approximately 50–80% of catechins transfer to the water from the dry product. The content of catechins determined in tea infusions is shown in Table S8.

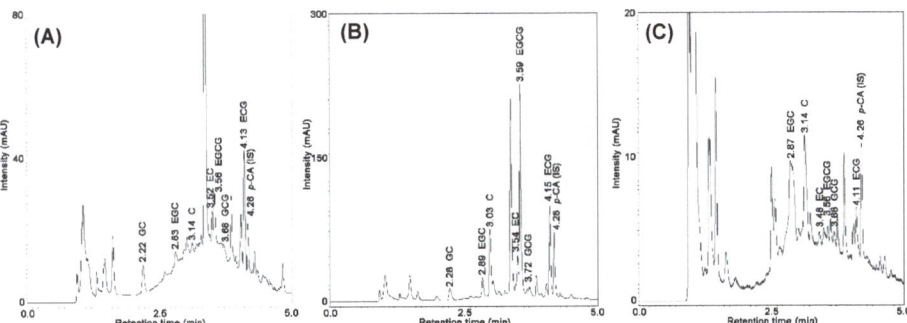

Figure 5. Representative chromatograms obtained for an extract of leaves of (**A**) black tea, (**B**) green tea, and (**C**) fruit tea using the proposed DES–solid–liquid extraction (SLE)–UHPLC–UV method.

3. Materials and Methods

3.1. Chemicals and Reagents

Analytical standards of CA, EC, EGC, ECG, EGCG, GC, GCG, *p*-coumaric acid (*p*-CA, used as internal standard) and Nile red were obtained from Sigma-Aldrich (Steinheim, Germany). Acetonitrile and trifluoroacetic acid (TFA) (HPLC grade) were from Merck (Darmstadt, Germany). Doubly distilled water was prepared using A Milli-Q water purification system (Merck Millipore, Bedford, MA, USA). Choline chloride, L-lactic acid, citric acid and malic acid were bought from Alfa Aesar (Lancashire, UK). Girard's reagent T (GrT) was obtained from Acros Organics (Geel, Belgium). The ILs, 1-butyl-3-methylimidazolium bromide [C_4MIM]Br, 1-butyl-3-methylimidazolium chloride [C_4MIM]Cl, 1-butyl-3-methylimidazolium nitrate [C_4MIM]NO_3, 1-butyl-3-methylimidazolium hydrogen sulphate [C_4MIM]HSO_4, 1-hexyl-3-methylimidazolium tetrafluoroborate [C_6MIM]BF_4, and 1-butyl-3-methylimidazolium hexafluorophosphate [C_4MIM]PF_6 were provided by Sigma-Aldrich (St. Louis, MO, USA). Methanol, hydrochloric acid, phosphoric acid and dimethylformamide (DMF) (analytical grade) were from Chempur (Piekary Śląskie, Poland).

Aqueous solutions of ILs (1 M) were prepared by dissolving a precise amount of each IL in deionized water.

A standard solution of each catechin (1 mg/mL) was obtained by dissolving a certain amount of the analytical standard in methanol (10.0 mL). A stock solution of 100 μg/mL, containing all analytes, was prepared by transferring 1.0 mL of each of the seven individual catechin standard solutions into a 10-mL volumetric flask, then adding methanol to the mark. The stock solution was stored in the refrigerator at 5 °C. The working solutions were prepared daily by dilution of the stock solution with the mobile phase (0.05% TFA in water:acetonitrile, 90:10 *v/v*).

3.2. Synthesis of Deep Eutectic Solvents

DESs based on Girard's reagent T were prepared according to our previously described ultrasound-assisted method [33].

Two or three components and a calculated amount of deionized water were added to a glass vial, sealed with a screw-cap and exposed to ultrasound (37 kHz, 30 W) at 60 °C until a homogeneous liquid was formed (10–30 min). Eight different DESs, including two or three components, were obtained and examined. The composition, molar ratios, and symbols of DESs used throughout this study are shown in Table S1.

3.3. Study of DES Properties

3.3.1. Polarity

The polarity of the obtained DESs was tested in solvatochromatic probe using Nile red (NR). λ_{max} was determined and used in the formula E_{NR} (kcal/mol) = $hc\lambda_{maxNA}$ = 28591/λ_{max} to obtain ENR [22].

The obtained results of DES polarity with 30% of water are summarized in Table S2. DES 8, DES 3 and DES 4 are the most polarized (ca. 47 kcal/mol), with polarities similar to water (46.9 kcal/mol). DES 5, DES 7 and DES 1 are the least polarized, with polarities similar to MeOH (51.8 kcal/mol).

3.3.2. Density and Viscosity

Viscosity tests of eight prepared DESs were carried out using a DV1 Viscometer (Brookfield, Middleboro, MA, USA) at 27 °C. A portable density meter, Densito 30PX (Mettler Toledo, Schwerzenbach, Switzerland), was used for density analysis at 27 °C. The obtained results are summarized in Table S2. No specific correlations between the efficiency of the extraction and the density or viscosity of the tested DESs are observed.

3.3.3. IR Spectroscopy

IR spectra were measured on the FT-IR spectrometer Nicolet 6700 at room temperature using the ATR method (Thermo Fisher Scientific, Waltham, MA, USA). The spectra of Girard's reagent, malic acid, their eutectic mixture and eutectic mixture with different water contents (10, 30, 50 and 75%) were recorded and compared.

3.4. Extraction of Catechins from Tea Samples

3.4.1. Comparison of Extraction Procedures

For the purpose of comparison, extractions employ: (1) conventional solvents (water, methanol and mixture water:methanol (1:1; *v/v*)); (2) ionic liquids ([C$_4$MIM]Br, [C$_4$MIM]Cl, [C$_4$MIM]NO$_3$, [C$_4$MIM]HSO$_4$, [C$_6$MIM]BF$_4$, [C$_4$MIM]PF$_6$); (3) DESs (Table S1) were performed. Briefly, the ground tea leaves (150 mg) were weighed, then the extraction solvent (1.5 mL) was added and the sample was stirred at 1100 rpm for 40 min at 40 °C in an Eppendorf tube using thermomixer comfort (Eppendorf AG, Hamburg, Germany). Then, the sample was centrifuged for 5 min at 2000× *g* (IKA mini G centrifuge, Staufen, Germany) and the liquid supernatant (600 µL) was transferred to another tube. The extract was diluted 1:1 with methanol. The final extract was filtered through a 0.2-µm nylon membrane syringe filter and transferred to a vial for UHPLC–UV analysis.

3.4.2. Optimization of DES–SLE Extraction Procedure

Firstly, during the optimization of the DES–SLE extraction procedure, the influence of water the content in DESs was investigated by considering four values: 10%, 30%, 50% and 75% of water in DES.

In the presented study, to optimize the extraction conditions (temperature, time of extraction and solid/liquid ratio), the response surface methodology (RSM) with three-factor and rotatable central composite design (CCD) was applied [34]. The design variables were extraction time (6.5–73.5 min; X_1), temperature (6.5–48.4 °C; X_2) and liquid-to-solid ratio: 1:2–1:12; mg/mL; X_3).

The generated runs are shown in Table S3. Randomizing the order of the experiments was applied to minimize the influences of unexplained variability in the observed response, caused extraneous factors. Multiple linear regression analysis was performed using Statistica 12 software (StatSoft, Krakow,

Poland). The experimental data were fitted to the second-order polynomial model (Equation (5)) and regression coefficients (βs) were obtained.

$$Y = \beta_0 + \sum_{i=1}^{k} \beta_i X_i + \sum_{i=1}^{k} \beta_{ii} X_i^2 + \sum_{\substack{i=1 \\ i<j}}^{k-1} \sum_{j=2}^{k} \beta_{ij} X_i X_j \tag{5}$$

where $X_1, X_2, \ldots, X_k$ are the independent variables affecting the responses Y's; β_0, β_i ($i = 1, 2, \ldots, k$), β_{ii} ($i = 1, 2, \ldots, k$), and β_{ij} ($i = 1, 2, \ldots, k; j = 1, 2, \ldots, k$) are the regression coefficients for intercept, linear, quadratic, and interaction terms, respectively; k is the number of variables.

All optimization procedures were carried out in triplicate. The selection of optimal conditions was based on the peak area obtained for selected catechins (EGC, EGCG, ECG, GCG).

3.5. Instrumentation and Chromatographic Conditions

A Hitachi Elite LaChrom UHPLC system coupled with a UV detector was used for chromatographic analysis (Merck Hitachi, Darmstadt, Germany). Chromatographic separation was achieved using Poroshell 120 SB-C18 (100 × 4.6 mm, 2.7 µm) from Agilent. A binary mobile phase was used for the chromatographic separation, comprised of 0.05% trifluoroacetic in water (solvent A) and acetonitrile (solvent B). The gradient elution started at 10% B and increased to 60% over 5.5 min, after which it decreased to 10% in 0.1 min, and was finally allowed to stabilize for 2.4 min; thus, the overall runtime was 8 min. The injection volume was 2 µL, the flow rate was fixed at 1.0 mL/min, and the separation was performed at 25 °C. The absorbance of all analytes was measured at $\lambda = 270$ nm.

Individual compounds were identified by comparing their retention times using the standard addition method.

A Dionex UHPLC system (Dionex Corporation, Sunnyvale, CA, USA) with an AB Sciex Q-Trap® 4000 mass spectrometer (Foster City, CA, USA) was used to confirm the presence of the determined catechins. The chromatographic conditions of UHPLC–MS/MS were the same as in the UHPLC–UV method, aside from the application of TFA in water as a component of the mobile phase. TFA was replaced by the 0.1% formic acid.

Electrospray ionization (ESI) conditions in the positive mode were first optimized with direct infusion into the mass spectrometer to select the precursor and product ions resulting from fragmentation, declustering potential (DP) and collision energy (CE) for each catechin (Table S4). The catechins were evaluated by employing multiple reaction monitoring (MRM) mode (Table S4). The other working parameters of the mass spectrometer were as follows: curtain gas 20 psi (nitrogen), collision gas-medium, ion spray voltage −4500 V, temperature 500 °C, ion source gases nebulizer gas 60 psi/auxiliary gas 50 psi (both nitrogen), entrance potential −5 V. Additionally, the dwell times of the analytes were set to 50 ms. Equipment control and data acquisition were performed with Analyst 1.5.1 software (Applied Biosystems, Foster City, CA, USA).

3.6. Analysis of Catechins in Tea Samples

In the quantitative analysis of tea, 9 samples of black tea (Assam), 9 samples of green tea (Long Jing, Bi Luo Chun, Yu Hua Cha, Jasmine, Sencha) and 2 samples of fruit tea were used in this study. All tea products were purchased from local markets in Poland. In total, 20 teas, cultivated in four countries (China, Japan, India, Vietnam), were used in this study. The origin of the teas was guaranteed by the seller. Blended products were used in this study (fruit tea).

Leaves of the teas were ground with a mill, 30 µL of the *p*-CA (internal standard; 50 µg/mL) was added and then samples were extracted using the following DES–SLE procedure: a 150-mg tea sample was added to a 2 mL Eppendorf tube and 1.5 mL of 30% DES-3 (malic acid:GrT; 2:1; *v/v*) was added. The sample was stirred at 50 °C for 50 min at 1100 rpm using thermomixer. Then, the sample was centrifuged at 1725 rpm for 5 min, the supernatant decanted, which was transferred to another tube, and 600 μL of the extract was diluted with 600 μL methanol. Thus, the mixture solvent was ready for UHPLC–UV analysis after being passed through a 0.22 μm nylon filter.

Additionally, catechins were extracted from tea infusions. A moderate to strong brew of black tea was prepared following the instructions provided with the 0.15 g tea by pouring 15 mL of boiling water into a beaker and dipping a teabag for 2–3 min. Green tea was prepared in water at a temperature of 80–90 °C. Next, all infusion tea samples were extracted based on the SPE procedure described in [35]. Samples were extracted using a solid-phase extraction (SPE) system (BAKERBOND spe-12G system, J.T. Baker Inc., Deventer, Netherlands). The cartridges (Oasis HLB, Waters) were conditioned with 3 mL of water (adjusted to pH = 3.5 with hydrochloric acid), then 3 mL 70% DMF with 0.1% phosphoric acid and 3 mL of water (adjusted to pH = 3.5 with hydrochloric acid). In the proposed method, 15 mL of a tea sample was adjusted to pH = 3.5 with hydrochloric acid and then loaded onto the cartridges. The catechins were eluted with 5 mL 70% DMF with 0.1% phosphoric acid. Finally, the eluate was injected into the UHPLC–UV system. The identity of the catechins in the leaves and infusion samples was confirmed by the UHPLC–MS/MS method in MRM mode.

4. Conclusions

The extraction of catechins from tea samples using a novel malic acid-based DES method demonstrated, for the first time, that DES-type solvents have promising prospects in the recovery of bioactive substances.

The extraction method using DES-3 (malic acid:GrT, 2:1) was significantly more efficient for extraction of seven main catechins than the previous time-consuming methods that employed organic solvents. The optimal performance was obtained at a temperature of 50 °C, a time of 50 min, the extraction solvent malic acid:GrT with a 2:1 M ratio and 30% water content, and the solid/liquid ratio 1:10 mg/mL. The effective UHPLC–UV method revealed excellent precision, accuracy, and recovery, and was applied to the determination of seven catechins in tea samples. We have shown the applicability of DES as an extraction solvent; however, due to its unique advantages, such as its environmentally benign behavior, low cost and non-toxicity, it holds great promise in other areas. In conclusion, DES–SLE provided a promising strategy to extract active compounds from tea samples for potential applications.

Supplementary Materials: The following are available online. Figure S1. UHPLC–UV chromatogram of a standard solution containing the analyzed catechins and IS, Figure S2. Effect of water content in DESs on extraction efficiency of catechins, Figure S3. The experimental data vs. predicted data for extraction efficiency of the conversion of (A) EGC, (B) EGCG, (C) GCG, (D) ECG, Figure S4. Pareto chart showing the values of effects from variables using the extraction efficiency of (A) EGC, (B) EGCG, (C) GCG, (D) ECG, Figure S5. Representative TIC-MRM chromatogram obtained for an extract of leaves of green tea after DES–SLE procedure using UHPLC–MS/MS method, Table S1. List of tested deep eutectic solvents (DES), Table S2. Polarity, viscosity and density of tested deep eutectic solvents (DESs), Table S3. Rotatable central composite design setting in the original and coded form of the independent variables (X_1, X_2 and X_3), Table S4. MRM transitions and mass spectrometer parameters, Table S5. Analysis of variance (ANOVA) for fit of extraction efficiency from central composite design, Table S6. Summary of calibration curve and linearity range of catechins ($n = 6$). Table S7. Summary of accuracy, precision and recovery of catechins ($n = 6$). Table S8. Content of catechins determined in infusion of tea.

Author Contributions: Conceptualization, S.B. and J.A.; formal analysis, S.B., J.A. and A.S.; methodology, S.B., J.A. and A.S.; supervision, S.B., writing—original draft, S.B. and J.A. All authors have read and agreed to the published version of the manuscript.

Funding: This project was supported by funds from the National Science Centre in the frame of the project SONATA No. 2014/13/D/ST4/01863 for the 2015–2018 period, Cracow, Poland. Publication partially supported by the Rector's grant in the field of scientific research and development works, Silesian University of Technology (Gliwice, Poland), grant number 04/010/RGJ20/0122.

Conflicts of Interest: The authors declare no conflict of interest.

References

1. Khan, N.; Mukhter, H. Tea and Health: Studies in Humans. *Curr. Pharm. Design* **2013**, *19*, 6141–6147. [CrossRef] [PubMed]
2. Cabrera, C.; Artacho, R.; Giménez, R. Beneficial Effects of Green Tea—A Review. *J. Am. Coll. Nutr.* **2006**, *25*, 79–99. [CrossRef] [PubMed]
3. Vauzour, D.; Rodriguez-Mateos, A.; Corona, G.; Oruna-Concha, M.J.; Spencer, J.P.E. Polyphenols and Human Health: Prevention of Disease and Mechanisms of Action. *Nutrients* **2010**, *2*, 1106–1131. [CrossRef]
4. Taylor, P.W.; Hamilton-Miller, J.M.T.; Stapleton, P.D. Antimicrobial properties of green tea catechins. *Food Sci. Technol. Bull.* **2005**, *2*, 71–81. [CrossRef]
5. Pandey, K.B.; Rizvi, S.I. Plant polyphenols as dietary antioxidants in human health and disease. *Ox. Med. Cell. Longevity* **2009**, *2*, 270–278. [CrossRef]
6. Banerjee, S.; Chatterjee, J. Efficient extraction strategies of tea (Camellia sinensis) biomolecules. *J. Food Sci. Technol.* **2015**, *52*, 3158–3168. [CrossRef] [PubMed]
7. Vuong, Q.V.; Golding, J.B.; Nguyen, M.; Roach, P.D. Extraction and isolation of catechins from tea. *J. Sep. Sci.* **2010**, *33*, 3415–3428. [CrossRef]
8. Perva-Uzunalić, A.; Škerget, M.; Knez, Ž.; Weinreich, B.; Otto, F.; Gruner, S. Extraction of active ingredients from green tea (Camellia sinensis): Extraction efficiency of major catechins and caffeine. *Food Chem.* **2006**, *96*, 597–605. [CrossRef]
9. Zuo, Y.; Chen, H.; Deng, Y. Simultaneous determination of catechins, caffeine and gallic acids in green, Oolong, black and pu-erh teas using HPLC with a photodiode array detector. *Talanta* **2002**, *57*, 307–316. [CrossRef]
10. Liang, H.; Liang, Y.; Dong, J.; Lu, J. Tea extraction methods in relation to control of epimerization of tea catechins. *J. Sci. Food Agricul.* **2007**, *87*, 1748–1752. [CrossRef]
11. Dong, J.J.; Ye, J.H.; Lu, J.L.; Zheng, X.Q.; Liang, Y.R. Isolation of antioxidant catechins from green tea and its decaffeination. *FBP* **2011**, *89*, 62–66. [CrossRef]
12. Sereshit, H.; Samandi, S. A rapid and simple determination of caffeine in teas, coffees and eight beverages. *Food Chem.* **2014**, *158*, 8–13. [CrossRef] [PubMed]
13. Rahim, A.A.; Nofrizal, S.; Saad, B. Rapid tea catechins and caffeine determination by HPLC using microwave-assisted extraction and silica monolithic column. *Food Chem.* **2014**, *147*, 262–268. [CrossRef]
14. Chang, C.J.; Chiu, K.L.; Chen, Y.L.; Chang, C.Y. Separation of catechins from green tea using carbon dioxide extraction. *Food Chem.* **2000**, *68*, 109–113. [CrossRef]
15. Sökmen, M.; Demir, E.; Alomar, S.Y. Optimization of sequential supercritical fluid extraction (SFE) of caffeine and catechins from green tea. *J. Supercrit. Fluids* **2018**, *133*, 171–176. [CrossRef]
16. Jun, X.; Shuo, Z.; Bingbing, L.; Rui, Z.; Ye, L.; Deji, S.; Guofeng, Z. Separation of major catechins from green tea by ultrahigh pressure extraction. *Int. J. Pharm.* **2010**, *386*, 229–231. [CrossRef] [PubMed]
17. Gadkari, P.V.; Balaraman, M. Catechins: Sources, extraction and encapsulation: A review. *Food Prod. Proces.* **2015**, *93*, 122–138. [CrossRef]
18. Pena-Pereira, F.; Namieśnik, J. Ionic Liquids and Deep Eutectic Mixtures: Sustainable Solvents for Extraction Processes. *Chem. Sus.* **2014**, *7*, 1784–1800. [CrossRef]
19. Ho, T.D.; Zhang, C.; Hantao, L.W.; Anderson, J.L. Ionic Liquids in Analytical Chemistry: Fundamentals, Advances, and Perspectives. *Anal. Chem.* **2014**, *86*, 262–285. [CrossRef]
20. Trujillo-Rodríguez, M.J.; Nan, H.; Varona, M.; Emaus, M.N.; Souza, I.D.; Anderson, J.L. Advances of Ionic Liquids in Analytical Chemistry. *Anal. Chem.* **2019**, *91*, 505–531. [CrossRef]
21. Smith, F.L.; Abbott, A.P.; Ryder, K.S. Deep Eutectic Solvents (DESs) and Their Applications. *Chem. Rev.* **2014**, *114*, 11060–11082. [CrossRef] [PubMed]
22. Ruβ, C.; König, B. Low melting mixtures in organic synthesis – an alternative to ionic liquids? *Green Chem.* **2012**, *14*, 2969–2982.
23. Shishov, A.; Bulatov, A.; Locatelli, M.; Carradori, S.; Andruch, V. Application of deep eutectic solvents in analytical chemistry. A review. *Microchem. J.* **2017**, *135*, 33–38. [CrossRef]
24. Zhang, H.; Tang, B.; Row, K. Extraction of Catechin Compounds from Green Tea with a New Green Solvent. *Chem. Rer. Chin. Univ.* **2014**, *30*, 37–41. [CrossRef]

25. Li, J.; Han, Z.; Zou, Y.; Yu, B. Efficient extraction of major catechins in Camellia sinensis leaves using green choline chloride-based deep eutectic solvents. *RSC Adv.* **2015**, *5*, 93937–93944. [CrossRef]

26. Jeong, K.M.; Ko, J.; Zhao, J.; Jin, Y.; Yoo, D.E.; Han, S.Y.; Lee, J. Multi-functioning deep eutectic solvents as extraction and storage media for bioactive natural products that are readily applicable to cosmetic products. *J. Clean. Prod.* **2017**, *151*, 87–95. [CrossRef]

27. Wang, M.; Wang, J.; Zhou, Y.; Zhang, M.; Xia, Q.; Bi, W.; Chen, D.D.Y. Ecofriendly Mechanochemical Extraction of Bioactive Compounds from Plants with Deep Eutectic Solvents. *ACD Sustainable Chem. Eng.* **2017**, *5*, 6297–6303. [CrossRef]

28. Ma, W.; Dai, Y.; Row, K.H. Molecular imprinted polymers based onmagnetic chitosan with different deepeutectic solvent monomers for the selectiveseparation of catechins in black tea. *Electrophoresis* **2018**, *39*, 2039–2046. [CrossRef]

29. Ma, W.; Row, K.H. Solid-Phase Extraction of Catechins from Green Tea with Deep Eutectic Solvent Immobilized Magnetic Molybdenum Disulfide Molecularly Imprinted Polymer. *Molecules* **2020**, *25*, 280. [CrossRef]

30. Magiera, S.; Baranowska, I.; Lautenszleger, A. UHPLC–UV method for the determination of flavonoids in dietary supplements and for evaluation of their antioxidant activities. *J. Pharm. Biomed. Anal.* **2015**, *102*, 468–475. [CrossRef]

31. Spáčil, Z.; Nováková, L.; Solich, P. Comparison of positive and negative ion detection of tea catechins using tandem mass spectrometry and ultra high performance liquid chromatography. *Food Chem.* **2010**, *123*, 535–541. [CrossRef]

32. Tao, W.; Zhou, Z.; Zhao, B. Simultaneous determination of eight catechins and four theaflavins in green, black and oolong tea using new HPLC-MS-MS method. *J. Pharm. Biomed. Anal.* **2016**, *131*, 140–145. [CrossRef] [PubMed]

33. Bajkacz, S.; Adamek, J. Development of a Method Based on Natural Deep Eutectic Solvents for Extraction of Flavonoids from Food Samples. *Food Anal. Meth.* **2018**, *11*, 1330–1344. [CrossRef]

34. Said, K.A.M.; Amin, M.A.M. Overview on the Response Surface Methodology (RSM) in Extraction Processes. *J. Appl. Sci. Process Eng.* **2015**, *2*, 8–17.

35. Unno, T.; Sagesaka, Y.M.; Kakuda, T. Analysis of Tea Catechins in Human Plasma by High-Performance Liquid Chromatography with Solid-Phase Extraction. *J. Agricul. Food Chem.* **2005**, *53*, 9885–9889. [CrossRef]

Sample Availability: Not available.

 *molecules*

Article

Multidimensional Liquid Chromatography Employing a Graphene Oxide Capillary Column as the First Dimension: Determination of Antidepressant and Antiepileptic Drugs in Urine

Edvaldo Vasconcelos Soares Maciel, Ana Lúcia de Toffoli, Jussara da Silva Alves and Fernando Mauro Lanças *

Institute of Chemistry of São Carlos, University of São Paulo, São Carlos, CEP 13566590, SP, Brazil; daltoniqsc@gmail.com (E.V.S.M.); ana_scalon@hotmail.com (A.L.d.T.); jussaraalves@usp.br (J.d.S.A.)
* Correspondence: flancas@iqsc.usp.br; Tel.: +55-163373-9984

Academic Editor: Victoria Samanidou
Received: 20 January 2020; Accepted: 13 February 2020; Published: 29 February 2020

Abstract: Human mental disorders can be currently classified as one of the most relevant health topics. Including in this are depression and anxiety, which can affect us at any stage of life, causing economic and social problems. The treatments involve cognitive psychotherapy, and mainly the oral intake of pharmaceutical antidepressants. Therefore, the development of analytical methods for monitoring the levels of these drugs in biological fluids is critical. Considering the current demand for sensitive and automated analytical methods, the coupling between liquid chromatography and mass spectrometry, combined with suitable sample preparation, becomes a useful way to improve the analytical results even more. Herein we present an automated multidimensional method based on high-performance liquid chromatography-tandem mass spectrometry using a lab-made, graphene-based capillary extraction column connected to a C8 analytical column to determined five pharmaceutical drugs in urine. A method enhancement was performed by considering the chromatographic separation and the variables of the loading phase, loading time, loading flow, and injection volume. Under optimized conditions, the study reports good linearity with $R^2 > 0.98$, and limits of detection in the range of 0.5–20 µg L^{-1}. Afterward, the method was applied to the direct analysis of ten untreated urine samples, reporting traces of citalopram in one of them. The results suggest that the proposed approach could be a promising alternative that provides direct and fully automated analysis of pharmaceutical drugs in complex biological matrices.

Keywords: liquid chromatography; mass spectrometry; sample preparation; automation; on-line; multidimensional; extraction column; urine; antidepressants; pharmaceutical drugs

---

1. Introduction

Diseases associated with human mental disorders can be currently classified as one of the most emergent topics in medicine. In this context are the widely known psychiatric illnesses called depression and anxiety. According to the World Health Organization, it is estimated that roughly 4.4% of the world population has already suffered from them. It is predicted that depression will be the second-most prevalent human disorder by 2030 [1].

In general, depression is considered a chronic disease that can arise in any stage of life, causing significant damage, including economic and social problems, and even leads to suicidal thoughts [2]. The most frequent symptoms of depression include unstable moods, fatigue, sadness, and insomnia. Additionally, anxiety can be considered another common type of psychiatric disorder that, when overlooked, leads to depression. In this case, arrhythmia, hyperventilation, sweating, racing thoughts, and insomnia indicate anxiety. Taking into account the similarities, there is presumably a direct correlation in terms of medical interventions. The most popular treatments involve cognitive psychotherapy, and mainly the use of pharmaceutical antidepressants (ADs) [3]. Therefore, considering the present panorama of mental disorders frequently reported in the 21st century, it is also expected that there will be an increase in antidepressant uptake by people in future.

Typically, these pharmaceutical drugs are divided into four main classes: tricyclic antidepressants (TCAs), selective serotonin reuptake inhibitor (SSRI), selective noradrenaline reuptake inhibitor (SNRI), and monoamine oxidase inhibitors (MOI) [4]. Although there are several different medicines commercially available, most of them have similar side effects (mainly in the early stages of administration), and a slow time to start acting on the human brain [5]. Besides these, other medications, such as antiepileptic drugs, can also be used to treat such disorders since they can act as mood stabilizers in some cases [6].

For these reasons, precise monitoring regarding their levels in the biological fluids is mandatory to guarantee therapeutic effectiveness and to diminish side effects. Moreover, the use of these drugs combined with other prescription medications may cause toxic problems, and, in the last few decades, their use for recreational purposes has concerned health organizations around the world [7,8]. Therefore, the development of analytical methods to determine the residues of ADs in human samples is very important in areas such as medicine and forensics. Several analytical techniques can be employed for these purposes, such as gas and liquid chromatography, capillary electrophoresis, and spectrophotometry, among others [1,9–11]. Considering the current demand for methods to be more sensitive and selective, the coupling between liquid chromatography and mass spectrometry becomes a useful way to improve the analytical results even more. Nonetheless, given the lower concentration levels of ADs and the complexity of biological samples, high-performance liquid chromatography-tandem mass spectrometry (HPLC-MS/MS) is not enough to achieve such results; hence, a previous step called sample preparation is often required [12].

Generally, these procedures are focused on removing interferents from the matrix, and on extracting/pre-concentrating target analytes [13]. The most common sample preparation techniques are conventional solid-phase extraction (SPE) and liquid–liquid extraction (LLE), which were proposed more than 50 years ago. These traditional approaches have many disadvantages, including laborious and time-consuming steps, large amounts of sample and solvent requirements, and disposable hardware (especially SPE), among other restrictions [14]. In order to overcome these shortcomings, modern sample preparation techniques based on the principles of the precursor solid-phase microextraction (SPME) began to appear in the early 1990s [14]. Consequently, the current trends are mainly based on miniaturization, automation, and high-throughput analysis, which point out automated methods that integrate sample preparation and HPLC-MS/MS as a suitable combination [15].

In this context, herein we propose an automated multidimensional method employing two columns, where the first one is specifically used for sample preparation and the second performs the chromatographic separation followed by tandem mass spectrometry detection. It is noteworthy that our capillary extraction column was packed with a lab-made extractive phase consisting of graphene oxide supported on an aminopropyl silica surface (GO-Sil). This column is much cheaper than the commercially available ones and has a reported excellent performance and robustness [16]. Additionally, the capillary dimensions of the extraction column (200-mm length and 508-μm i.d.) allow for economies in quantities of solvent, sample, and extractive phase, which are under the principles of green chemistry, which is so important nowadays. Its excellent extractive performance is attributed mainly to the high surface area of the graphene oxide, together with the delocalized π-electron system, which suggests a good affinity with molecules containing aromatic rings like the pharmaceutical drugs herein analyzed. In this case, the π-π interaction is the main interaction mechanism responsible for selective extraction. Aiming to evaluate the system performance, we selected four antidepressant drugs (ADs) as chemical probes, namely carbamazepine, citalopram, clomipramine, and desipramine, and one anticonvulsant AC, namely sertraline.

2. Results and Discussion

2.1. Method Enhancement

2.1.1. Chromatographic Separation

During the early stages of this work, experiments were performed that aimed to optimize the analytes' chromatographic separation. Figure 1 illustrates the main results obtained by varying the mobile phase composition. As can be seen, our first attempt using isocratic mode (Figure 2E) reported a lower chromatographic resolution. However, as we were evaluating different combinations of mobile phases (D → B), improvements on the resolution were achieved. Finally, Figure 2A shows the best conditions regarding the separation of the five target analytes. In this case, an elution gradient employing ultrapure water and acetonitrile, both acidified with 0.2% formic acid, reported the best results. These gains in the resolution using the elution gradient might be due to the similarities in the analytes' chemical structure, which required subtle variations on the mobile phase elution strength, in order to separate one from another compound. Additionally, as our mass spectrometer operated in electrospray (ESI) positive mode, which is known to suffer from a matrix effect that might lead to ion suppression or enhancement, the acidification of the mobile phases could aid the analytes to be more ionizable, increasing the analytical signal.

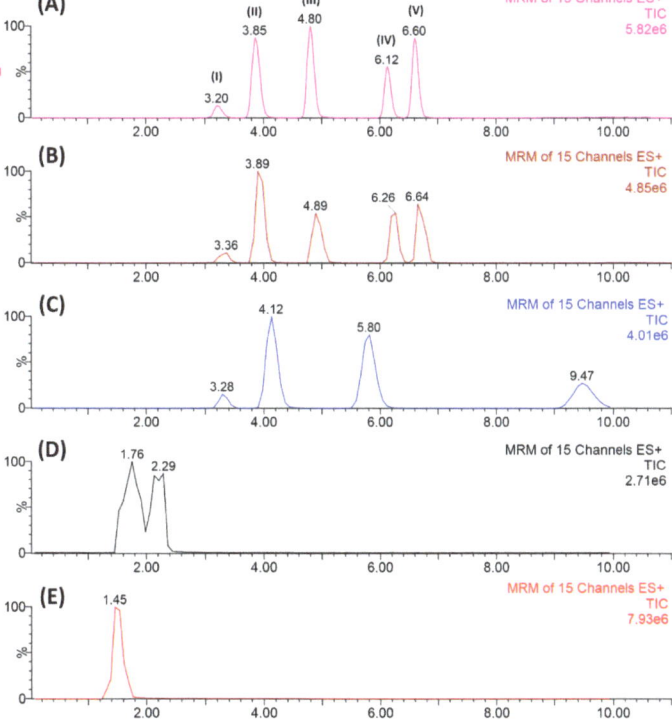

Figure 1. Representation of the chromatographic separation enhancement from E → A: (**A**) best condition applying elution gradient (H_2O/ACN + 0.2% formic acid), (**B**) satisfactory separation but the dwell-time was not adjusted, (**C–E**) mobile phase without acidification and mobile flow rate not adjusted. Elution order: (I) carbamazepine, (II) citalopram, (III) desipramine, (IV) sertraline, and (V) clomipramine.

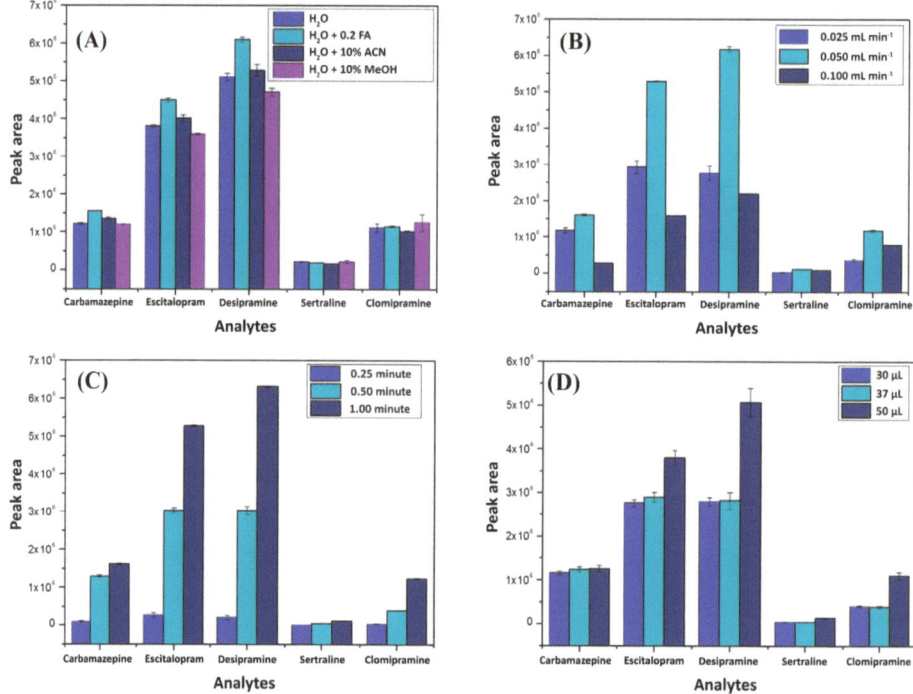

Figure 2. Method enhancement parameters obtained by univariate experiments considering the following parameters: (**A**) loading phase, (**B**) loading flow, (**C**) loading time, and (**D**) injection volume.

2.1.2. Multidimensional Automated Procedure

In the sequence, a batch of experiments aiming to achieve an ideal analytical condition for all other influential parameters was conducted. Figure 2 depicts the results obtained for each investigated variable through univariate experiments by considering the area under the chromatographic peak as the response variable. All parameters were studied using triplicate injections. It is important to emphasize that when a parameter was not being evaluated, it was kept in the following standard analytical conditions: loading phase, H_2O; loading flow, 0.05 mL min^{-1}; loading time, 0.5 min; and injection volume, 50 μL.

First, the best composition of the loading mobile phase was evaluated. As can be seen in Figure 2A, the best extraction performance was reported using ultrapure water with formic acid (0.2%). This behavior can be explained due to the lower pH ($\approx$3.2) obtained when formic acid (FA) is used, which can favor the interactions between the analytes and the sorbent phase. In this pH range, most molecules are charged and consequently have more affinity for the polar oxygen groups present on the graphene oxide surface [17,18]. Apart from that, using methanol and acetonitrile in the loading phase is expected to produce a higher elution strength, which makes the sorption of the analytes in the extraction column difficult; they pass directly through it, going to waste. Sequentially, the loading flow was investigated using univariate experiments with three different values: 0.025, 0.050, and 0.100 mL min^{-1}. Figure 2B depicts the results using 0.050 mL min^{-1}, reporting the best performance for the majority of the analytes. As can be seen, the intermediate value had the best performance when comparing it with 0.025 mL min^{-1}. This fact can be explained by considering that the lower flow rate value might not be enough to ensure that all analytes had passed through the extraction column at the time the valve was switched to the elution position, causing analytes not to be sorbed into the extraction columns. Conversely, when considering 0.05 mL min^{-1}, a higher flow hampered the analytes since they were desorbed due to a more diluted condition or due to the higher force that pushed them inside the extraction column, resulting in lower extraction performance.

After determining the best characteristic of the loading phase composition and flow rate, the other parameters were studied. Figure 2C shows that by increasing the loading time in which the analytes were pumped inside the extraction column, a better extraction performance was achieved. This effect is reasonable since a greater loading time implies more interaction between the analytes and the sorbent phase. Therefore, 1 min was fixed as the selected loading time. Furthermore, the volume of the sample injected into the system was varied to include these three values: 30, 37, and 50 µL. As can be expected, the larger sample volume (50 µL) resulted in better extraction performance since this is directly proportional to the number of analytes available to interact with the extraction column. For this reason, 50 µL was fixed as the injection volume.

2.2. Figures of Merit

The figures of merit herein evaluated were determined according to the International Conference on Harmonization (ICH) guidelines [19].

First, the method selectivity was evaluated by analyzing a sample obtained from a pool formed by blank urines, collected from consenting volunteers, which were compared with those obtained from the same sample after being spiked with a mixture containing the target analytes. As no peaks were observed in the multiple reaction monitoring (MRM) ion transition for each compound, the method was considered as being selective (Figure 3). In the sequence, the limits of detection and quantification were determined via successive injections of spiked urine samples until observing a signal to noise ratio near to 3:1 and 10:1, for LOD and LOQ, respectively. Therefore, the limits of detection ranged from 0.01–2.0 µg L^{-1} and the limits of quantification from 0.5–20 µg L^{-1}. The method linearity was determined considering six different concentration levels, with each one being evaluated on triplicate injections. The linear interval for each analyte was: 1–200 µg L^{-1} for carbamazepine, citalopram, and desipramine, and 20–200 µg L^{-1} for sertraline and clomipramine. As shown in Table 1, the method presented good linearity with correlation coefficients (R^2) higher than 0.985.

Table 1. Method linearity characteristics and its limits of detection (LOD) and quantification (LOQ).

Analytes	Linear Equation	R^2	LOD (µg L^{-1})	LOQ (µg L^{-1})
Carbamazepine	y = 1681.7 + 2835.6x	0.999	0.01	0.5
Citalopram	y = −2542.9 + 6904.3x	0.997	0.04	0.5
Clomipramine	y = −14200.5 + 1167.5x	0.985	0.5	25
Desipramine	y = −1593.3 + 7048.4x	0.994	0.01	0.5
Sertraline	y = −1614.1 + 128.4x	0.985	2.0	20

Afterward, the method accuracy, precision, and enrichment factor were all determined by considering three concentration levels (low, medium, and high) evaluated using injection triplicates. As can be seen in Table 2, the method presented good accuracy, with the values being between 83.2 and 117.6, which is considered acceptable according to the ICH guidelines (80–120%). Sequentially, the intra-day precision was determined on the same day of those other validation parameters, while the inter-day precision was evaluated on a subsequent day. Table 2 shows the obtained relative standard deviation (RSD) values, ranging from 1.4–13.6%, which were also per the ICH guidelines. Finally, as our analytical method was based on a multidimensional automated approach, it was essential to study the enrichment factor obtained by pushing the analytes through the extraction column before chromatographic analysis. In general, an increase in the analytical signal is expected when a pre-concentration step is carried out. Table 2 shows the obtained results for it, highlighting a good enrichment factor for all target compounds providing a signal enhancement varying from 4.7 to 59.4 when compared to the direct injection approach. Therefore, these results support the choice for a multidimensional and automated method to perform sample preparation and determination of pharmaceutical drugs in complex samples as urine. Furthermore, it must be underscored that the exceptional robustness of the in-house prepared extractive phase GO-Sil packed into the capillary extraction column was used for more than 250 urine injections without losing its original performance.

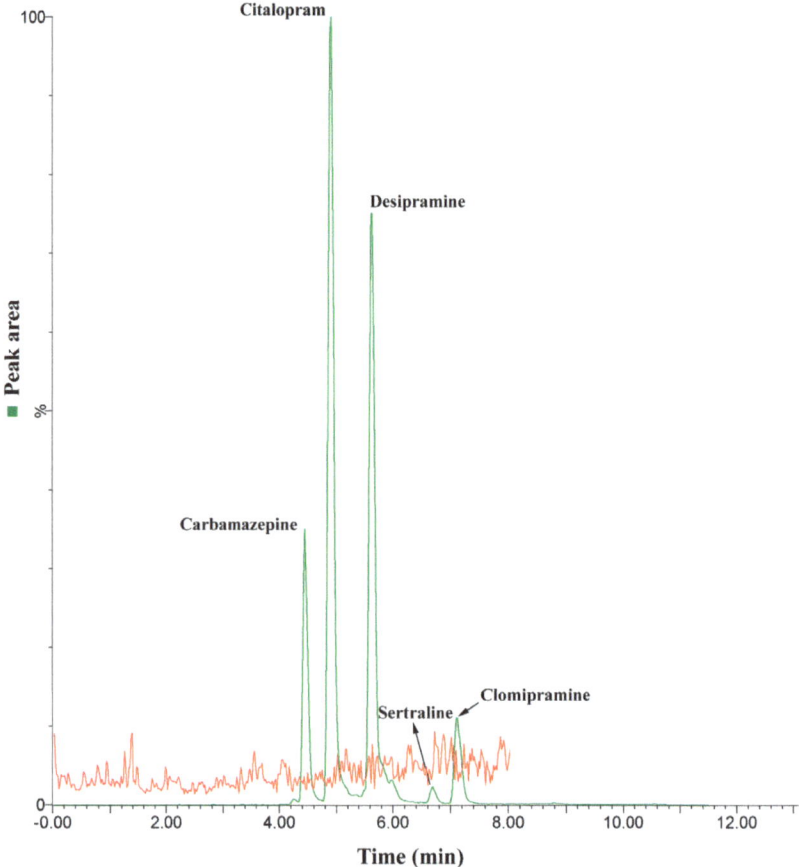

Figure 3. Chromatograms obtained by comparing a 100 µg L^{-1} spiked urine sample with an unspiked blank one in order to verify the selectivity of the proposed analytical method.

Table 2. Additional figures of merit including the method enrichment factor, accuracy, and precision. RSD: Relative Standard Deviation.

Analytes	Enrichment Factor			Accuracy (%)			Precision (% RSD)					
							Intra-Day			Inter-Day		
	L	M	H	L	M	H	L	M	H	L	M	H
Carbamazepine	4.7	5.3	5.1	83.2	95.8	98.8	12.3	2.3	3.6	13.6	2.1	1.4
Citalopram	6.8	7.6	7.0	125.3	89.7	99.1	2.0	1.9	3.2	6.8	2.9	5.5
Clomipramine	17.3	18.1	17.4	98.7	117.6	102.4	5.2	6.1	4.0	9.2	3.2	4.8
Desipramine	18.2	16.4	15.0	105.8	114.8	102.3	6.9	3.5	11.8	11.2	4.5	2.4
Sertraline	21.2	59.4	13.1	98.7	117.6	102.4	12.8	4.5	6.5	8.1	4.1	1.4

2.3. Overall Method Performance

When looking to compare our obtained results with other published papers in the literature, we can underscore some advantages, as well as limitations. First, as our paper presents the use of a synthesized graphene-based sorbent packed into a capillary extraction column, its robustness is noteworthy, as just described, given that it was applied to more than 250 injections. As examples, other recent works pinpoint their lab-made extractive hardware being re-used five and seventy times without losing its efficiency, respectively [20,21]. Likewise, our developed extraction column surpasses by far the commercially available SPE cartridges, which can be ideally used only once. Furthermore, considering our automated multidimensional approach using two columns, the system required only 50 µL of urine with reduced reagent consumption and consequent waste generation [4,22,23]. The lack of steps demanding operator intervention due to the automation can lead to remarkable gains in analysis time (≈8 min), while it also diminishes analytical errors resulting from sample handling [4,23]. Another great quality of it is the capacity to perform the analysis of antidepressants and antiepileptics in undiluted and unprecipitated urine. As highlighted by Cai et al. [24], several methods developed to analyze ADs in urine have been carried out by considering a dilution step due to the high complexity of the samples. Finally, the LODs and LOQs of the proposed approach are in a similar range with most published works; although some methods can be more sensitive, our results provide a suitable range for its main goal [25,26]. From the authors' point of view, the major limitation of this proposed methodology is in its system configuration, since it demands an auxiliary pump and a switching valve, which might consist of a restriction for some laboratories.

2.4. Method Application

Separately from the pool of blank samples used during the development step, the analytical methodology herein described was applied to the analysis of other urine samples collected from consenting volunteers. From ten samples analyzed for the target compounds, one presented traces of citalopram in a concentration estimated to be in the order of 150 µg L^{-1}. This result is probably due to the considerably widespread use of citalopram (SSRI) at present since it has a broad spectrum of action, treating not only depression, but also obsessive-compulsive disorder, panic disorder, and social phobia [26]. Figure 4 shows the results comparing the referred sample (red line) with a blank one fortified with the analytes in a concentration range that resulted in an area similar to that obtained for the unknown sample. As can be seen, the signals for citalopram were in similar magnitude; the MRM transitions, the relative ratio between the monitored ions, and similarity of the retention time verifies the observed results.

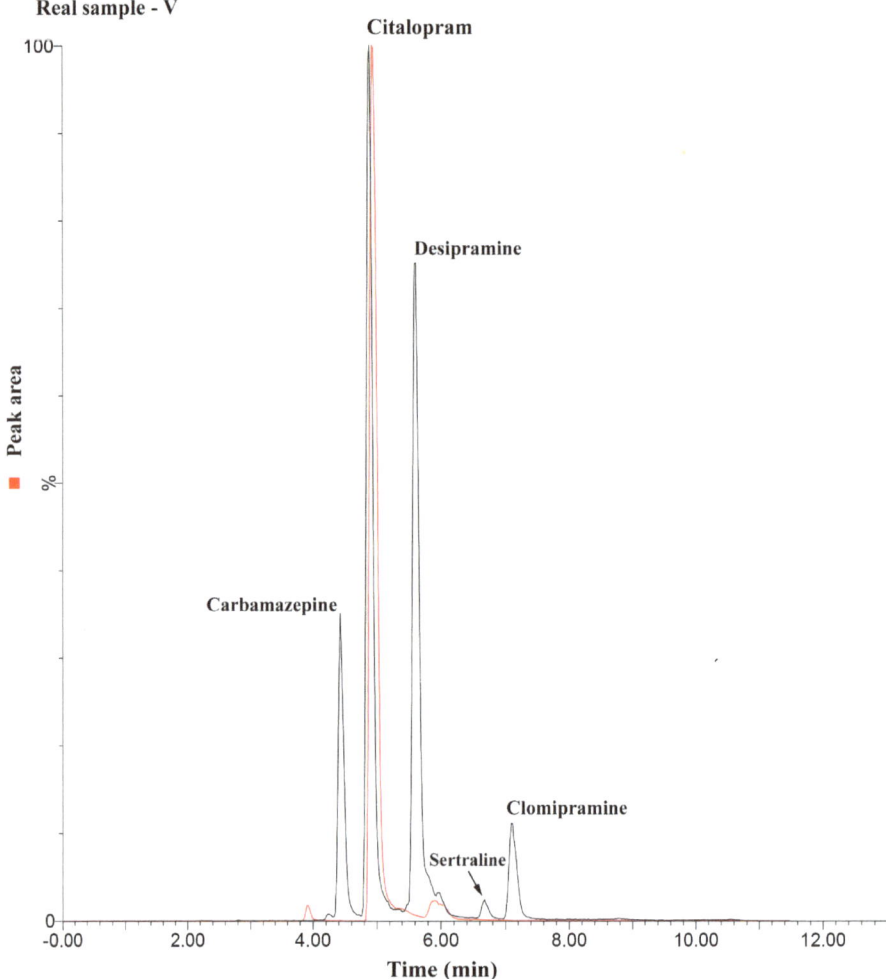

Figure 4. Chromatogram comparing a 150 μg L^{-1} spiked urine with a sample from a volunteer (red line) in which traces of citalopram were found.

3. Experimental

3.1. Reagents and Standard Solutions

High purity (99%) analytical standards of carbamazepine, citalopram, clomipramine, desipramine, and sertraline were all acquired from Fluka Analytical (St Louis, MO, USA). The analytes' stock solutions were all prepared in methanol at a concentration of 1000 mg L^{-1}, and subsequently diluted to 100 mg L^{-1}. The work solutions were prepared from the stock ones in a proper concentration by considering the goal of each experiment to be performed. It should be highlighted that all standard solutions were temperature-controlled (−30 °C) inside the amber flasks.

The HPLC grade solvents acetonitrile (ACN) and methanol (MeOH) were purchased from TEDIA (Farfield, OH, USA) and the ultrapure water was produced at our laboratory using a MILLI-Q purification system from Millipore (Burlington, MA, USA). Furthermore, MS grade formic acid (FA) acquired from Sigma-Aldrich (St Louis, MO, USA) was used to acidify the chromatographic mobile phases. The GO-Sil extractive phase was synthesized and had already been used in previous works published by our research group [16,18].

3.2. Extraction Column Preparation

As our extraction column possessed capillary physical dimensions (200-mm length and 508-μm i.d.), our best choice to produce it was using the slurry packing procedure. In short, this consisted of using a high-pressure pump to push a suspension containing the stationary phase inside the column tubing, similar to that utilized in the production of HPLC and U-HPLC analytical columns. Therefore, the slurry packing system mainly consisted of a packing solvent, a slurry solvent to dissolve the stationary phase, a reservoir where the suspension was kept, and the column hardware often placed in the inferior part of the system.

In this work, a Haskell DSFH-300 hydropneumatic pump acquired from Haskel (Burbank, CA, USA) was employed as the pushing pump, while ultrapure water was used as the packing solvent. The suspension consisted of 10 mg of GO-Sil extractive phase dissolved in 700 μL of the slurry solvent (isopropanol/tetrahydrofuran; 6:1 *v/v*). The packing pressure was maintained at ≈600 bar during the procedure (≈60 min) in order to fill the column tubing. For more detailed information about the extraction column production, as well as for the GO-Sil extractive phase characterization assays (SEM and FTIR), please refer to a recent manuscript published by our research group [16].

3.3. Instrumentation

The analytical system was composed of an Acquity UPLC liquid chromatograph equipped with a binary solvent manager, and a sample manager coupled to a Xevo TQ S mass spectrometer using electrospray ionization, all from Waters (Milford, MA, USA). Moreover, a Shimadzu LC 10Ai equipped with a degasser 10A from Shimadzu (Kyoto, JAP), and an electronically assisted switching valve from Supelco (St. Louis, MO, USA) were used to carry out the automated sample loading step, transferring the sample from its original vial to inside the first (extraction) column.

The chromatographic separations were achieved using a Poroshell 120 SB-C8 analytical column from Agilent (Santa Clara, CA, USA) (100 mm × 2.1 mm × 2.7 μm d_p) at a temperature of 40 °C. The mobile phase consisted of ultrapure water and acetonitrile (both acidified with 0.2% formic acid) at a flow rate of 0.20 mL min^{-1}, and the loading phase contained acidified ultrapure water (0.2% formic acid) at a flow rate of 0.05 mL min^{-1}.

The mass spectrometry parameters were optimized via direct infusion of each analyte in standard solutions at a concentration of 0.5 mg mL^{-1}, assisted by the IntelliStart optimization software (4.1) from Waters (USA). Under the optimized conditions, the detection method included a positive ESI, capillary voltage of 3.9 kV, source temperature of 150 °C, desolvation gas (N$_2$) temperature of 650 °C and flow of 1000 L h^{-1}, and collision gas (Ar) flow of 0.15 mL min^{-1}. In order to enhance the method selectivity, the MS/MS configuration operation in the multiple reaction monitoring (MRM) was chosen to be used. All the analytes' transitions used for identification/quantification, as well as its main detection parameters, can be found in Table 3.

3.4. Multidimensional Analytical Method

The multidimensional analytical method was composed of two columns (extraction and analytical) connected using the switching valve, which was responsible for steering the flow depending on the purpose. Figure 5 illustrates the configuration assembled to perform the automated analysis.

Before starting any analysis, the urine samples were simply filtered through a 0.22-μm cellulose membrane to avoid clogging the whole system.

During each analysis, the autosampler was responsible for controlling the chromatographic injection and the valve positions. This was done through a sequence of events scheduled in the software. First, the sample injection was performed with the valve set at the loading position (valve ports connected through the purple line; see Figure 5). Therefore, the LC 10Ai auxiliary pump carried the sample through the capillary extraction column, at a flow of 0.05 mL min^{-1}, in order to retain the analytes while the majority of interferents went to waste. Meanwhile, the HPLC binary solvent pump conditioned the analytical column with the initial composition of the elution gradient. After 1 min, the valve was switched to the eluting position (valve ports connected through the red dotted lines; see Figure 5). Thus, the chromatographic mobile phase was pumped inside the extraction column, at a flow rate of 0.2 mL min^{-1}, to desorb the analytes, shifting them to the analytical column and further to the mass spectrometer. In the sequence, the multidimensional system was washed and conditioned again to be ready for the next injection. Table 4 summarizes the main steps regarding the described analytical procedure.

Table 3. Analytes' multiple reaction monitoring (MRM) precursor and product ions and its main detection parameters.

Analyte	Precursor Ion (*m/z*)	Product Ion (*m/z*)	Cone Voltage (V)	Collision Energy (V)	Dwell Time (ms)
Carbamazepine	253	152	24	42	0.075
		167	24	44	0.075
		180	24	32	0.075
Desipramine	267	72	22	14	0.075
		193	22	42	0.075
		208	22	24	0.075
Sertraline	306	123	16	48	0.075
		159	16	30	0.075
		275	16	14	0.075
Clomipramine	315	58	24	30	0.075
		86	24	18	0.075
		227	24	42	0.075
Citalopram	325	109	32	30	0.075
		234	32	26	0.075
		262	32	20	0.075

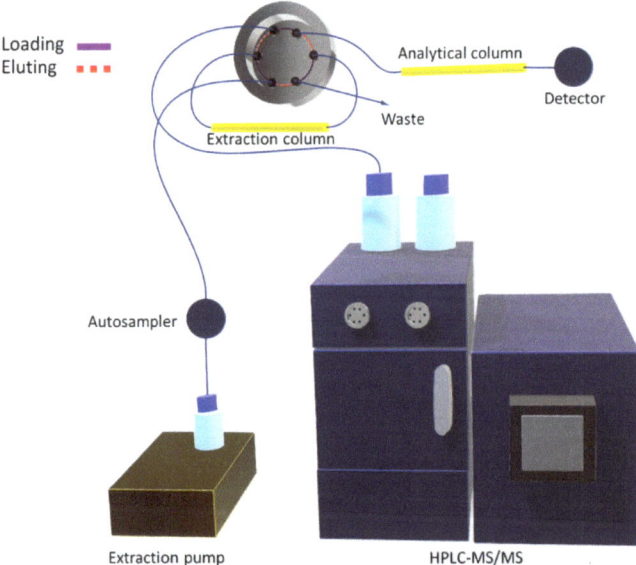

Figure 5. Illustrative drawing representing the multidimensional configuration, including the extraction column (first dimension) connected to the analytical column (second dimension) via a switching valve placed before the HPLC-MS/MS instrument.

Table 4. Analytical steps involved in the automated multidimensional extraction/determination of the analytes.

Event	Time (min)	Solvent Composition (Extraction Column)	Solvent Composition (Analytical Column)
Loading	0.00–1.00	H_2O + 0.2% FA	H_2O (A)/ACN (B) * (30%:70%)
	1.00–3.00	H_2O (A)/ACN (B) * (30%:70% → 35%:65%)	H_2O (A)/ACN (B) * (30%:70% → 35%:65%)
Eluting	3.00–6.00	H_2O (A)/ACN (B) * (35%:65% → 40%:60%)	H_2O (A)/ACN (B) * (35%:65% → 40%:60%)
	5.00–6.00	H_2O (A)/ACN (B) * (40%:60%)	H_2O (A)/ACN (B) * (40%:60%)
	6.00–7.00	H_2O (A)/ACN (B) * (40%:60% → 50%:50%)	H_2O (A)/ACN (B) * (40%:60% → 50%:50%)
Cleaning	7.00–7.66	H_2O (A)/ACN (B) * (50%:50% → 10%:90%)	H_2O (A)/ACN (B) * (50%:50% → 10%:90%)
	7.66–8.60	H_2O + 0.2% FA	H_2O (A)/ACN (B) * (10%:90%)
Conditioning	8.60–11.50	H_2O + 0.2% FA	H_2O (A)/ACN (B) * (30%:70%)

* Both mobile phases acidified with 0.2% formic acid.

3.5. Method Enhancement

In order to achieve a satisfactory sample clean-up (eliminating the majority of endogenous urine compounds) combined with a good chromatographic resolution and MS detectability, a batch of univariate experiments were performed. Therefore, the influences of the elution gradient, injection volume, loading flow, loading time, and loading phase composition were all investigated. These experiments were performed via injection of triplicates of blank urine samples spiked at 100 μg L^{-1}.

First, the chromatographic separation was studied by changing the mobile phase solvent composition as well as the pH. Three solvents were tested (MeOH, ACN, and H_2O), and formic acid was added to modify the pH. Sequentially, four parameters directly related to the extraction column were considered: (i) the loading phase composition: H_2O, H_2O (0.2% FA), H_2O/ACN, and H_2O/MeOH; (ii) the loading flow: 0.025, 0.05, and 0.1 mL min^{-1}; (iii) the loading time: 0.25, 0.5, and 1.0 min; and (iv) the injection volume: 30, 37, and 50 μL. The parameters and its evaluation conditions were chosen by considering our experience with such types of multidimensional configurations [16,25].

3.6. Figures of Merit

Afterward, a systematic study regarding the analytical figures of merit commonly considered for validation procedures was performed according to international guidelines [19]. Therefore, individual experiments were carried out by contemplating six different variables: linearity, accuracy, precision, limits of quantification and detection, pre-concentration factor, and selectivity. It is essential to highlight that the pool of urine samples used in this step was collected from consenting volunteers and previously tested to verify the absence of the analytes such that they could be considered blank samples that would not interfere with the spiked concentration levels.

The method linearity was studied through the matrix-matched calibration method by spiking urine samples at six different concentration levels: 1, 25, 50, 75, 100, 150, and 200 μg L^{-1} for carbamazepine, citalopram, and desipramine; 20, 40, 80, 100, 150, and 200 μg L^{-1} for sertraline; and 25, 50, 75, 100, 150, and 200 μg L^{-1} for clomipramine. Each concentration level was evaluated using triplicate extractions with the automated multidimensional approach. The limits of detection (LODs) and quantification (LOQs) were determined via comparison of the signal to noise ratio in blank samples and those spiked at known concentration levels. Determination of the LOD was chosen at a signal to noise ratio of 3:1, while for LOQ, a signal to noise ratio of 10:1 was considered. The selectivity was investigated via comparing the pool of "blank" urine with those spiked at known concentration levels to verify the absence of interferent signal on the compounds' retention time or MRM transitions. First, the accuracy was determined in three different concentrations via measuring the actual value obtained from the linearity equation (C_r) and comparing it with the theoretical concentration value of each spiking level on the analytical curve (C_t). Sequentially, precision was studied in terms of the relative standard deviation (RSD %) at three different levels of concentration, repeated in two consecutive days (intra- and inter-day assays). Finally, the pre-concentration factor (or enrichment factor) was evaluated by performing several injections of spiked urine samples via employing the multidimensional system (passing through the extraction column), which were compared with those similarly spiked and were directly injected into the analytical column.

3.7. Method Application

Urine samples used in this work were collected from consenting volunteers. Part of it was prior analyzed for the presence of the target drugs; in its absence, they formed a pool of samples used as "blank samples" during all stages of the study development. Additionally, the other samples not tested were used to verify the method's applicability after the determination of the figures of merit. All aliquots were only filtered through 0.22 μm cellulose membrane prior injection into the automated multidimensional system.

4. Conclusions

Herein an online automated analytical method based on multidimensional liquid chromatography coupled to tandem mass spectrometry was developed to extract and determine four antidepressants and one antiepileptic drug in human urine. The approach was based on the interconnection between two columns being the first accountable to perform the analytes' extraction (first dimension) while the second worked as a chromatographic analytical column (second dimension). Our capillary extraction column was packed with a synthesized graphene-based sorbent that exhibits excellent extraction performance and robustness being used for more than 250 injections. The method takes roughly 8 min and used 50 μL of undiluted and unprecipitated urine, demanding only a simple filtration step before injection into the multidimensional system. Besides, essential parameters were investigated to find out an ideal analytical condition allowing the determination of some validation figures of merit: linearity, accuracy, precision, selectivity, enrichment factor, LOD, and LOQ. Afterward, all ten urine samples collected from the consenting individuals in the study were analyzed to verify the proposed procedure. The presence of citalopram residues at a concentration level of around 150 μg L^{-1} was found in one of the ten analyzed samples. Therefore, based on the results obtained and reported in this manuscript, the proposed multidimensional analytical method was revealed to be a promising way to perform rapid and effective trace analysis of antidepressant and antiepileptic drugs in urine that easily adaptable to work with other biological complex matrices, such as saliva and plasma, among others.

Author Contributions: E.V.S.M. wrote this version of the manuscript. E.V.S.M. and A.L.d.T. performed the synthesis of the extractive phase/produced the extraction column, supervised the method enhancement, and performed the validation/urine applications. A.L.d.T. processed the data. J.d.S.A. performed the method enhancement and wrote a Portuguese version of this manuscript. F.M.L. conceptualized and supervised the whole research project, provided all required facilities, and reviewed/edited this manuscript. All authors have read and agreed to the published version of the manuscript.

Funding: The authors are grateful to The São Paulo Research Foundation (FAPESP—grants 2017/02147-0, 2015/15462-5, and 2014/07347-9) and the National Council for Scientific and Technological Development (CNPq—307293/2014-9) for the financial support provided. This research project was financed in part by the Coordenação de Aperfeiçoamento de Pessoal de Nível Superior—Brasil (CAPES), Finance Code 001.

Conflicts of Interest: The authors declare no conflict of interest.

Abbreviations

ACN, acetonitrile; ADs, antidepressants; ESI, electrospray ionization; FA, formic acid; GO-Sil, graphene oxide supported onto aminosílica; HPLC, High-performance liquid chromatography; i.d., inner diameter; ICH, International Conference on Harmonization; LC, liquid chromatography; LLE, liquid–liquid extraction; LOD, limit of detection; LOQ, limit of quantification; MeOH, methanol; MRM, multiple reaction monitoring; MS, mass spectrometry; RSD, relative standard deviation; SPE, solid-phase extraction; SPME, solid-phase microextraction; SSRI, selective serotonin reuptake inhibitor; UPLC, ultra-performance liquid chromatography.

References

1. Murtada, K.; de Andrés, F.; Zougagh, M.; Ríos, Á. Strategies for antidepressants extraction from biological specimens using nanomaterials for analytical purposes: A review. *Microchem. J.* **2019**, *150*, 104193. [CrossRef]

2.	Gotlib, I.H.; Joormann, J. Cognition and Depression: Current Status and Future Directions. *Annu. Rev. Clin. Psychol.* **2010**, *6*, 285–312. [CrossRef] [PubMed]

3.	Beck, A.T.; Haigh, E.A.P. Advances in Cognitive Theory and Therapy: The Generic Cognitive Model. *Annu. Rev. Clin. Psychol.* **2014**, *10*, 1–24. [CrossRef] [PubMed]

4.	Alves, V.; Gonçalves, J.; Conceição, C.; Câmara, H.M.T.; Câmara, J.S. An improved analytical strategy combining microextraction by packed sorbent combined with ultra high pressure liquidchromatography for the determination of fluoxetine, clomipramineand their active metabolites in human urine. *J. Chromatogr. A* **2015**, *1408*, 30–40. [CrossRef] [PubMed]

5.	Dome, P.; Tombor, L.; Lazary, J.; Gonda, X.; Rihmer, Z. Natural health products, dietary minerals and over-the-counter medications as add-on therapies to antidepressants in the treatment of major depressive disorder: A review. *Brain Res. Bull.* **2019**, *146*, 51–78. [CrossRef] [PubMed]

6.	Ghoraba, Z.; Aibaghi, B.; Soleymanpour, A. Application of cation-modified sulfur nanoparticles as an efficient sorbent for separation and preconcentration of carbamazepine in biological and pharmaceutical samples prior to its determination by high-performance liquid chromatography. *J. Chromatogr. B* **2017**, *1063*, 245–252. [CrossRef]

7.	Dear, J.W.; Bateman, D.N. Antidepressants. *Medicine* **2016**, *44*, 135–137. [CrossRef]

8.	Liveri, K.; Constantinou, M.A.; Afxentiou, M.; Kanari, P. A fatal intoxication related to MDPV and pentedrone combined with antipsychotic and antidepressant substances in Cyprus. *Forensic Sci. Int.* **2016**, *265*, 160–165. [CrossRef]

9.	Zheng, M.M.; Wang, S.T.; Hu, W.K.; Feng, Y.Q. In-tube solid-phase microextraction based on hybrid silica monolith coupled to liquid chromatography-mass spectrometry for automated analysis of ten antidepressants in human urine and plasma. *J. Chromatogr. A* **2010**, *1217*, 7493–7501. [CrossRef]

10.	Rodríguez-Flores, J.; Salcedo, A.M.C.; Fernández, L.M. Rapid quantitative analysis of letrozole, fluoxetine and their metabolites in biological and environmental samples by MEKC. *Electrophoresis* **2009**, *30*, 624–632. [CrossRef]

11.	Truta, L.; Castro, A.L.; Tarelho, S.; Costa, P.; Sales, M.G.F.; Teixeira, H.M. Antidepressants detection and quantification in whole blood samples by GC–MS/MS, for forensic purposes. *J. Pharm. Biomed. Anal.* **2016**, *128*, 496–503. [CrossRef] [PubMed]

12.	Mohebbi, A.; Farajzadeh, M.A.; Yaripour, S.; Mogaddam, M.R.A. Determination of tricyclic antidepressants in human urine samples by the three–step sample pretreatment followed by hplc–uv analysis: An efficient analytical method for further pharmacokinetic and forensic studies. *EXCLI J.* **2018**, *17*, 952–963. [PubMed]

13.	Niu, Z.; Zhang, W.; Yu, C.; Zhang, J.; Wen, Y. Recent advances in biological sample preparation methods coupled with chromatography, spectrometry and electrochemistry analysis techniques. *TrAC Trends Anal. Chem.* **2018**, *102*, 123–146. [CrossRef]

14.	Maciel, E.V.S.; de Toffoli, A.L.; Neto, E.S.; Nazario, C.E.D.; Lanças, F.M. New materials in sample preparation: Recent advances and future trends. *TrAC Trends Anal. Chem.* **2019**, *119*, 115633. [CrossRef]

15.	Maciel, E.V.S.; de Toffoli, A.L.; Lanças, F.M. Current status and future trends on automated multidimensional separation techniques employing sorbent-based extraction columns. *J. Sep. Sci.* **2019**, *42*, 258–272. [CrossRef] [PubMed]

16.	De Toffoli, A.L.; Maciel, E.V.S.; Lanças, F.M. Evaluation of the tubing material and physical dimensions on the performance of extraction columns for on-line sample preparation-LC–MS/MS. *J. Chromatogr. A* **2019**, *1597*, 18–27. [CrossRef]

17.	De Toffoli, A.L.; Maciel, E.V.S.; Fumes, B.H.; Lanças, F.M. The role of graphene-based sorbents in modern sample preparation techniques. *J. Sep. Sci.* **2018**, *41*, 288–302. [CrossRef]

18.	Vasconcelos Soares Maciel, E.; Henrique Fumes, B.; Lúcia de Toffoli, A.; Mauro Lanças, F. Graphene particles supported on silica as sorbent for residue analysis of tetracyclines in milk employing microextraction by packed sorbent. *Electrophoresis* **2018**, *39*, 2047–2055. [CrossRef]

19.	Ich Validation of Analytical Procedures: Text and Methodology Q2(R1) 2005, 1–17. Available online: https://database.ich.org/sites/default/files/Q2_R1__Guideline.pdf (accessed on 17 February 2020).

20.	Murtada, K.; de Andrés, F.; Ríos, A.; Zougagh, M. Determination of antidepressants in human urine extracted by magnetic multiwalled carbon nanotube poly(styrene-co-divinylbenzene) composites and separation by capillary electrophoresis. *Electrophoresis* **2018**, *39*, 1808–1815. [CrossRef]

Molecules **2020**, *25*, 1092

21. Fuentes, A.M.A.; Fernández, P.; Fernández, A.M.; Carro, A.M.; Lorenzo, R.A. Microextraction by packed sorbent followed by ultra high performance liquid chromatography for the fast extraction and determination of six antidepressants in urine. *J. Sep. Sci.* **2019**, *42*, 2053–2061. [CrossRef]
22. Farsimadan, S.; Goudarzi, N.; Chamjangali, M.A.; Bagherian, G. Optimization of ultrasound-assisted dispersive liquid-liquid microextraction based on solidification of floating organic droplets by experimental design methodologies for determination of three anti-anxiety drugs in human serum and urine samples by high p. *Microchem. J.* **2016**, *128*, 47–54. [CrossRef]
23. Resende, S.; Deschrijver, C.; van de Velde, E.; Verstraete, A. Development and validation of an analytical method for quantification of 15 non-tricyclic antidepressants in serum with UPLC-MS/MS. *Toxicol. Anal. Clin.* **2016**, *28*, 294–302. [CrossRef]
24. Cai, J.; Zhu, G.T.; He, X.M.; Zhang, Z.; Wang, R.Q.; Feng, Y.Q. Polyoxometalate incorporated polymer monolith microextraction for highly selective extraction of antidepressants in undiluted urine. *Talanta* **2017**, *170*, 252–259. [CrossRef] [PubMed]
25. Ferreira, D.C.; de Toffoli, A.L.; Maciel, E.V.S.; Lanças, F.M. Online fully automated SPE-HPLC-MS/MS determination of ceftiofur in bovine milk samples employing a silica-anchored ionic liquid as sorbent. *Electrophoresis* **2018**, *39*, 2210–2217. [CrossRef] [PubMed]
26. El Sherbiny, D.; Wahba, M.E.K. Micellar liquid chromatographic method for the simultaneous determination of citalopram hydrobromide with its two demethylated metabolites. Utility as a diagnostic tool in forensic toxicology. *J. Pharm. Biomed. Anal.* **2019**, *164*, 173–180. [CrossRef] [PubMed]

Sample Availability: Not available.

MDPI

St. Alban-Anlage 66

4052 Basel

Switzerland

Tel. +41 61 683 77 34

Fax +41 61 302 89 18

www.mdpi.com

Molecules Editorial Office

E-mail: molecules@mdpi.com

www.mdpi.com/journal/molecules